Reinforced Concrete Design for Buildings

Other books by Paul Rogers:

Tables and Formulas for Fixed End Moments

Steel Columns Eccentrically Loaded

Reinforced Concrete Design for Buildings

PAUL ROGERS
Fellow, American Society of Civil Engineers
California State University, Los Angeles
Consulting Structural Engineer

with the assistance of

MICHAEL L. BALTAY
Fellow, American Society of Civil Engineers
Structural Engineer

VNR
VAN NOSTRAND REINHOLD COMPANY
New York Cincinnati Toronto London Melbourne

Van Nostrand Reinhold Company Regional Offices:
New York Cincinnati Chicago Millbrae Dallas

Van Nostrand Reinhold Company International Offices:
London Toronto Melbourne

Library of Congress Catalog Card Number: 73-6914
ISBN: 0-442-27018-6

Manufactured in the United States of America

Published by Van Nostrand Reinhold Company
450 West 33rd Street, New York, N.Y. 10001

Published simultaneously in Canada by Van Nostrand Reinhold Ltd.

15 14 13 12 11 10 9 8 7 6 5 4 3 2 1

Library of Congress Cataloging in Publication Data

Rogers, Paul, 1909–
Reinforced concrete design for buildings.

1. Reinforced concrete construction. I. Title.
TA683.2.R63 693.5′4 73-6914
ISBN 0-442-27018-6

Preface

The recently published Standard Building Code of the American Concrete Institute, ACI 318-71, contains several changes of major importance. These changes obviate most of the existing textbooks on reinforced concrete design and create an urgency to provide up-to-date information to the design profession.

This book is intended to assist the structural engineer in the design of reinforced concrete buildings. In order to avoid repetitions, information available through the ACI 318-71 Code is not repeated here but is referred to in examples and explanations. Also omitted are details of design for prestressed concrete, shells and folded plates, etc. The inclusion of these would substantially increase the volume of this book without assisting the designer of building structures.

Strength Design (formerly Ultimate Strength Design, USD) is employed exclusively in this book. Since most of the notations have changed, both the new and the old are presented. The reader will find many computer outputs in this text for the solution of *intermediate* steps, and for which longhand solutions are very time-consuming.

Perhaps the most important part of this book is the inclusion of complete work sheets for the design of beams and columns. The step-by-step procedure permits a complete longhand computation; on the other hand, computerization of this procedure is not difficult.

Certain deviation from the ACI 318-71 Code is recommended. The two-way slab provisions are unduly complicated and time-consuming, and the seismic specifications are, in the author's opinion, not conservative enough. There is a movement presently on the West Coast to demand dynamic response analyses, based on time history or on spectrum, for every structure six stories or higher.

The excellent assistance of Michael L. Baltay, S.E., in developing the numerous computer solutions, is gratefully acknowledged. Credit is also due to Henry Vogl, C.E., for his conscientious review and checking of text and examples. The author appreciates the excellent artwork prepared by John Németh and Elizabeth Gerlóczy.

The author hopes that his more than thirty years of experience as a consulting structural engineer, educator, and longtime member of the American Concrete Institute's Standard Building Code Committee 318 will enable him to convey to the reader a practical approach to the design of reinforced concrete building structures.

Los Angeles, California PAUL ROGERS, S.E.

Notations

The Code in Appendix B has a new set of notations which is mostly in accordance with the internationally used concrete design term. Some old notations remained without change, some were subjected to change.

In order to facilitate the interpretation of the new Code notations, the reader will find on the next few pages a copy of Appendix B, completed with the notations formerly used.

	Formerly	
a	a	= depth of equivalent rectangular stress block, defined by Section 10.2.7—Chapters 8, 10, and 12
a	a	= shear span, distance between concentrated load and face of support—Chapter 11
a	A	= maximum deflection under test load of a member relative to a line joining the ends of the span, or of the free end of a cantilever relative to its support, in.—Chapter 20
A	A	= effective tension area of concrete surrounding the main tension reinforcing bars and having the same centroid as that reinforcement, divided by the number of bars, sq in. When the main reinforcement consists of several bar sizes the number of bars shall be computed as the total steel area divided by the area of the largest bar used—Chapter 10
A	A_t	= area of that part of the cross section between the flexural tension face and the center of gravity of the gross section—Chapter 18
A_b	a_s	= area of an individual bar, sq in.—Chapter 12
A_c	A_c	= area of core of spirally reinforced column measured to the outside diameter of the spiral, sq in.—Chapter 10 and Appendix A
A_c	A_c	= area of concrete at the cross section considered—Chapter 18
A_{ch}	A_c'	= area of rectangular core of column measured out-to-out of hoop—Appendix A

	Formerly	
A_g	A_g	= gross area of section, sq in.—Chapters 8, 9, 10, 11, 14, and Appendix A
A_h	A_h	= area of shear reinforcement parallel to the main tension reinforcement, sq in.—Chapter 11
A_l	A_l	= total area of longitudinal reinforcement to resist torsion, sq in.—Chapter 11
A_{ps}	A_s	= area of prestressed reinforcement in tension zone—Chapters 11 and 18
A_s	A_s	= area of nonprestressed tension reinforcement, sq in.—Chapters 8, 9, 10, 11, 12, 18, and Appendix A
A_s'	A_s'	= area of compression reinforcement, sq in.—Chapters 8, 9, 18, and Appendix A
A_{sh}	A_{sh}	= area of transverse hoop bar (one leg)—Appendix A
A_t	A_t	= area of structural steel or tubing in a composite section—Chapter 10
A_t	A_o	= area of one leg of a closed stirrup resisting torsion within a distance s, sq in.—Chapter 11
A_v	A_v	= area of shear reinforcement within a distance s, or area of shear reinforcement perpendicular to main reinforcement within a distance s for deep beams, sq in.—Chapter 11
A_v	A_v	= area of shear reinforcement within a distance s—Chapter 12 and Appendix A
A_{vf}	A_v	= area of shear-friction reinforcement, sq in.—Chapter 11
A_{vh}	A_{vh}	= area of shear reinforcement parallel to the main tension reinforcement within a distance s_2, sq in.—Chapter 11
A_w	A_w	= area of a deformed wire, sq in.—Chapters 7 and 12
A_1	A_b	= loaded area—Chapter 10
A_2	A_b'	= maximum area of the portion of the supporting surface that is geometrically similar to and concentric with the loaded area—Chapter 10
b	b	= width of compression face of member—Chapters 8, 10, 11, and 18
b_o	b_o	= periphery of critical section for slabs and footings—Chapter 11
b_v	b'	= the width of the cross section being investigated for horizontal shear—Chapter 17
b_w	b'	= web width, or diameter of circular section, in.—Chapters 11 and 12
c	c	= distance from extreme compression fiber to neutral axis—Chapter 10
c_1	c_1	= size of rectangular or equivalent rectangular column, capital, or bracket measured in the direction in which moments are being determined—Chapter 11
c_2	c_2	= size of rectangular or equivalent rectangular column, capital, or bracket measured transverse to the direction in which moments are being determined—Chapters 11 and 13

	Formerly	
C	C	= cross-sectional constant to define the torsional properties. See Eq. (13-7)–Chapter 13
C_m	C_m	= a factor relating the actual moment diagram to an equivalent uniform moment diagram–Chapter 10
d	d	= distance from extreme compression fiber to centroid of tension reinforcement, in.–Chapters 7, 8, 10, 11, 12, 17, and Appendix A
d	d	= distance from extreme compression fiber to centroid of prestressing steel, or to combined centroid when nonprestressing tension reinforcement is included, in.–Chapter 18
d'	d'	= distance from extreme compression fiber to centroid of compression reinforcement, in.–Chapter 9
d_b	D	= nominal diameter of bar, wire, or prestressing strand, in.–Chapters 7 and 12
d_c	t_b	= thickness of concrete cover measured from the extreme tension fiber to the center of the bar located closest thereto–Chapter 10
d_p	D	= diameter of the pile at footing base–Chapter 15
d_s	d''	= distance from centroid of tension reinforcement to the tensile face of the member, in.–Chapter 9
D	D	= dead loads, or their related internal moments and forces–Chapters 9 and 20
e	e	= eccentricity of design load parallel to axis measured from the centroid of the section. It may be calculated by conventional methods of frame analysis–Chapter 10
e		= base of Napierian logarithms–Chapter 18
E	E	= load effects of earthquake, or their related internal moments and forces–Chapter 9
E_c	E_c	= modulus of elasticity of concrete, psi. See Section 8.3.1–Chapters 8, 9, and 10
E_{cb}	E_{cb}	= modulus of elasticity for beam concrete–Chapter 13
E_{cc}	E_{cc}	= modulus of elasticity for column concrete–Chapter 13
E_{cs}	E_{cs}	= modulus of elasticity for slab concrete–Chapter 13
EI	EI	= flexural stiffness of compression members. See Eq. (10-8) and Eq. (10-9)–Chapter 10
E_s	E_s	= modulus of elasticity of steel, psi. See Section 8.3.2–Chapters 8 and 10
f'_c	f'_c	= specified compressive strength of concrete, psi–Chapters 4, 7, 8, 9, 10, 11, 12, 14, 18, 19, and Appendix A
$\sqrt{f'_c}$	$\sqrt{f'_c}$	= square root of specified compressive strength of concrete, psi–Chapters 7, 9, 11, 12, 15, and 19
f'_{ci}	f'_{ci}	= compressive strength of concrete at time of initial prestress–Chapter 18
f_{ct}	f'	= average splitting tensile strength of lightweight aggregate concrete, psi–Chapters 4, 9, 11, and 12

	Formerly	
f_d	f_d	= stress due to dead load, at the extreme fiber of a section at which tensile stresses are caused by applied load, psi—Chapter 11
f_h	f_h	= tensile stress developed by standard hook, psi—Chapter 12
f_{pc}	f_{pc}	= compressive stress in the concrete, after all prestress losses have occurred, at the centroid of the cross section resisting the applied loads or at the junction of the web and flange when the centroid lies in the flange, psi. (In a composite member, f_{pc} will be the resultant compressive stress at the centroid of the composite section, or at the junction of the web and flange when the centroid lies within the flange, due to both prestress and to bending moments resisted by the precast member acting alone)—Chapter 11
f_{pe}	f_{pe}	= compressive stress in concrete due to prestress only after all losses, at the extreme fiber of a section at which tensile stresses are caused by applied loads, psi—Chapter 11
f_{ps}	f_{su}	= calculated stress in prestressing steel at design load, psi—Chapters 12 and 18
f_{pu}	f_s'	= ultimate strength of prestressing steel, psi—Chapters 3, 11, and 18
f_{py}	f_{sy}	= specified yield strength of prestressing steel, psi—Chapter 18
f_r	f_r	= modulus of rupture of concrete, psi—Chapter 9 and Appendix A
f_s	f_s	= calculated stress in reinforcement at service loads, ksi—Chapter 10
f_{se}	f_{se}	= effective stress in prestressing steel, after losses, psi—Chapters 12 and 18
f_y	f_y	= specified yield strength of nonprestressed reinforcement, psi—Chapters 3, 7, 8, 9, 10, 11, 12, 18, 19, and Appendix A
F	--	= lateral or vertical pressure of liquids, or their related internal moments and forces—Chapter 9
h	t beams: t_b slabs: t_s	= overall thickness of member, in.—Chapters 7, 8, 9, 10, 11, 13, 14, 19, 20, and Appendix A
h_v	h	= total depth of shearhead cross section—Chapter 11
h_w	H	= total height of wall from its base to its top—Chapter 11
H	Q	= lateral earth pressure, or its related internal moments and forces—Chapter 9
I	I	= moment of inertia of section resisting externally applied design loads—Chapter 11
I_b	I_b	= moment of inertia about centroidal axis of gross section of a beam as defined in Section 13.1.5—Chapter 13
I_c	I_c	= moment of inertia of gross cross section of columns—Chapter 13
I_{cr}	I_{cr}	= moment of inertia of cracked section transformed to concrete—Chapter 9

	Formerly	
I_e	I_{eff}	= effective moment of inertia for computation of deflection—Chapter 9
I_g	I_g	= moment of inertia of gross concrete section about the centroidal axis, neglecting the reinforcement—Chapters 9 and 10
I_s	I_s	= moment of inertia about centroidal axis of gross section of slab
		= $h^3/12$ times width of slab specified in definitions of α and β_t—Chapter 13
I_{se}	I_s	= moment of inertia of reinforcement about the centroidal axis of the member cross section—Chapter 10
I_t	I_t	= moment of inertia of structural steel or tubing in a cross section about the centroidal axis of the member cross section—Chapter 10
k	k	= effective length factor for compression members—Chapter 10
K	K	= wobble friction coefficient per foot of prestressing steel—Chapter 18
K_b	K_b	= flexural stiffness of beam; moment per unit rotation—Chapter 13
K_c	K_c	= flexural stiffness of column; moment per unit rotation—Chapter 13
K_{ec}	K_{ec}	= flexural stiffness of an equivalent column; moment per unit rotation. See Eq. (13-5)—Chapter 13
K_s	K_s	= flexural stiffness of slab; moment per unit rotation—Chapter 13
K_t	K_t	= torsional stiffness of torsional member; moment per unit rotation—Chapter 13
l	l	= span length of beam or one-way slab, as defined in Section 8.5.2; clear projection of cantilever, in.—Chapter 9
l	L	= length of prestressing steel element from jacking end to any point x—Chapter 18
l_a	L_a	= additional embedment length at support or at point of inflection, in.—Chapter 12
l_c	h	= height of column, center-to-center of floors or roof—Chapter 13
l_c	h	= vertical distance between supports—Chapter 14
l_d	L_d	= development length, in. See Chapter 12—Chapters 7 and 12
l_e	L_e	= equivalent embedment length, in.—Chapter 12
l_h	h'	= maximum unsupported length of rectangular hoop measured between perpendicular legs of the hoop or supplementary crossties—Appendix A
l_n	l_c	= clear span for positive moment or shear and the average adjacent clear spans for negative moment—Chapter 8
l_n	l_c	= length of clear span in long direction of two-way construction, measured face-to-face of columns in slabs without beams and face-to-face of beams or other supports in other cases—Chapter 9

	Formerly	
l_n	l_c	= length of clear span, in the direction moments are being determined, measured face-to-face of supports—Chapter 13
l_n	l_c	= clear span measured face-to-face of supports—Chapter 11 and Appendix A
l_t	l	= span of member under load test (the shorter span of flat slabs supported on four sides). The span, except as provided in Section 20.4.6(c) is the distance between the centers of the supports or the clear distance between supports plus the depth of the member, whichever is smaller, in.—Chapter 20
l_u	h	= unsupported length of compression member—Chapter 10
l_v	L_s	= length of shearhead arm from centroid of concentrated load or reaction—Chapter 11
l_w	L_o	= total lengths of wire extending beyond outermost cross wires, for each pair of spliced wires, in.—Chapter 7
l_w	--	= horizontal length of wall—Chapters 8 and 11
l_1	l_1	= length of span in the direction moments are being determined measured center-to-center of supports—Chapter 13
l_2	l_2	= length of span transverse to l_1, measured center-to-center of supports—Chapter 13
L	L	= live loads, or their related internal moments and forces—Chapters 9 and 20
M_a	M_{max}	= maximum moment in member at stage for which deflection is being computed—Chapter 9
M_c	M	= moment to be used for design of compression member—Chapter 10
M_{cr}	M_{cr}	= cracking moment. See Section 9.5.2.2—Chapters 9 and 11
M_m	M'	= modified bending moment—Chapter 11
M_{max}	M_1	= maximum bending moment due to externally applied design loads—Chapter 11
M_o	M_o	= total static design moment—Chapter 13
M_p	M_p	= required full plastic moment of shearhead cross section—Chapter 11
M_t	M_{uo}	= theoretical moment strength, in.-lb, of a section $= A_s f_y \left(d - \frac{a}{2}\right)$—Chapters 8 and 12
M_u	M_u	= applied design load moment at a section, in.-lb—Chapter 11
M_v	M_s	= moment resistance contributed by shearhead reinforcement—Chapter 11
M_1	M_1	= value of smaller end moment on compression member calculated from a conventional elastic frame analysis, positive if member is bent in single curvature, negative if bent in double curvature—Chapter 10
M_2	M_2	= value of larger end moment on compression member calculated from a conventional elastic frame analysis, always positive—Chapter 10

	Formerly	
n	N	= number of pairs of cross wires in splice—Chapter 7
n	n	= modular ratio = E_s/E_c—Chapter 8
n	N	= number of cross wires in anchorage zone of welded deformed wire fabric—Chapter 12
N_c	T_c	= tensile force in the concrete under load of $D + 1.2L$—Chapter 18
N_u	T_u	= design tensile force on bracket or corbel acting simultaneously with V_u—Chapter 11
N_u	N_u	= design axial load normal to the cross section occurring simultaneously with V_u to be taken as positive for compression, negative for tension, and to include the effects of tension due to shrinkage and creep—Chapters 8 and 11
P_b	P_b	= axial load capacity at simultaneous assumed ultimate strain of concrete and yielding of tension steel (balanced conditions)—Chapter 9 and Appendix A
P_c	P_c	= critical load. See Section 10.11.5—Chapter 10
P_e	P_u'	= maximum design axial load acting on a column or wall during an earthquake—Appendix A
P_s	T_o	= steel force at jacking end—Chapter 18
P_u	P_u	= axial design load in compression member—Chapters 8, 9, 10, and 14
P_x	T_x	= steel force at any point x—Chapter 18
r	r	= radius of gyration of the cross section of a compression member—Chapter 10
s	s	= tie spacing, in.—Chapter 7
s	s	= shear or torsion reinforcement spacing in a direction parallel to the longitudinal reinforcement—Chapter 11
s	s	= spacing of stirrups, in.—Chapter 12
s	s	= shear reinforcement spacing in the direction of the longitudinal reinforcement—Appendix A
sh	--	= center-to-center spacing of hoops—Appendix A
s_w	s_1	= spacing of deformed wires, in.—Chapters 7 and 12
s_1	s_v	= spacing of vertical reinforcement in a wall—Chapter 11
s_2	s_h	= shear or torsion reinforcement spacing in a direction perpendicular to the longitudinal reinforcement—or spacing or horizontal reinforcement in a wall—Chapter 11
T	T	= cumulative effect of temperature, creep, shrinkage, and differential settlement—Chapter 9
T_u	M_{tu}	= design torsional moment—Chapter 11
U	U	= required strength to resist design loads or their related internal moments and forces—Chapter 9
v_c	v_c	= nominal permissible shear stress carried by concrete—Chapters 8 and 11
v_{ci}	v_{ci}	= shear stress at diagonal cracking due to all design loads, when

	Formerly	
		such cracking is the result of combined shear and moment—Chapter 11
v_{cw}	v_{cw}	= shear stress at diagonal cracking due to all design loads, when such cracking is the result of excessive principal tensile stresses in the web—Chapter 11
v_{dh}	v_h	= design horizontal shear stress at any cross section, psi—Chapter 17
v_h	v_t	= permissible horizontal shear stress, psi—Chapter 17
v_{tc}	τ_c	= nominal permissible torsion stress carried by concrete—Chapter 11
v_{tu}	τ_u	= nominal total design torsion stress—Chapter 11
v_u	v_u	= nomial total design shear stress—Chapter 11
V_d	V_d	= shear force at section due to dead load—Chapter 11
V_1	V_1	= shear force at section occurring simultaneously with M_{max}—Chapter 11
V_p	V_p	= vertical component of the effective prestress force at the section considered—Chapter 11
V_u	V_u	= total applied design shear force at section—Chapters 8, 11, 12, and 17
w	w	= design load per unit length of beam or per unit area of slab—Chapter 8
w	w	= weight of concrete, lb per cu ft—Chapters 8 and 9
w	w	= design load per unit area—Chapter 13
w_d	w_D	= design dead load per unit area—Chapter 13
w_l	w_L	= design live load per unit area—Chapter 13
W	W	= wind load, or its related internal moment and forces—Chapter 9
x	x	= shorter overall dimension of a rectangular part of a cross section—Chapters 11 and 13
x_1	x_1	= shorter center-to-center dimension of a closed rectangular stirrup—Chapter 11
y	y	= longer overall dimension of a rectangular part of a cross section—Chapters 11 and 13
y_t	y_t	= distance from the centroidal axis of gross section, neglecting the reinforcement, to the extreme fiber in tension—Chapters 9 and 11
y_1	y_1	= longer center-to-center dimension of a closed rectangular stirrup—Chapter 11
z	Z	= a quantity limiting distribution of flexural reinforcement. See Section 10.6—Chapter 10
α (alpha)	H	= ratio of flexural stiffness of beam section to the flexural stiffness of a width of slab bounded laterally by the center line of the adjacent panel, if any, on each side of the beam $= \frac{E_{cb} I_b}{E_{cs} I_s}$—Chapters 9 and 13

	Formerly	
α	α	= angle between inclined web bars and longitudinal axis of member—Chapter 11
α	α	= total angular change of prestressing steel profile in radians from jacking end to any point x—Chapter 18
α_c	K'	= ratio of flexural stiffness of the columns above and below the slab to the combined flexural stiffness of the slabs and beams at a joint taken in the direction moments are being determined $= \dfrac{\Sigma K_c}{\Sigma(K_s + K_b)}$—Chapter 13
α_{ec}	K_e'	= ratio of flexural stiffness of the equivalent column to the combined flexural stiffness of the slabs and beams at a joint taken in the direction moments are being determined $= \dfrac{K_{ec}}{\Sigma(K_s + K_b)}$—Chapter 13
α_m	H_{av}	= average value of α for all beams on the edges of a panel—Chapter 9
α_{min}	K_r'	= minimum α_c to satisfy Section 13.3.6.1(a)—Chapter 13
α_t	Ω	= a coefficient as a function of y_1/x_1. See Section 11.8.2—Chapter 11
α_v	K	= ratio of stiffness of shearhead arm to surrounding composite slab section. See Section 11.11.2—Chapter 11
α_1	H_1	= α in the direction of l_1—Chapter 13
α_2	H_2	= α in the direction of l_2—Chapter 13
β (beta)	S	= ratio of clear spans in long to short direction of two-way construction—Chapter 9
β	S	= ratio of long side to short side of a footing—Chapter 15
β_a	α	= ratio of dead load per unit area to live load per unit area (in each case without load factors)—Chapter 13
β_b	r_b	= ratio of area of bars cut off to total area of bars at the section—Chapter 12
β_d	R_m	= the ratio of maximum design dead load moment to maximum design total load moment, always positive—Chapter 10
β_s	R_a	= ratio of length of continuous edges to total perimeter of a slab panel—Chapter 9
β_t	R	= ratio of torsional stiffness of edge beam section to the flexural stiffness of a width of slab equal to the span length of the beam, center-to-center of supports $= \dfrac{E_{cb}C}{2E_{cs}I_s}$—Chapter 13
β_1	k_1	= a factor defined in Section 10.2.7—Chapters 8 and 10
δ (delta)	--	= moment magnification factor for columns. See Section 10.11.5—Chapter 10
δ_s	--	= factor defined by Eq. (13-4). See Section 13.3.6.1—Chapter 13

	Formerly	
μ (mu)	μ	= coefficient of friction. See Section 11.15–Chapter 11
μ	μ	= curvature friction coefficient–Chapter 18
ξ (xi)	K	= constant for standard hook–Chapter 12
ρ (rho)	p	= ratio of nonprestressed tension reinforcement = A_s/bd–Chapters 8, 10, 11, 18, and Appendix A
ρ'	ρ'	= A_s'/bd–Chapters 8 and 18
ρ_b	p_b	= reinforcement ratio producing balanced conditions. See Section 10.3.3.–Chapters 8 and 10
ρ_h	p_h	= the ratio of horizontal shear reinforcement area to the gross concrete area of a vertical section–Chapter 11
ρ_{min}	p_{min}	= minimum reinforcement ratio–Chapter 10
ρ_n	p_n	= the ratio of vertical shear reinforcement area to the gross concrete area of a horizontal section–Chapter 11
ρ_p	p^*	= ratio of prestressed reinforcement = A_{ps}/bd–Chapter 18
ρ_s	p_s	= ratio of volume of spiral reinforcement to total volume of core (out-to-out of spirals) of a spirally reinforced concrete or composite column–Chapter 10
ρ_s	p_s	= ratio of volume of spiral reinforcement to total volume of core (out-to-out of spirals)–Appendix A
ρ_v	p_v	= $(A_s + A_h)/bd$–Chapter 11
ρ_w	p_w	= $A_s/b_w d$–Chapter 11
ϕ (phi)	ϕ	= capacity reduction factor. See Section 9.2–Chapters 8, 9, 10, 11, 14, 15, 17, 18, and 19
ω (omega)	q	= $\rho f_y/f_c'$–Chapter 18
ω'	q'	= $\rho' f_y/f_c'$–Chapter 18
ω_p	q^*	= $\rho_p f_{ps}/f_c'$–Chapter 18
ω_w, ω_{pw}, ω_w'	q_w, q_w^*, q_w'	= reinforcement indices for flanged sections computed as for ω, ω_p, ω' except that b shall be the web width, and the steel area shall be that required to develop the compressive strength of the web only–Chapter 18

Metric Equivalents

Metric equivalents, using the units of the technical system (mks, meter-kilogram-second) are given below.

To convert to Système International (SI) units, use the following conversion factors:

1 kilogram (kg) = 9.806 newtons (N)
1 kg/sq cm = 0.9806 bars

A newton ($kg/m/sec^{-2}$) is a force which, when applied to a body having a mass of 1 kilogram, gives the body an acceleration of 1 $meter/sec^{-2}$. In the SI system the term "mass" (unit: kg) or "unit" is used to indicate the amount of matter in the object. The term "weight" (unit: newton) should be used only when the gravitational forces are specified. These terms contradict the present U.S. usage.

The relation between the mks and SI systems could be visualized by converting the acceleration due to gravity (which varies due to geographical location) to mks units.

Assume: $g = 386.1\ in./sec^{-2}$
then: $g = 32.2\ ft/sec^{-2} = 9.806\ m/sec^{-2}$

MULTIPLE AND SUBMULTIPLE UNITS IN THE METRIC SYSTEM

Multiplication factor	*Prefix*	*Symbol*
$1\,000 = 10^3$	kilo	k
$100 = 10^2$	hecto	h
$10 = 10^1$	deca	da
$0.1 = 10^{-1}$	deci	d
$0.01 = 10^{-2}$	centi	c
$0.001 = 10^{-3}$	milli	m
$0.000\,001 = 10^{-6}$	micro	μ
10^{-9}	nano	n

METRIC UNITS

micron (μ)	*millimeter* (*mm*)	*centimeter* (*cm*)	*decimeter* (*dm*)	*meter* (*m*)	*kilometer* (*km*)
1000	1	0.1	0.01	0.001	–
–	10	1	0.1	–	
10^6	1000	100	10	1	0.001
–	–	–	–	1000	1

gram	*kilogram*	*ton*
1	0.001	–
1000	1	0.001
–	1000	1

	liter	dm^3	m^3
	1	1	0.001
	1000	1000	1
weight of water	1 kg	1 kg	1 ton

ENGLISH–MKS (SI) CONVERSION FACTORS

to convert	*to*	*multiply by*
LENGTH		
inch	millimeter (mm)	25.40
inch	centimeter (cm)	2.54
inch	meter (m)	0.0254
feet	meter (m)	0.3048
mile	kilometer (km)	1.61
yard	meter (m)	0.91
AREA		
sq inch	cm^2	6.45
sq feet	m^2	0.093
acre	m^2	4047.0
sq mile	km^2	2.59
VOLUME		
cubic inch	cm^3	16.4
cubic feet	m^3	0.028
cubic yard	m^3	0.765
U.S. gallon	liter (dm^3)	3.8
WEIGHT ("mass" in SI)		
pound (lb)	kilogram (kg)	0.453
U.S. ton	ton	0.9072
pound/feet	kilogram/meter	1.49
FORCE ("weight" in SI)		
pound force (lbf)	newton (N)	4.45
kilogram force (kgf)	newton (N)	9.81
STRESS or PRESSURE		
MKS units:		
psi	kg/cm^2	0.07

to convert	*to*	*multiply by*
U.S. ton/sq ft	ton/m^2	9.76
psf	kg/m^2	4.882
SI units:		
psi	kilonewton/m^2 (kN/m^2)	6.9
psf	newton/m^2 (N/m^2)	47.9
newton/m^2 (N/m^2)	pascal (Pa)	1.0
UNIT WEIGHT ("unit" in SI)		
pound/cu ft (lb/cu ft)	kg/m^3	16.02
MOMENT OF INERTIA ("moment of section" in SI)		
in.^{4}	cm^4	41.62
MOMENT		
foot-pound	m-kg	0.138

METRIC EQUIVALENTS OF DIMENSIONAL UNITS

English	*Metric*	
Length		
1/4 inch	6.35	mm
3/8 inch	9.53	mm
1/2 inch	1.27	cm
5/8 inch	1.59	cm
3/4 inch	1.91	cm
1 inch	2.54	cm
6 inch	15.24	cm
1 ft	30.50	cm
3 ft	0.915	m
10 ft	3.05	m

English	*Metric*
Stress (pressure)	
1 psi	0.07 kg/cm^2
100	77.03
3000	211.0
4000	281.0
5000	352.0
20000	1406.0
60000	4219.0
29000000	2039000.0

English	*Metric*	
Weight ("mass" in SI)		
62.4 lbs/cu ft	1	kg/dm^3
62.4	1	ton/m^3
145.0	2.323	ton/m^3

Standard Reinforcing Bars

Bar size	Diameter mm	Area cm^2
# 3	9.52	0.71
# 4	12.70	1.29
# 5	15.88	2.00
# 6	19.05	2.84
# 7	22.22	3.87
# 8	25.40	5.10
# 9	28.65	6.45
#10	32.26	8.19
#11	35.81	10.06
#14	43.00	14.52
#18	57.33	25.81

Standard cylinder (concrete testing)

6 × 12 inch — 15 × 30 cm

Temperature

degree F	*degree C*
0	-18
32	0
40	4
95	35
150	65
212	100

Comparisons of ACI 318-71 and ACI 318-63

Outline of Major Changes

1. *Strength Design* (formerly Ultimate Strength Design, USD) is now exclusively used (8.1). Working Stress Design (WSD) is treated as an alternate, special case (8.10).
2. Quality control—more restricted proportions and acceptance requirements than in previous Code (4.2, 4.3).
3. Accepting 80 ksi reinforcing bars, and detailed specific strain requirements. Added specifications for welded wire fabric and tubing (3.5, 9.4).
4. Deflection control—more restricted than before, and provides specific method for deflection calculations (9.5).
5. Small decrease of *load factor "U"*; added recommendations for earth-pressure, liquids and impact (9.3).
6. Detailed guidelines for *EI and stiffness factors* for beams and columns (8.5, 9.5).
7. *Crack-control* provisions introduced the first time (10.6).
8. *Deep beams*—first time covered in the Code (10.7, 11.9).
9. *Column limitations*—minimum dimensions eliminated, but a 1 inch eccentricity is required (10.3). Minimum bar size was eliminated (10.9). Spiral or tie diameter to be a minimal 3/8 in.
10. Slenderness effect for columns by *moment magnification* method similar to structural steel design. Magnification factor to be computed for *all* the columns per floor (10.11).
11. *Composite column* concept redefined and design method outlined (10.15).
12. *Shear and torsion* section has several innovations: *Full length* web reinforcement for beams (11.1). The *shear friction* concept introduced (11.15).

Brackets and *corbels* specially considered (11.14). *Deep beams* are first time included (11.9). *Shear head* reinforcement (structural steel) is now permitted. Provisions for design and reinforcement for *torsion* (11.7, 11.8). Moment transfer to columns (11.13). Provisions for *shear walls* (11.16).

13. *Development* of reinforcement. Flexural bond computations are omitted. *Anchorage length* concept is emphasized. *Tension splices* are classified and detailed requirement added (12).
14. *Slab systems* comprise an entirely new method. *Flat slabs*, *flat plates*, and *two-way slabs* are all on the same basis. The *Direct Design Method* is the substitute to the previous *Empirical Method*; the *Equivalent Frame Method* is comparable to the previous *Elastic Analysis*; the new two-way slab provisions, however, are entirely new (13).
15. *Walls*, with the load within the kern, may be designed by a new empirical method (14).
16. *Footings* to be designed for *factored loads and factored soil reactions*. *Dowel* requirements are relaxed (15).
17. Prestressed concrete: higher allowable tensile stresses (18).
18. *Shells and folded plates*—new chapter (19).
19. Seismic design—new chapter (Appendix A).

Contents

Reinforced Concrete Design for Buildings

The "Carlyle" Apartments, Chicago, Illinois. Forty stories, flat plate, shear wall. Architects: Hirshfeld, Pawlan, Reinheimer. Structural Engineers: Paul Rogers & Associates.

1

Properties of Reinforced Concrete

Concrete is a mixture of portland cement, coarse and fine aggregate, and water. After proper curing and hardening, the concrete becomes artificial stone. As such it has excellent qualities for resisting compressive forces, but weak in resisting tensile forces. In order to compensate for this deficiency, reinforcement is added which will assist in resisting flexural and diagonal tension forces. Reinforcement is also beneficial against temperature stresses; it increases the capacity of compression members, assists in producing ductile members, and reduces the flow-effect of concrete, called creep.

Since detailed specifications for materials, concrete quality, mixing and placing of concrete, formwork, embedded pipes, construction joints, details of reinforcement, etc., are extensively covered in the ACI 318-71 Code, these are not repeated in this book.

1150 Lake Shore Dr., Chicago, Illinois. Twenty-five stories, flat plate, shear wall. Architects: Hausner & Macsai, Chicago. Structural Engineers: Paul Rogers & Associates.

2

Analysis and Design

Required loadings are, as a rule, prescribed in applicable local codes and ordinances. The supporting structure must resist any and all of the prescribed loads. These should be increased, if warranted, by the structural engineer, architect, or building official.

Lateral loads (wind, earthquake, blast, etc.,) should be assessed according to the latest information, which may be more conservative than existing codes. For structures which are to resist seismic forces, and buildings of great heights and/or importance, dynamic response analyses are recommended.

Analysis of structural members and frames is based, presently, on the elastic theory. Certain redistribution of negative support moments are permitted; otherwise the bending moments and shears may be obtained from accepted structural analyses.

For gravity loads on continuous beams and frames which satisfy certain prescribed restrictions, ACI Code 318-71 permits the use of some approximate coefficients.

Limitations for use of coefficients:

(1) Number of spans: $\geqslant 2$

(2) $l_{n_2} \leqslant 1.2\, l_{n_1}, l_{n_3}$, etc.

(3) $w_D \leqslant 3w_L$.

Positive moment:

End spans, if discontinuous end is unrestrained	$wl_n^2/11$
End spans, if discontinuous end is integral with the support	$wl_n^2/14$
Interior spans	$wl_n^2/16$

Negative moment at exterior face of 1st interior support, Two spans	$wl_n^2/9$
More than two supports	$wl_n^2/10$
Negative moment at other faces of interior supports	$wl_n^2/11$
Negative moment at face of all supports for (a) slabs with spans not exceeding 10 ft, and (b) beams and girders where ratio of sum of column stiffnesses to beam stiffness exceeds 8 at each end of the span	$wl_n^2/12$

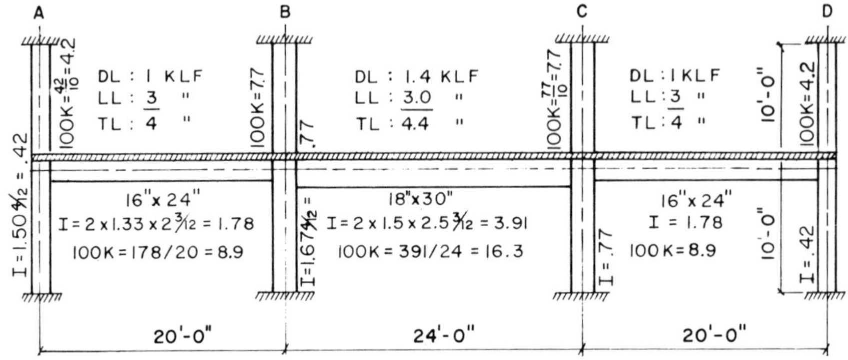

Fig. 2-1 Example of continuous moment distribution.

Fixed End Moments

DL:	$\frac{20^2}{12} \times 1 = 33'^k$	$\frac{24^2}{12} \times 1.4 = 67'^k$	$\frac{20^2}{12} \times 1 = 33'^k$
LL:	$\frac{20^2}{12} \times 3 = 100$	$\frac{24^2}{12} \times 3.0 = 144$	$\frac{20^2}{12} \times 3 = 100$
TL:	$\frac{20^2}{12} \times 4 = 133'^k$	$\frac{24^2}{12} \times 4.4 = 211'^k$	$\frac{20^2}{12} \times 4 = 133'^k$

Procedure of Continuous Moment Distribution:

Step 1) Start with DL Fixed End Moments on spans A-B and C-D, and Full Load (DL + LL) Fixed End Moment on span B-C. This loading produces the maximum Positive Moment on span B-C. Complete moment distribution.

Step 2) *Add* LL Fixed End Moments on span A-B only, and continue moment distribution. This loading produces the maximum Negative Moment at B.

Step 3) *Add* LL Fixed End Moments on span C-D and *deduct* the LL Fixed End Moments on span B-C; continue moment distribution. This loading produces the maximum Positive Moments in spans A-B and C-D.

Step 4) Draw envelope of maximum moments.

Negative moment at interior faces of exterior supports for members built integrally with their supports:

- Where the support is a spandrel beam or girder — $wl_n^2/24$
- Where the support is a column — $wl_n^2/16$
- Shear in end members at face of 1st interior support — $1.15wl_n/2$
- Shear at face of all other supports — $wl_n/2$

The above coefficients, although not entirely correct, have withstood the test of time.

For frames that do not satisfy the above restrictions, more detailed analyses are needed. For gravity loads only, each floor may be considered separately with the columns above and below fixed at their far ends. The following method was first presented by the author in the October 1949 issue of the *ACI Journal*.

A			B				C				D			JOINT
↓	↑	A-B	B-I	↓	↑	B-C	C-B	↓	↑	C-D	D-C	↓	↑	MEMBER
4.2	4.2	8.9	8.9	7.7	7.7	16.3	16.3	7.7	7.7	8.9	8.9	4.2	4.2	STIFFNESS, K
24	24	52	22	19	19	40	40	19	19	22	52	24	24	K/ΣK
		+ 33	− 33			+211	−211			+ 33	− 33			F.E.M.
− 8	− 8	− 17	− 9							+ 9	+ 17	+ 8	+ 8	
		− 19	− 37	−32	−32	− 68	− 34							
+ 5	+ 4	+ 10	+ 5			+ 40	+ 81	+39	+39	+ 44	+ 22			
		− 5	− 9	− 9	− 9	− 18	− 9			− 6	− 11	− 5	− 5	
+ 1	+ 2	+ 2	+ 1			+ 3	+ 6	+ 3	+ 3	+ 3	+ 2			
			− 1	− 1	− 1	− 1					− 1	− 1		
− 2	− 2	+ 4	− 83	−42	−42	+167	−167	+42	+42	+ 83	− 4	+ 2	+ 2	
		+100	−100											
−24	−24	+ 52	− 26											
		+ 14	+ 28	+24	+24	+ 50	+ 25							
− 3	− 3	− 8	− 4			− 5	− 10	− 5	− 5	− 5	− 3			
			+ 2	+ 1	+ 1	+ 5	+ 2				+ 1	+ 1	+ 1	
							− 1			− 1				
−29	−29	+ 58	−183	−17	−17	+217	−151	+37	+37	+ 77	− 6	+ 3	+ 3	
						−144	+144			+100	−100			
										+ 26	+ 52	+24	+24	
						− 54	−109	−51	−51	− 59	− 30			
		+ 21	+ 43	+38	+38	+ 79	+ 40			+ 8	+ 16	+ 7	+ 7	
− 5	− 5	− 11	− 6			− 10	− 20	− 9	− 9	− 10	− 6			
		+ 3	+ 5	+ 2	+ 2	+ 7	+ 4			+ 1	+ 4	+ 1	+ 1	
− 1	− 1	− 1					− 3			− 2				
−35	−35	+ 70	−141	+23	+23	+95	−95	−23	−23	+141	− 70	+35	+35	

Fig. 2-2 Moment distribution.

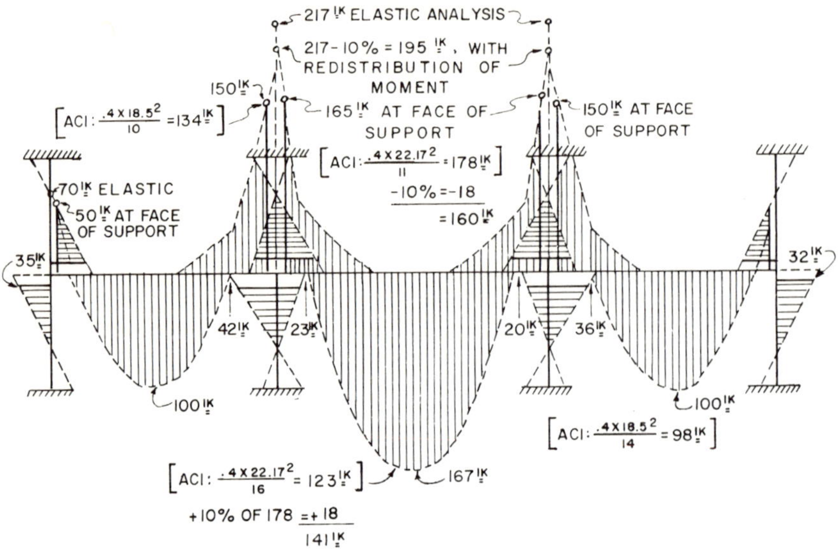

Fig. 2-3 Final moment diagram.

Comments: The author's simplified continuous moment distribution method is easy to apply and it furnishes accurate and speedy results. A comparison with ACI coefficients indicates good rapport except for Positive Moments in spans where 10% redistribution of Negative Moments were employed. The ACI formula $wl_n^2/16$ gives an unconservative result.

In the above example a 10 percent redistribution of the negative support moments was applied. Larger redistribution, $20\left(1 - \dfrac{\rho - \rho'}{\rho_b}\right)\%$, is permitted in the ACI 318-71 Code. For this application, however, actual reinforcement ratios are needed. Consequently, an accurate redistribution may be employed as a refinement after preliminary design is completed.

While the above example is at the utmost limit where the Code coefficients may be used, analysis indicates that for the interior positive moment the Code coefficient is unconservative.

With the advent of high-speed electronic computers and the availability of programs, theoretically accurate analyses can be obtained at reasonable cost. For multistory buildings, subject to wind and/or seismic forces, accurate analyses are strongly recommended.

TEE-BEAM CONSTRUCTION

Since most reinforced concrete construction is monolithic, slabs are cast as integral parts of the beams.

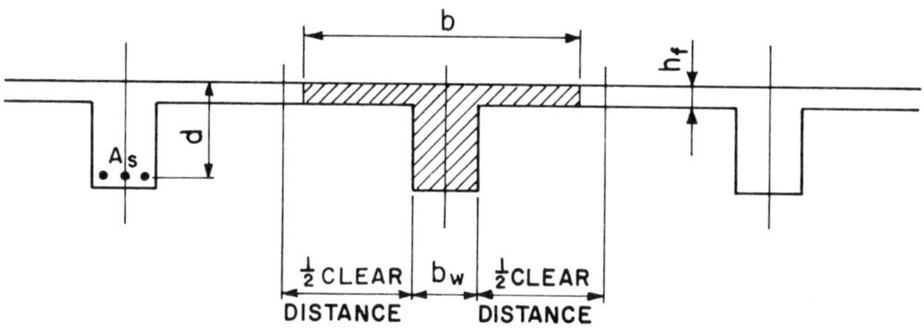

Fig. 2-4 Limits of a tee-beam.

As the beam deflects under the effect of loads, it pulls the slab downward. Thus a tee section is created which provides additional resistance to compression forces. (Note that no such assistance is available at supports where compression is at the bottom of beams.) The ACI 318-71 Code, similarly to previous Codes, prescribes the following limitations:

$$\left.\begin{aligned} b_{max} &= b_w + 16h_f \leqslant \text{center to center dimension of slabs} \\ &= \tfrac{1}{4}\, l \ (l = \text{span length of beam}) \end{aligned}\right\} \text{Flange of thickness } h_f \text{ on both sides}$$

$$\left.\begin{aligned} b_{max} &= b_w + 6h_f \leqslant \text{one half of clear span of slab} \\ &= \tfrac{1}{12}\, l \end{aligned}\right\} \text{Flange on one side only}$$

$$b_{max} = 4b_w, \text{ and } h_f \leqslant b_w/2 \qquad \text{For isolated beams only}$$

JOIST FLOOR CONSTRUCTION

While joist construction certainly features tee-beam design, certain additional restrictions are to be followed, as shown in Fig. 2-5.

STRENGTH AND SERVICEABILITY

Unlike the ACI Code 318-63, the ACI 318-71 does not include extensive formulas for the design of members. Thus the ACI 318-63 Code still forms the basis for calculations. The load factors have been reduced as follows:*

I. $U = 1.4\,D + 1.7\,L$ (ACI-9-1)

II. $U = 0.75(1.4\,D + 1.7\,L + 1.7\,W)$ (ACI-9-2)

III. $U = 0.75(1.4\,D + 1.7\,L + 1.87\,E)$ (ACI-9-2)

IV. $U = 0.90\,D - 1.3\,W$ } Reversals due to uplift (ACI-9-3)

V. $U = 0.90\,D - 1.43\,E$ }

*Local authorities may require larger load factors than prescribed by the ACI 318-71 Code. Thus, designers are cautioned to use the legally specified values pertaining to their areas. For seismic areas, the author recommends higher load factors and urges the designers to consult regional ordinances and specifications.

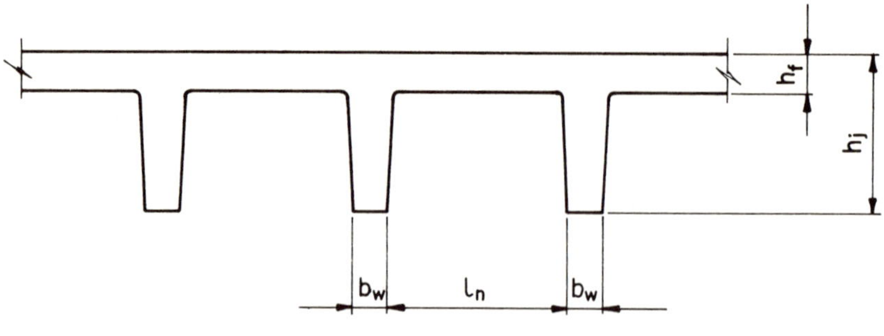

$l_n \leq$ 30 in.

$b_w \geq$ 4 in.

$h_j \leq 3.5\, b_w$

$h_f \geq$ { 2 in. (w/FILLERBLOCKS: 1½) ; $l_n/12$; O.D. CONDUITS OR PIPES + 1" }

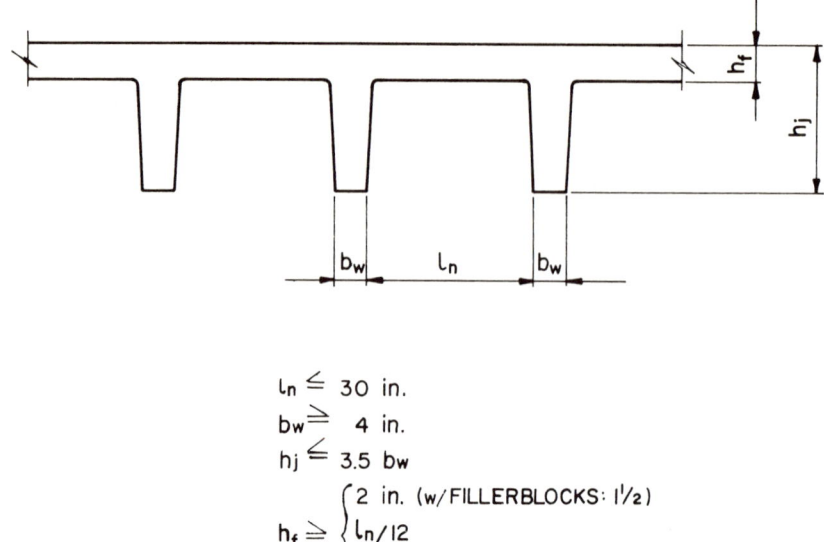

Fig. 2-5 Concrete joist floor construction.

If lateral earth pressure H is included in the design,

VI. $U = 1.4\,D + 1.7\,L + 1.7\,H$

but where D or L reduce the effect of H,

VII. $U = 0.90\,D + 0.00\,L - 1.7\,H$

If lateral liquid pressure F is included in the design,

VIII. $U = 1.4\,D + 1.7\,L + 1.4\,F$,

but where D or L reduce the effect of F,

IX. $U = 0.90\,D + 0.00\,L - 1.4\,F$

X. $U = 0.75(1.4\,D + 1.7\,L)$ where D includes the effect of creep, settlement, shrinkage, temperature

The above values of U will further increase indirectly, using the ϕ *Capacity Reduction Factor* in the computation of the strength capacity of the resisting members. The following values for ϕ are to be used:

Bending in reinforced concrete, and for axial tension	0.90
Shear and torsion	0.85
Axial compression or axial compression combined with bending:	
(a) Reinforced members with spiral reinforcement	0.75
(b) Other reinforced members	0.70
Bearing on concrete	0.70
Bending in plain concrete	0.65

IMPORTANT NOTE: All the tables of this text already include the effect of ϕ; thus ϕ should not be considered while using the tables.

Strength Design Method (Formerly Ultimate Strength Design)

The ACI 318-71 Code is based on the *Strength Design Method.* Working Stress Design (WSD) is permitted under Alternate Design Method, but use of it is not encouraged. This book presents the Strength Design Method only. Working Stress Design is well known to practicing engineers and several excellent texts are available; thus repetition of WSD is not warranted.

The elastic theory assumes that stresses and strains are proportional. However, it was recognized, for a long time that, as stresses in reinforced concrete become of higher intensity, these do not remain on a straight line but assume a curved shape. The *Strength Design Method* (SDM) is based on the curvilinear shape of the compression stress block, simplified by being represented by an equivalent rectangular stress block. Deflections of flexural members consist of two parts: the deflection that occurs immediately on application of load, and an additional long-term deflection.

The immediate deflection can be computed by the usual elastic methods. The modulus of elasticity, E, is to be taken as:

$$E_c = w^{1.5}\, 33\sqrt{f'_c} \text{ or } 57000\sqrt{f'_c} \text{ (psi)}$$

for normal weight concrete. The effective moment of inertia, I_e, to be used is as follows:

$$I_e = (M_{cr}/M_a)^3 \times I_g + [1 - (M_{cr}/M_a)^3]\, I_{cr} \qquad \text{(ACI-9.4)}$$

where

$$M_{cr} = f_r I_g / y_t, \qquad \text{(ACI-9.5)}$$

and

$$f_r = 7.5\sqrt{f'_c}$$

but f_r should be multiplied by 0.75 for "all-lightweight" concrete, and by 0.85 for "sand-lightweight" concrete.

The *additional* long-time deflection may be computed by comprehensive analysis or by multiplying the immediate deflection caused by the sustained load with $[2 - 1.2\,(A'_s/A_s)] \geqslant 0.6$.

However, checking by the above deflection formulas is not required if the minimum thicknesses of Table 2-1 are employed.

Table 2.1. Maximum span/depth ratio for slabs and beams.

Member, length, l, (in inches)	*Simply supported*	*One end continuous*	*Both ends continuous*	*Cantilever*
Solid one-way slabs	1/20	1/24	1/28	1/10
Beams or ribbed one-way slabs	1/16	1/18.5	1/21	1/8

It is the author's practice to camber all flexural members for dead loads and permanent installation loads. It is essential, in case of cambering, that the *top* also be cambered. Horizontal screeding merely reduces the thickness at midpoint.

Minimum thicknesses of two-way slabs, flat slabs, etc., are to be established as follows:

$$h = \frac{l_n (800 + 0.005 f_y)}{36{,}000 + 5000\beta\,[\alpha_m - 0.5\,(1 - \beta_s)(1 + 1/\beta)]} \qquad \text{(ACI-9-6)}$$

but not less than:

$$h = \frac{l_n (800 + 0.005 f_y)}{36{,}000 + 5000\beta\,(1 + \beta_s)} \qquad \text{(ACI-9-7)}$$

The thickness need not be more than:

$$h = \frac{l_n (800 + 0.005 f_y)}{36{,}000} \qquad \text{(ACI-9-8)}$$

However, minimum thicknesses to be as follows:

h = for slabs without beams or drop panels: 5 in.
h = for slabs without beams but with drop panels as prescribed in the Code: 4 in.
h = for slabs with beams: 3½ in.

For slabs with drop panels of adequate size and thickness, the Code formulas of (9-6), (9-7), and (9-8) may be reduced by 10 percent.

Harbor House Apartments, Chicago, Illinois. Twenty-nine stories, depressed two-way slabs. Architects: Hausner & Macsai, Chicago. Structural Engineers: Paul Rogers & Associates.

3

Bending

A primitive structure consists of vertical load-bearing elements (walls, piers, posts) and horizontal flexing elements (slabs, beams, girders.) A more sophisticated modern structure still retains the above basic elements with the exception that most, if not all, of its elements are interacting. Such a behavior, in case of reinforced concrete construction, is brought about by cast-in-place methods, by appropriate connections, by post-tensioning, or by other means.

The result of monolithic construction is that most axially loaded members are also subject to bending forces, while many flexural members are subject to additional axial and torsional forces. The American Concrete Institute's latest Standard Building Code, 318-71, actually groups together flexure and axial loads.

In customary office practice, however, the design of flexural members is separated from axially loaded members (columns). The working drawings include, as a rule, separate beam and column schedules. This chapter follows this routine and is devoted to bending and shear only.

It is assumed that readers are acquainted with basic statics and strength of materials. Consequently, only such basic information will be repeated here as is believed to be of assistance. The following is a brief review of the elastic theory.

Basic Laws of Equilibrium:

1) The sum of horizontal forces, including reactions, must equal zero.
2) The sum of vertical forces, including reactions, must equal zero.
3) The sum of moments about any arbitrary point must equal zero.

HOOKE'S LAW

Within the *elastic* range of materials:

$$\frac{\Delta L}{L} = \frac{f}{E} = \frac{P}{AE}$$

where

L = original length
ΔL = deformation
P = applied load
A = cross-sectional area of member
f = P/A, unit stress due to load P
E = modulus of elasticity of the material [to be established experimentally; for concrete: $E_c = w^{1.5} \cdot 33\sqrt{f'}$ (psi), & for steel: $E_s = 29 \cdot 10^6$ (psi)].
$\Delta b = b - b'$
$\frac{\Delta b}{b} = \frac{f}{E} \cdot m$ (m = Poisson's Ratio)

If dA is an infinitely thin layer and f is the stress acting on it:

1) $\int fdA = 0; \quad \frac{f}{f_1} = \frac{c}{c_1}; \quad f = \frac{cf_1}{c_1}; \quad \frac{f_1}{c_1}\int cdA = 0; \quad \text{or} \int cdA = 0$

The expression $\int cdA$ is the first or *STATICAL MOMENT*.

2) $P \cdot x = M_x$ = force × distance = moment, $M_x = \int fdAc = \int \frac{cf_1}{c_1} cdA = \frac{f_1}{c_1}\int c^2 dA$

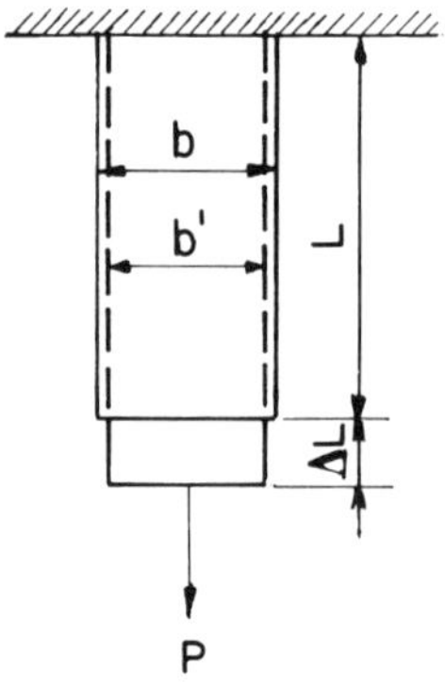

Fig. 3-1 Elastic deformation of an axially loaded member.

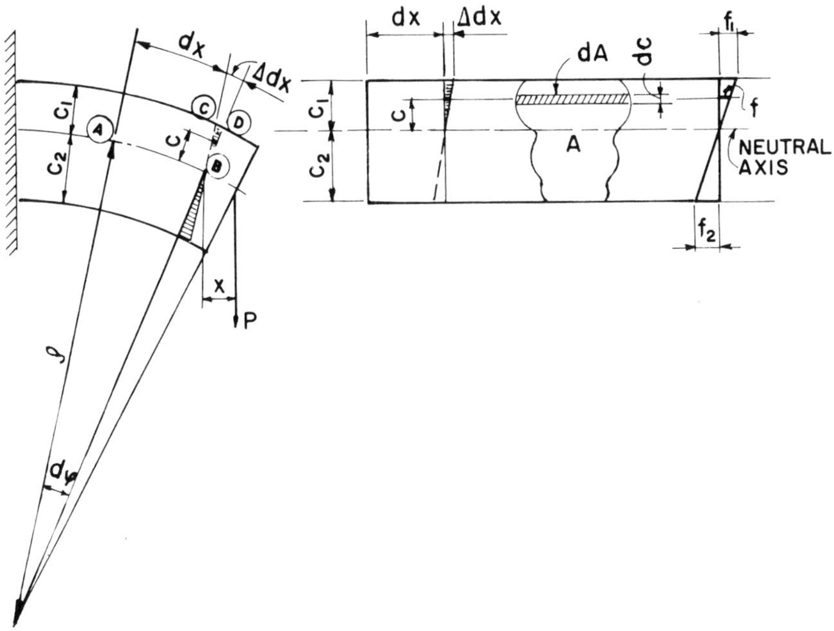

Fig. 3-2 Elastic deformation and stresses of a flexural member.

The expression $\int c^2 dA = I$ is called the *MOMENT OF INERTIA*. The expression $\frac{I}{c} = S$ is called the *SECTION MODULUS;* $f_1 = \frac{M_x c_1}{I} = \frac{M_x}{S_1}$.

According to differential calculus, the radius of curvature ρ (Fig. 3-2) is:

$$\rho = \frac{[1 + dy/dx)^2]^{3/2}}{d^2y/dx^2}.$$

For small angles $(dy/dx)^2$ is approaching zero and can be neglected. Thus:

$$\rho = \frac{1}{d^2y/dx^2};$$

or employing c instead of y,

$$\rho = \frac{1}{d^2c/dx^2}.$$

From the geometry of deflection:

$$\frac{\rho}{dx} = \frac{c_1}{\Delta dx}; \quad \rho = \frac{c_1 dx}{\Delta dx} = \frac{c_1 E}{f_1};$$

but

$$f_1 = \frac{M_x \cdot c_1}{I} \quad \text{and} \quad \rho = \frac{c_1 EI}{M_x c_1} = \frac{EI}{M_x} = \frac{1}{d^2c/dx^2};$$

finally:

$$\frac{d^2c}{dx^2} = \frac{M}{EI};$$

or reverting to y instead of c:

$$\frac{d^2y}{dx^2} = \frac{M}{EI}$$

Thus, the Basic Laws of Equilibrium (cont'd.):

4) The sum of loads on a flexural member is equal to the shear forces: $\Sigma P = V$;

5) The sum of shear on a flexural member is equal to the bending moments: $\Sigma V = M$;

6) The sum of moments on a flexural member, divided by EI, is equal to the angle change: $\Sigma M/EI = \theta$.

7) The sum of θ is equal to the deflection: $\Sigma\theta = \Delta$.

It is advisable to remember, particularly in the present time of high-speed electronic computers and sophisticated analyses, that the laws of equilibrium must be maintained and all effects on the structural system must be compatible.

The *Strength Design Method* assumes that ultimate flexural compression capacity of concrete is reached when the strain $\epsilon_c = 0.003$, and that balanced condition is present when the reinforcement, the steel cross-sectional area $A_s = \rho_b bd$, with $\epsilon_s = f_y/E_s$.

$$\frac{c}{d} = \frac{0.003}{0.003 + f_y/E_s} = \frac{0.003\,E_s}{0.003\,E_s + f_y} = \frac{87{,}000}{87000 + f_y} = \frac{87}{87 + f_y\ (\text{ksi})}$$

$$T = \rho_b bdf_y = C = b\beta_1 c\, 0.85\, f'_c = 0.85\, f'_c b\beta_1 d \frac{87}{87 + f_y}$$

solving:

$$\rho_b = \frac{\beta_1\, 0.85\, f'_c}{f_y} \cdot \frac{87}{87 + f_y}$$

In order to avoid brittle and sudden failure, the Code limits the maximum reinforcement to three-fourths of ρ_b. The value of

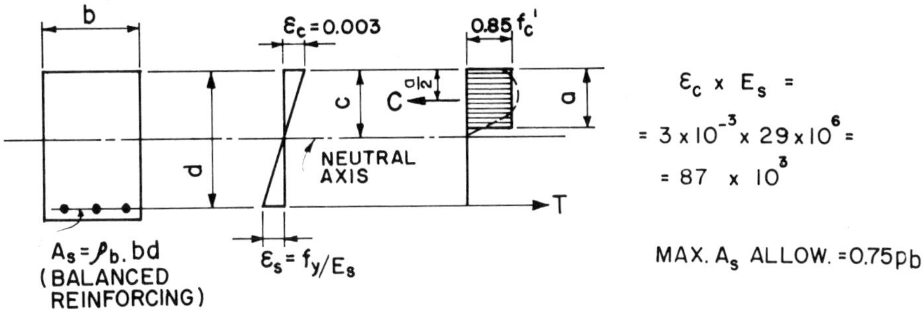

Fig. 3-3 Balanced condition.

$$\beta_1 = 0.85 - 0.05\,(f'_c - 4) \leqslant 0.85.$$

$$\rho_{\max} = 0.75\,\rho_b = \frac{0.75 \times 0.85\,\beta_1 f'_c}{f_y} \cdot \frac{87}{87 + f_y}$$

$$\rho_{\min} = 200/f_y; \quad \frac{c}{d}\max = \frac{3}{4} \cdot \frac{87}{87 + f_y}$$

The depth of the compression block may be computed from the formula:

$$a = A_s f_y / 0.85\, f'_c b, \text{ or } a/d = \rho\,\frac{f_y}{0.85\, f'_c}$$

Table 3-1. Maximum and minimum percentage of reinforcement for beams (investigate deflection whenever ρ exceeds the value of 0.018).

f'_c (psi) \ f_y (psi)	A_s	40,000	50,000	60,000	70,000	80,000	Important Note
3,000	ρ_{max}	0.02784	0.02065	0.01604	0.01287	0.01059	For f_y = 70,000 & 80,000 full scale tests are required for acceptable crack widths.
	ρ_{min}	0.0050	0.0040	0.0033	0.0029	0.0025	
4,000	ρ_{max}	0.03712	0.02753	0.02138	0.01716	0.01411	
	ρ_{min}	0.0050	0.0040	0.0033	0.0029	0.0025	
5,000	ρ_{max}	0.04367	0.03239	0.02515	0.02019	0.01661	
	ρ_{min}	0.0050	0.0040	0.0033	0.0029	0.0025	
6,000	ρ_{max}	0.04913	0.03644	0.0283	0.02271	0.01868	
	ρ_{min}	0.0050	0.0040	0.0033	0.0029	0.0025	

Table 3-2. Coefficient *a/d* for maximum depth of compression block.

f'_c (psi) \ f_y (psi)	40,000	50,000	60,000	70,000	80,000	Formula
3,000	0.4367	0.4049	0.3774	0.3533	0.3322	$a/d_{max} =$
4,000	0.4367	0.4049	0.3774	0.3533	0.3322	$\rho_{max} f_y / 0.85 f'_c$
5,000	0.4110	0.3811	0.3551	0.3325	0.3126	
6,000	0.3853	0.3573	0.3329	0.3117	0.2930	

The maximum capacity of a rectangular beam without compression reinforcement may be computed from the formula:

$$M_{u_{max}} \text{ (in. kips)} = \phi\, 0.85 f'_c bd^2 \left(1 - 0.5 \frac{a}{d} \max\right), \quad \text{or}$$

$$K_{u_{max}} = \frac{M_{u_{max}}}{bd^2} = 0.765 f'_c \frac{a}{d} \max \left(1 - 0.5 \frac{a}{d} \max\right) \quad (\phi = 0.90)$$

The reinforcement

$$A_s = \frac{M_{u_{actual}} \text{ (ft. kips)}}{a_u d}$$

where

$$a_u = \phi f_y \left(1 - 0.5 \frac{a}{d}\right) \Big/ 12{,}000$$

Table 3-3. Coefficient $K_{u_{max}} = M_{u_{max}}$ (in. kips)/bd^2.

f'_c (psi) \ f_y (psi)	40,000	50,000	60,000	70,000	80,000	Formula
3,000	0.783	0.741	0.703	0.668	0.636	$K_{u_{max}} = \dfrac{M_{u_{max}} \text{ (in. k.)}}{bd^2}$
4,000	1.044	0.988	0.937	0.890	0.847	$= 0.765 f'_c \times a/d \max$
5,000	1.249	1.180	1.117	1.060	1.009	$\times \left(1 - .5 \dfrac{a}{d} \max\right)$
6,000	1.428	1.347	1.274	1.208	1.148	

For the maximum allowable steel percentages, a_u can be tabulated as follows:

Table 3-4. Coefficients a_u for maximum allowable ρ.

f'_c (psi) \ f_y (psi)	40,000	50,000	60,000	70,000	80,000	Formula
3,000	2.345	2.991	3.651	4.322	5.003	$a_{u\,max} = \dfrac{\phi f_y \left(1 - .5\,\dfrac{a}{d}\,max\right)}{12{,}000}$
4,000	2.345	2.991	3.651	4.322	5.003	
5,000	2.384	3.035	3.701	4.377	5.062	
6,000	2.422	3.080	3.751	4.432	5.121	In general: $A_s = \dfrac{M_u}{a_u d}$

Example: Rectangular beam with tension reinforcement only.

Bending moment M_u from frame analysis, multiplied by the proper load factors: 200 ft kips; f'_c: 4 ksi; f_y: 60 ksi; $bd^2 = 10 \times 20^2 = 4000$; $K_{u_{max}}$: 0.936, (Table 3-3); $a_{u_{max}}$: 3.65, (Table 3-4);

$$M_{u_{max\ conc.}} = K_{u_{max}} \times bd^2/12 = 312 \text{ ft kips (O.K.)}$$

$$A_{s_{max}} = 0.0214 \times 10 \times 20 = 4.28 \text{ sq in. (Table 3-1)}$$

$$A_{s_{min}} = 0.0033 \times 10 \times 20 = 0.66 \text{ sq. in. (Table 3-1)}$$

The required amount of reinforcement cannot be established directly without design aids. A trial-and-error method is presented first. Two useful direct design methods will be presented afterward; one was developed by Frederick P. Wiesinger and the other by Alfred Zweig.

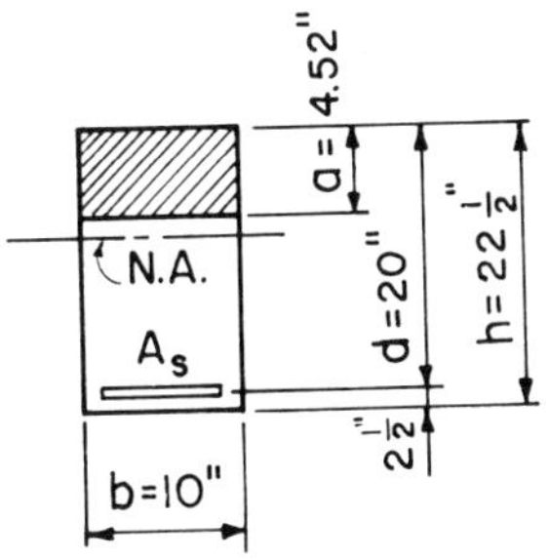

Fig. 3-4.

$$M_{u_{\text{max steel}}} = A_{s_{\text{max}}} \times a_{u_{\text{max}}} \times d \times \phi$$

$$= 4.28 \times 3.65 \times 20 \times 0.90 = 281 \text{ ft kips} > 200 \text{ ft kips}$$

First approximation for actual $A_s = (200/281) \times 4.28 = 3.05$ sq in.

$$a = A_s f_y / 0.85 f'_c b = 3.05 \times 60/0.85 \times 4 \times 10 = 3.05 \times 1.765 = 5.38 \text{ in.}$$

$$M_{u_{\text{steel (3.05)}}} = A_s f_y (d - a/2)\, \phi/12 = 3.05 \times 60\,(20 - 2.69)0.90/12$$

$$= 238 \text{ ft kips}$$

Second approximation for actual $A_s = (200/238)\ 3.05 = 2.56$ sq in.

$$a = 2.56 \times 1.765 = 4.52 \text{ in.}$$

$$M_{u_{\text{steel (2.56)}}} = 2.56 \times 60\,(20 - 2.26)\ 0.90/12 = 204 \text{ ft kips}$$

$$> 200 \text{ ft kips (O.K.)}$$

The above example is clear evidence that direct design, without aids, is time-consuming and unrealistic in a practicing engineer's office.

BEAMS WITH COMPRESSION REINFORCEMENT

It is seldom that compression reinforcement is really needed with the ACI 318-71 *Strength Design Method* because the concrete can absorb much higher forces than was possible before; also, excessively shallow beams are undesirable. The compression reinforcement may be present, however (e.g., bottom reinforcement passing through the interior support, seismic design, etc.). In such a case the compression reinforcement serves as an additional equivalent concrete compression block. Strain compatibility, of course, must be maintained, and ACI 318-71 safeguards must be complied with.

From Fig. 3-5:

$$\frac{c - d'}{c} = \frac{\epsilon'_s}{\epsilon_c} = \frac{f'_s/29{,}000}{0.003} = \frac{f'_s}{87}$$

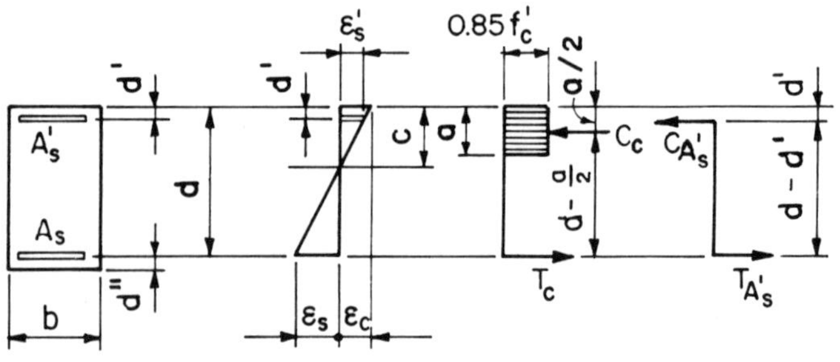

Fig. 3-5 Balanced condition compression reinforcement.

limiting $\rho_{max} = 0.75\ \rho_b$ and substituting $\frac{c}{d}\max = \frac{3}{4}\cdot\frac{87}{87+f_y}$:

$$f'_s \max = 87 - \frac{4}{3}(87+f_y)\frac{d'}{d} \leqslant f_y$$

it follows that:

$$f'_s \max = f_y \quad \text{if} \quad \frac{d'}{d} \leqslant \frac{3}{4}\cdot\frac{87-f_y}{87+f_y}$$

Table 3-5. Ratio of d'/d to develop f_y in compression reinforcing.

$f'_{s\,max}$ \ f_y	40,000	50,000	60,000	70,000	80,000
$\frac{d'}{d}$ max	0.278	0.202	0.138	0.081	0.031

The influence of the compression reinforcement can most conveniently be ascertained by computing the value of the internal couple due to the compression reinforcement, A'_s, and due to the corresponding additional tension reinforcement, $A_{s_{s'}}$.

Since the compression reinforcement displaces a concrete area, A'_s and $A_{s_{s'}}$ are not equal.

$$M'_{u\,\text{comp.}} = \frac{\phi A'_s (f_y - 0.85 f'_c)(d - d')}{12} \text{ (ft kips)}$$

$$M'_{u\,\text{tension}} = \frac{\phi A_{s_{s'}} f_y (d - d')}{12} \text{ (ft kips)}$$

Tables 3-6 and 3-7 present the values of $M'_{u\,\text{comp.}}$ and $M'_{u\,\text{tension}}$ due to 1 sq in. of reinforcement for slabs and beams of 12 in. of width, and for a variety of depths. It is to be noted that the restriction of $\rho \leqslant 0.75\ \rho_b$ does not apply to the additional tensile reinforcement $A_{s_{s'}}$ which balances the compression reinforcement A'_s.

TEE-BEAMS

As the beam deflects under the effect of loads, it pulls the slab downward. Thus a tee section is created which provides additional resistance to compression forces. (Note that no such assistance is available at the supports where compression is at the bottom of the beams.) The ACI 318-71 Code, similarly to previous Codes, prescribes certain restrictions.

Table 3-6

RECTANGULAR SECTION PROPERTIES FOR SLABS,
DESIGN METHOD, ACI CODE 318-71, f_y=60 ksi,
Resisting Moments(ft.kips), & Shears(kips).
The effect of capacity reduction Ø included.

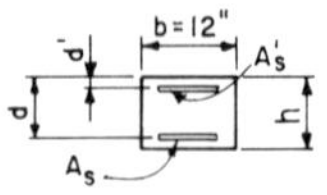

		f'_c = 3ksi				f'_c = 4 ksi				f'_c = 5 ksi			
d'	d	$\bar{M}_u$ conc.	$\bar{M}'_t$ tensn	$\bar{M}'_c$ compn	$\bar{V}_c$ conc.	$\bar{M}_u$ conc.	$\bar{M}'_t$ tensn	$\bar{M}'_c$ compn	$\bar{V}_c$ conc.	$\bar{M}_u$ conc.	$\bar{M}'_t$ tensn	$\bar{M}'_c$ compn	$\bar{V}_c$ conc.
1"	2.5"	4.4			2.8	5.9			3.2	7.0			3.6
	3.0"	6.3			3.4	8.4			3.9	10.1			4.3
	3.5"	8.6			3.9	11.5			4.5	13.7			5.0
	4.0"	11.2			4.5	15.0			5.2	17.9			5.8
	4.5"	14.2	11.4	10.7	5.0	19.0	11.4	10.5	5.8	22.6	11.4	10.3	6.5
	5.0"	17.6	14.3	13.6	5.6	23.4	14.3	13.3	6.5	27.9	14.3	13.1	7.2
	5.5"	21.3	18.9	17.9	6.1	28.3	18.9	17.6	7.1	33.8	18.9	17.3	7.9
1.25"	6.0"	25.3	16.5	15.6	6.7	33.7	16.5	15.3	7.7	40.2	16.5	15.0	8.7
	6.5"	29.7	19.4	18.4	7.3	39.6	19.4	18.1	8.4	47.2	19.4	17.8	9.4
	7.0"	34.4	22.4	21.2	7.8	45.9	22.4	20.9	9.0	54.7	22.4	20.6	10.1
	7.5"	39.6	25.4	24.2	8.4	52.7	25.4	23.8	9.7	62.8	25.4	23.4	10.8
	8.0"	45.0	28.6	27.3	8.9	60.0	28.6	26.8	10.3	71.5	28.6	26.4	11.5
	8.5"	50.8	31.6	30.3	9.5	67.7	31.6	29.8	11.0	80.7	31.6	29.3	12.3
	9.0"	56.9	34.8	33.2	10.1	75.9	34.8	32.8	11.6	90.5	34.8	32.3	13.0
1.5"	9.5"	63.6	33.6	32.1	10.6	84.5	33.6	31.5	12.3	100.8	33.6	31.1	13.7
	10.0"	70.2	36.7	35.0	11.2	93.7	36.7	34.5	12.9	111.7	36.7	33.9	14.4
	10.5"	77.5	39.8	38.1	11.7	103.3	39.8	37.5	13.5	123.2	39.8	37.0	15.1
	11.0"	85.0	42.9	41.1	12.3	113.3	42.9	40.5	14.2	135.2	42.9	39.9	15.9
	11.5"	92.9	46.0	44.0	12.8	123.9	46.0	43.4	14.8	147.7	46.0	42.8	16.6
	12.0"	101.2	49.2	47.2	13.4	134.9	49.2	46.6	15.5	160.7	49.2	45.9	17.3

	f'_c = 3ksi	f'_c = 4 ksi	f'_c = 5 ksi
$K_{u\ max}$	0.703	0.937	1.117
ρ max	0.016	0.0214	0.0252
a/d max	0.377	0.377	0.355

$\bar{M}_u$ conc.= Ultimate Moment capacity of 1 ft. wide concrete section, ft.kips.
$\bar{M}'_t$ tensn= Resisting Moment of 1 sq.in. of tension reinf. required to balance the force in compression reinforcement, ft.kips.
$\bar{M}'_c$compn = Resisting Moment of 1 sq.in. compression reinforcement, ft.kips.
$\bar{V}_c$ con. = Shear capacity of 1 ft. wide concrete section, kips.

If the depth a is within the flange, the design is as for a rectangular beam. If a is deeper than h, then the web compression below the flange has to be considered.

Design Examples

1) *Rectangular Beam, Simple Bending, No Compression Reinforcing.*
$M_{u_{req'd}}$: 200 ft kips; f'_c: 4 ksi; f_y: 60 ksi; $b = 10''$, $d = 20''$.

Table 3-7

RECTANGULAR SECTION PROPERTIES FOR BEAMS,
DESIGN METHOD, ACI CODE 318-71, f_y=60 ksi,
Resisting Moments (ft.kips) & Shears (kips).
The effect of capacity reduction Ø included.

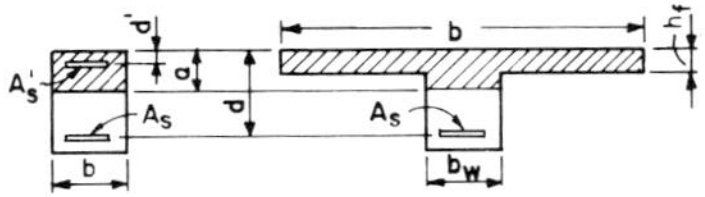

		f'_c = 3 ksi				f'_c = 4 ksi				f'_c = 5 ksi			
d'	d	$\bar{M}_u$	$\bar{M}'_t$	$\bar{M}'_c$	$\bar{V}_c$	$\bar{M}_u$	$\bar{M}'_t$	$\bar{M}'_c$	$\bar{V}_c$	$\bar{M}_u$	$\bar{M}'_t$	$\bar{M}'_c$	$\bar{V}_c$
		concrete	tensn	compn	conc.	conc.	tensn	compn	conc.	conc.	tensn	compn	conc
2.5"	10"	70.3	33.7	20.0	11.2	93.7	33.7	19.6	12.9	111.7	33.7	19.1	14.4
	10.5"	77.5	36.0	22.8	11.7	103.3	36.0	22.3	13.5	123.1	36.0	21.8	15.1
	11"	85.1	38.2	25.5	12.3	113.4	38.2	25.0	14.2	135.2	38.2	24.4	15.9
	11.5"	93.0	40.5	28.3	12.7	123.9	40.5	27.7	14.8	147.7	40.5	27.2	16.6
	12"	101.2	42.7	31.2	13.4	134.9	42.7	30.6	15.5	160.8	42.7	29.9	17.3
	12.5"	109.7	45.0	34.0	14.0	146.4	45.0	33.4	16.1	174.5	45.0	32.7	18.0
	13"	118.8	47.2	36.9	14.5	158.4	47.2	36.2	16.8	188.8	47.2	35.5	18.8
	13.5"	128.1	49.5	39.8	15.1	170.8	49.5	39.1	17.4	203.6	49.5	38.4	19.5
	14"	137.8	51.7	42.7	15.6	183.7	51.7	42.0	18.1	218.9	51.7	41.5	20.2
	14.5	147.8	54.0	45.7	16.2	197.0	54.0	44.9	18.7	234.8	54.0	44.2	20.9
	15"	158.2	56.2	48.7	16.8	210.8	56.2	47.8	19.4	251.3	56.2	47.0	21.6
	15.5"	168.9	58.5	51.5	17.3	225.1	58.5	50.7	20.0	268.3	58.5	50.0	22.6
	16"	178.0	60.7	55.3	17.9	239.9	60.7	54.4	20.6	286.0	60.7	53.6	23.1
	16.5"	191.4	63.0	57.6	18.4	255.1	63.0	56.7	21.3	304.1	63.0	55.8	23.8
	17"	203.2	65.2	60.7	19.0	270.8	65.2	59.7	21.9	322.8	65.2	58.8	24.5
	17.5"	215.3	67.5	63.7	19.6	286.9	67.5	62.7	22.6	342.1	67.5	61.7	25.2
	18"	227.8	69.7	66.7	20.1	303.6	69.7	65.6	23.2	361.9	69.7	64.7	26.0
	18.5"	240.6	72.0	68.9	20.7	320.7	72.0	67.9	23.9	382.3	72.0	66.9	26.7
	19"	253.8	74.2	71.1	21.2	338.2	74.2	70.0	24.5	403.2	74.2	69.0	27.4
	19.5"	267.3	76.5	73.2	21.8	356.3	76.5	72.2	25.2	424.7	76.5	71.1	28.1

Table 3-7 (continued)

		f'_c = 3 ksi				f'_c = 4 ksi				f'_c = 5 ksi			
d'	d	$\bar{M}_u$	$\bar{M}'_t$	$\bar{M}'_c$	$\bar{V}_c$	$\bar{M}_u$	$\bar{M}'_t$	$\bar{M}'_c$	$\bar{V}_c$	$\bar{M}_u$	$\bar{M}'_t$	$\bar{M}'_c$	$\bar{V}_c$
		concrete	tensn	compn	conc.	conc.	tensn	compn	conc.	conc.	tensn	compn	conc.
2.5"	20"	281.2	78.7	75.4	22.3	374.8	78.7	74.3	25.8	446.8	78.7	73.2	28.9
	21"	310.0	83.2	79.7	23.5	413.2	83.2	78.5	27.1	492.6	83.2	77.4	30.3
	22"	338.8	87.7	84.0	24.6	451.6	87.7	83.8	28.4	538.4	87.7	81.5	31.7
	23"	371.9	92.2	88.3	25.7	495.7	92.2	87.0	29.7	590.9	92.2	85.7	33.2
	24"	404.9	96.7	92.6	26.8	539.7	96.7	91.3	31.0	643.4	96.7	89.9	34.6
	25"	439.4	101.2	96.9	27.9	585.6	101.2	95.5	32.3	698.1	101.2	94.1	36.1
	26"	475.2	105.7	101.2	29.1	633.4	105.7	99.8	33.5	755.1	105.7	98.3	37.5
	27"	512.5	110.2	105.6	30.2	683.1	110.2	104.0	34.8	814.3	110.2	102.4	38.9
	28"	551.2	114.7	109.9	31.3	734.6	114.7	108.2	36.1	875.7	114.7	106.6	40.4
	29"	591.2	119.2	114.2	32.4	788.0	119.2	112.5	37.4	939.4	119.2	110.8	41.8
	30"	632.7	123.7	118.5	33.5	843.3	123.7	116.7	38.7	1005.3	123.7	115.0	43.3
	31"	675.6	128.2	122.8	34.6	900.5	128.2	121.0	40.0	1073.4	128.2	119.2	44.7
	32"	719.9	132.7	127.1	35.8	959.5	132.7	125.2	41.3	1143.8	132.7	123.3	46.2
	33"	765.6	137.2	131.4	36.9	1020.4	137.2	129.5	42.6	1216.4	137.2	127.5	47.6
	34"	812.7	141.7	135.7	38.0	1083.2	141.7	133.7	43.9	1291.3	141.7	131.7	49.0
	35"	861.2	146.2	140.0	39.1	1147.8	146.2	138.0	45.2	1368.3	146.2	135.9	50.5
	36"	911.1	150.7	144.3	40.2	1214.4	150.7	142.2	46.4	1447.6	150.7	140.1	51.9
	37"	962.4	155.2	148.7	41.3	1282.8	155.2	146.4	47.7	1529.2	155.2	144.3	53.4
	38"	1015.1	159.7	153.0	42.5	1353.0	159.7	150.7	49.0	1612.9	159.7	148.4	54.8
	39"	1069.3	164.2	157.3	43.6	1425.2	164.2	154.9	50.3	1699.0	164.2	152.6	56.3
	40"	1124.8	168.7	161.6	44.7	1499.2	168.7	159.2	51.6	1787.2	168.7	156.8	57.7
	42"	1240.1	177.7	170.2	46.9	1652.9	177.7	167.7	54.2	1970.4	177.7	165.2	60.6
	44"	1361.0	186.7	178.8	49.2	1814.0	186.7	176.2	56.8	2162.5	186.7	173.5	63.5
	46"	1487.5	195.7	187.4	51.4	1982.7	195.7	184.7	59.3	2363.6	195.7	181.9	66.4

Table 3-7 (continued)

d'	d	$f'_c = 3$ ksi $\bar{M}_u$ conc.	$\bar{M}'_t$ tensn	$\bar{M}'_c$ compn	$\bar{V}_c$ conc.	$f'_c = 4$ ksi $\bar{M}_u$ conc.	$\bar{M}'_t$ tensn	$\bar{M}'_c$ compn	$\bar{V}_c$ conc.	$f'_c = 5$ ksi $\bar{M}_u$ conc.	$\bar{M}'_t$ tensn	$\bar{M}'_c$ compn	$\bar{V}_c$ conc.
2.5"	48"	1619.7	204.7	196.0	53.6	2158.8	204.7	193.1	61.9	2573.4	204.7	190.2	69.2
	50"	1757.5	213.7	204.7	55.9	2342.5	213.7	201.6	64.5	2792.5	213.7	198.6	72.1
	52"	1900.9	222.7	213.3	58.1	2533.6	222.7	210.1	67.1	3020.4	222.7	207.0	75.0
	54"	2049.9	231.7	221.9	60.3	2732.3	231.7	218.6	69.7	3257.2	231.7	215.3	77.9
	56"	2204.6	240.7	230.5	62.6	2938.4	240.7	227.1	72.2	3502.9	240.7	223.7	80.8
	58"	2364.9	249.7	239.1	64.8	3152.1	249.7	235.6	74.8	3757.6	249.7	232.1	83.7
	60"	2530.8	258.7	247.8	67.0	3373.2	258.7	244.1	77.4	4021.2	258.7	240.4	86.6
	62"	2702.3	267.7	256.4	69.3	3601.8	267.7	252.6	80.0	4293.7	267.7	248.8	89.4
	64"	2879.5	276.7	265.0	71.5	3838.0	276.7	261.1	82.6	4575.2	276.7	257.1	92.3
	66"	3062.3	285.7	273.6	73.7	4081.6	285.7	269.6	85.1	4865.7	285.7	265.5	95.2
	68"	3250.7	294.7	282.2	76.0	4332.7	294.7	278.0	87.7	5165.0	294.7	273.9	99.2
	70"	3444.7	303.7	290.8	78.3	4591.3	303.7	286.6	90.3	5473.3	303.7	282.2	101.0
	72"	3644.3	312.7	299.5	80.5	4857.4	312.7	295.0	92.9	5790.5	312.7	290.6	103.8

	$f'_c = 3$ ksi	$f'_c = 4$ ksi	$f'_c = 5$ ksi
K_u max	0.703	0.937	1.117
ρ max	0.016	0.0214	0.0252
a/d max	0.377	0.377	0.355

$\bar{M}_u$ conc. = Ultimate Moment capacity of 1 ft. wide concrete section, ft.kips.

$\bar{M}'_t$ tensn = Resisting Moment of 1 sq.in. of tension reinforcement required to balance the force in compression reinforcement, ft.kips.

$\bar{M}'_c$ compn = Resisting Moment of i sq. in. of compression reinforcement, ft.kips.

$\bar{V}_c$ = Shear capacity of 1 ft. wide concrete section, kips.

Table 3-8

ZWEIG COEFFICIENTS : "K"

$\bar{\rho}$	f_y	40 ksi			50 ksi			60 ksi		
	f'_c	3 ksi	4 ksi	5 ksi	3 ksi	4 ksi	5 ksi	3 ksi	4 ksi	5 ksi
	C	7.86	5.90	4.72	9.83	7.30	5.90	11.80	8.85	7.08
0.0050		1.042692	1.031369	1.024780	1.054575	1.039874	1.031369	1.066980	1.048599	1.038128
0.0055		1.047372	1.034727	1.027395	1.060701	1.044219	1.034727	1.074691	1.053990	1.042268
0.0060		1.052133	1.038128	1.030038	1.066963	1.048636	1.038128	1.082600	1.059488	1.046472
0.0065		1.056977	1.041573	1.032707	1.073364	1.053124	1.041573	1.090722	1.065095	1.050743
0.0070		1.061907	1.045064	1.035404	1.079909	1.057687	1.045064	1.099063	1.070814	1.055081
0.0075		1.066924	1.048599	1.038128	1.086599	1.062325	1.048599	1.107630	1.076643	1.059488
0.0080		1.072031	1.052181	1.040881	1.093440	1.067040	1.052181	1.116429	1.082600	1.063965
0.0085		1.077228	1.055811	1.043662	1.100435	1.071834	1.055811	1.125465	1.088671	1.068513
0.0090		1.082518	1.059488	1.046472	1.107586	1.076708	1.059488	1.134746	1.09865	1.073134
0.0095		1.087905	1.063214	1.049312	1.114898	1.081664	1.063214	1.144276	1.101184	1.077829
0.0100		1.093384	1.066989	1.052181	1.122375	1.086703	1.066080	1.154063	1.107630	1.082600
0.0105		1.098963	1.070814	1.055081	1.130020	1.091826	1.070814	1.164112	1.114207	1.087447
0.0110		1.104642	1.074691	1.058011	1.137836	1.097935	1.074691	1.174430	1.120917	1.092372
0.0115		1.110423	1.078619	1.060972	1.145826	1.102332	1.078619	1.185023	1.127762	1.097377
0.0120		1.116308	1.082600	1.063965	1.153996	1.107719	1.082600	1.195897	1.134746	1.102463
0.0125		1.122298	1.086634	1.066989	1.162347	1.113195	1.086634	1.207058	1.141870	1.107630
0.0130		1.128396	1.090722	1.070045	1.170885	1.118764	1.090722	1.218512	1.149137	1.112881
0.0135		1.134602	1.094865	1.073134	1.170611	1.124427	1.094865		1.156551	1.118217
0.0140		1.140920	1.099063	1.076256	1.188531	1.130185	1.099063		1.164112	1.123639
0.0145		1.147350	1.103318	1.079411	1.197646	1.136039	1.103318		1.171825	1.129148
0.0150		1.153895	1.107630	1.082600	1.206962	1.141992	1.107630		1.179692	1.134746
0.0155		1.160556	1.112000	1.086823	1.210481	1.148044	1.112000		1.187715	1.140434
0.0160		1.167336	1.116429	1.089080	1.226207	1.154198	1.116429		1.195697	1.146213
0.0165		1.174236	1.120917	1.092372	1.236144	1.160454	1.120917		1.204240	1.152085
0.0170		1.181257	1.125465	1.095700		1.166814	1.125465		1.212748	1.158051
0.0175		1.188402	1.130075	1.099063		1.173280	1.130075		1.221422	1.164112
0.0180		1.195673	1.134746	1.102463		1.179854	1.134746			1.170271
0.0185		1.203071	1.139479	1.105899		1.186536	1.139479			1.176527
0.0190		1.210598	1.144276	1.109371		1.193328	1.144276			1.182882
0.0195		1.218256	1.149137	1.112881		1.200232	1.140137			1.189339
0.0200		1.226047	1.154063	1.116429		1.207249	1.154063			1.195897
0.0205			1.159055	1.120015		1.214381	1.159055			1.202559
0.0210			1.164112	1.123639			1.164112			
0.0215				1.127302						

$A_s = K \; \bar{A}_s$;

$\bar{A}_s = M_u \,/\, ad$;

$a = \phi \; f_y/12000$;

$K = 1 + \bar{p}c + 2(\bar{p}c)^2 + 5(\bar{p}c)^3$

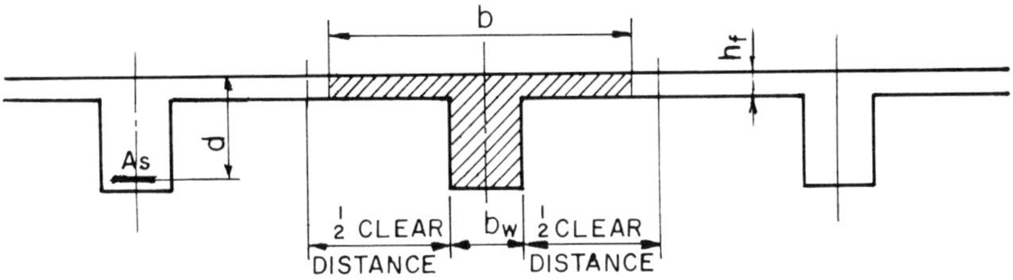

Fig. 3-6 Limits of a tee-beam.

From Table 3-7, $\overline{M}_{u\,\text{conc.}} = \dfrac{10}{12}\,374.8 = 312$ ft kips > 200 ft kips (O.K.).

Ratio of $\dfrac{M_{u\,\text{req'd}}}{\overline{M}_{u\,\text{conc.}}} = \dfrac{200}{312} = 0.64$; from Fig. 3-8, $a_u = 4.0$; $A_s = \dfrac{M_{u\,\text{req'd}}}{a_u d} = \dfrac{200}{4 \times 20} = 2.50$ sq in. (Wiesinger Method)

From Table 3-8, $\overline{A}_s = \dfrac{M_{u\,\text{req'd}}}{ad} = \dfrac{200}{4.5 \times 20} = 2.22$; $a = \dfrac{\phi f_y}{12{,}000} = \dfrac{0.9 \times 60}{12{,}000} = 4.5$

$A_s = \overline{A}_s \cdot K = 2.22 \times 1.121 = 2.50$ sq in. (Zweig Method)

A_s min $= \dfrac{200}{60{,}000} \times 10 \times 20 = 0.66$ sq in. (O.K.) ($\overline{\rho} = \dfrac{2.22}{10 \times 20} = 0.0111$; $K = 1.121$)

Either method is applicable.

2) *Rectangular Beam, Simple Bending, With Compression Reinforcement.*

$M_{u\,\text{req'd}}$: 360 ft kips.; $f'_c = 4$ ksi; $f_y = 60$ ksi; $b = 10''$; $d = 20''$; $d' = 2.5''$.

From Table 3-7, $\overline{M}_{u\,\text{conc.}} = \dfrac{10}{12}\,374.8 = 312$ ft kips < 360 ft kips (compression reinforcement is required).

$M' = M_{u\,\text{req'd}} - M_{u\,\text{conc.}} = 360 - 312 = 48$ ft kips

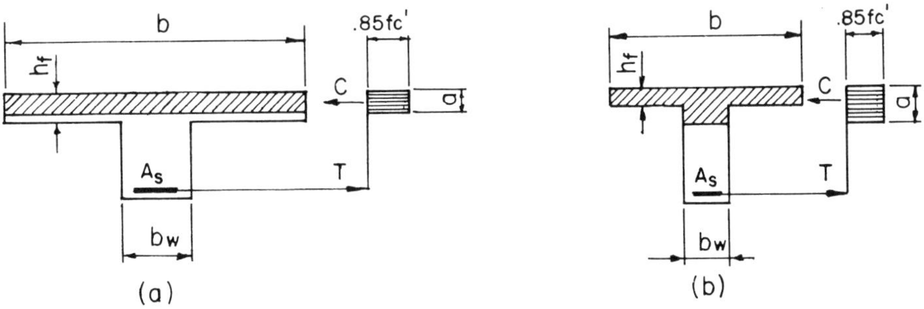

Fig. 3-7 Compression stresses in tee-beams.

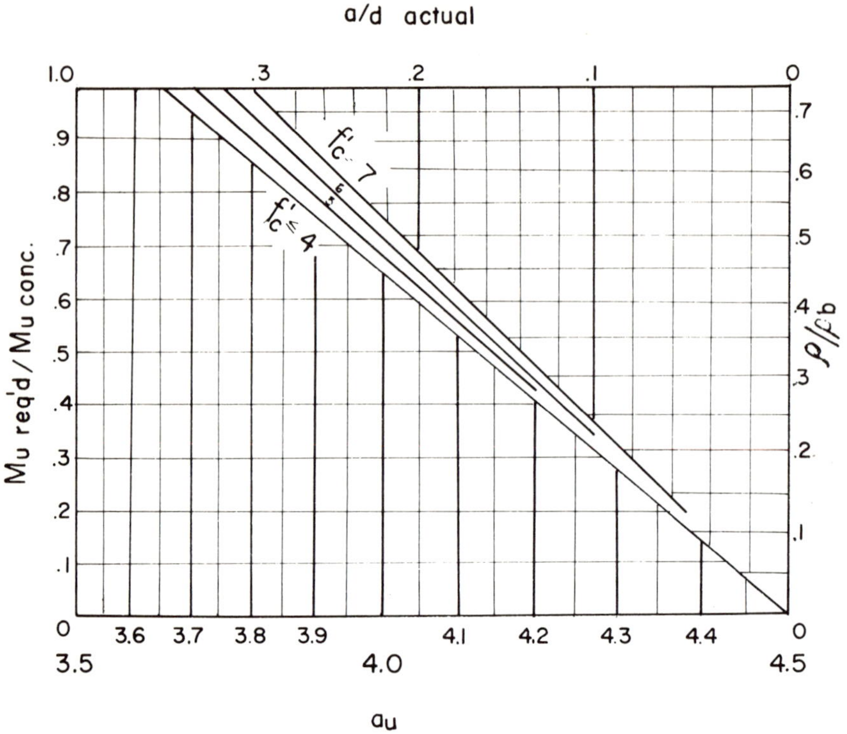

Fig. 3-8 Chart to obtain the value "a_u".

$$A'_s = \frac{M'}{\overline{M}'_{\text{comp.}}} = \frac{48}{74.3} = 0.65 \text{ sq in. (Wiesinger Method)}$$

To balance concrete: (Table 3-1):

$$\rho_{max}\, bd \quad = 0.0214 \times 10 \times 20 = 4.28 \text{ sq in;}$$

To balance compr. reinf. (Table 3-7):

$$M'/\overline{M}'_{\text{tens.}} = 48/78.7 \qquad = 0.61 \text{ sq in.}$$

$$\text{Total, } A_s = 4.89 \text{ sq in.}$$

(Note: The Zweig Method does not provide for compression reinforcement.)

3) *Tee-Beam, Compression Block Within the Flange.*

$M_{u_{\text{req'd}}}$ = 400 ft kips; f'_c = 4 ksi; f_y = 60 ksi; b_w = 10″; d = 20″; h_f = 6″; span: 30′-0″; beams @ 10′ o.c.

$$b = \begin{cases} l/4 = 30/4 = 7'\text{-}6'' = 90''; \textit{USE} \\ b_w + 16h_f = 10'' + (16 \times 6) = 106''; \\ \text{center to center of beams} = 120''; \end{cases}$$

From Table 3-7, $M_{u\,\text{conc.}} = \frac{90}{12}\,374.8 = 2810$ ft kips > 400 ft kips; from Fig. 3-8, for $\frac{M_{u\,\text{req'd}}}{M_{u\,\text{conc.}}} = \frac{400}{2810} = 0.1425$,

$$\frac{a}{d}\text{ actual} = 0.049 < \frac{h_f}{d}\left(\frac{6}{20} = 0.30\right)\ \text{(O.K.)}$$

$a = 0.049 \times 20 \approx 1.00;\ a_u = 4.4;$

$$A_s = \frac{M_{u\,\text{req'd}}}{a_u d} = \frac{400}{4.4 \times 20} = 4.55 \text{ sq in. (Wiesinger Method).}$$

From Table 3-8, $\overline{A}_s = \frac{M_{u\,\text{req'd}}}{ad}$

$$A_s = \overline{A}_s K = 4.44 \times 1.025 = 4.55 \text{ sq in.} \left(a = \frac{\phi f_y}{12000} = \frac{0.9 \times 60}{12} = 4.5;\ \overline{\rho} = \frac{4.44}{90 \times 20} = 0.0025;\ K = 1.025\right) \text{(Zweig Method)}$$

Check: $C = 0.85\,f'_c bt\phi = 3.4 \times 90 \times 1 \times 0.90 = 275$ kips;
$T = C = 275$ kips; $A_s = T/f_y = 275/60 = 4.57$ sq in. (O.K.)

4) *Tee-Beam, Compression Block Deeper Than Flange.*

$M_{u\,\text{req'd}} = 220$ ft kips; $f'_c = 3$ ksi; $f_y = 60$ ksi; $b_w = 6''$; $d = 16''$; $h_f = 2.5''$; span: 24′, joists @ 26″ o.c.

$$b = \begin{cases} l/4 = 24/4 = 6'\text{-}0'' & = 72'' \\ b_w + 16h_f = 6 + 16 \times 2.5 = 46'' & \\ \text{c.l. to c.l. of joists} & = 26''\ \textit{USE} \end{cases}$$

From Table 3-7, $M_{u\,\text{conc.}}$ 26/12 · 178 = 386 ft kips > 220 ft kips; from Fig. 3-8, for $M_{u\,\text{req'd}}/M_{u\,\text{conc.}} = 200/386 = 0.517$;

$$\frac{a}{d}\text{ actual} = 0.175 > \frac{h_f}{d}\left(\frac{2.5}{16} = 0.156\right)$$

Compression stress block extends below the flange.

$$M_{u\,\text{max web}} = K_{u\,\text{max}} b_w d^2/12 = 0.703 \times 6 \times 16^2/12 = 89 \text{ ft kips}$$

$$M_{u\,\text{max flange}} = \frac{0.85\,f'_c(b - b_w)\,h_f\phi\left(d - \frac{h_f}{2}\right)}{12} = \frac{\overbrace{2.55 \times 20 \times 2.5}^{C\,=\,127.5\text{ kips}} \times 0.9 \times 14.75}{12} = 141 \text{ ft kips}$$

Total = 230 ft kips > 220 ft kips

From Fig. 3-8, for $\dfrac{220 - 141}{89} = 0.887; \dfrac{a}{u} = 3.76$

$$A_{s_{web}} = \frac{220 - 141}{3.76 \times 16} = 1.31 \text{ sq in.}$$

$$A_{s_{flange}} = \frac{C\,(=T)}{f_y} = \frac{127.5}{60} = \underline{2.13 \text{ sq in.}}$$

Total A_s = 3.44 sq in. (Wiesinger Method)

$$\rho_{web} = \frac{1.31}{6 \times 16} = 0.0137 < 0.01604 \text{ (Table 3-1) (O.K.)}$$

From Table 3-8:

$$\bar{A}_s = \frac{M_{u_{req'd}}}{ad} = \frac{220}{4.5 \times 16} = 3.06$$

$$A_s = \bar{A}_s K = 3.06 \times 1.11 = 3.40 \text{ sq in.} \left(a = \frac{\phi f_y}{12} = \frac{0.9 \times 60}{12} = 4.5; \quad \bar{\rho} = \frac{3.06}{26 \times 16} = 0.007; K = 1.11 \right) \text{ (Zweig Method)}$$

CRACK CONTROL

With the use of higher-strength steel and *Strength Design*, the service load stresses can be quite high. While this is perfectly safe, visible cracks are bound to develop. In order to minimize the width of the cracks, the ACI 318-71 has introduced a new formula:

$z = f_s \sqrt[3]{d_c A} \leqslant 175$ kips per inch for interior exposure
$\leqslant 145$ kips per inch for experior exposure, where

f_s (ksi) = stress in reinforcing steel at service load, or assuming $f_s = 0.60 f_y$
d_c (in.) = concrete cover to the centerline of the reinforcing bar
A (sq in.) = effective tension area of concrete = $2 d_c s_b$
s_b (in.) = reinforcing bar spacing
z/f_s = $\sqrt[3]{2 d_c^2 s_b}$, and
s_b (in.) = $1/2 d_c^2 (z/f_s)^3$

The maximum spacing of reinforcing bars in slabs is shown in Table 3-9 for several slab thicknesses, bar size combinations, and interior or exterior locations. The first column shows the requirements for shrinkage and temperature reinforcements per Section 7.13 of ACI Code, which applies to one-way slabs perpendicular to the main reinforcement as well as to the two-way slab systems

Table 3-9

$F_y = 60.0$			Maximum Spacing of Reinforcement in Slabs					
			Interior (cover ¾")			Exterior (cover 1½")		
Slab thickness (inches)	Bar size	Tempera. reinf. (one-way)	One-way slab	2-way or flat slb.	Crack contr.	One-way	2-way or flat slb.	Crack contr.
4	3	15.5	11.0	8.0*	65.5	12.0*	8.0*	11.5
	4	18.0*	12.0*	8.0*	57.5	12.0*	8.0*	10.5
	5	18.0	12.0*	8.0*	51.0	12.0*	8.0*	10.0
5	3	12.0	8.0	10.0*	65.5	10.0	10.0*	11.5
	4	18.0*	15.0	10.0*	57.5	15.0*	10.0*	10.5
	5	18.0*	15.0*	10.0*	51.0	15.0*	10.0*	10.0
	6	18.0*	15.0*	10.0*	45.5	15.0*	10.0*	6.0
6	4	18.0*	12.0	12.0*	57.5	14.0	12.0*	10.5
	5	18.0*	18.0*	12.0*	51.0	18.0*	12.0*	10.0
	6	18.0*	18.0*	12.0*	45.5	18.0*	12.0*	6.0
8	4	14.0	8.5	14.0	57.5	9.5	14.0	10.5
	5	18.0*	13.5	16.0*	51.0	15.0	16.0*	10.0
	6	18.0*	18.0*	16.0*	45.5	18.0*	16.0*	6.0
	7	18.0*	18.0*	16.0*	40.5	18.0*	16.0*	5.5
9	4	12.5	7.5	12.5	57.5	8.5	12.5	10.5
	5	18.0*	11.5	18.0*	51.0	13.0	18.0*	10.0
	6	18.0*	17.0	18.0*	45.5	18.0*	18.0*	6.0
	7	18.0*	18.0*	18.0*	40.5	18.0*	18.0*	5.5
	8	18.0*	18.0*	18.0*	37.0	18.0*	18.0*	5.0
10	5	17.0	10.5	17.0	51.0	11.5	17.0	10.0
	6	18.0*	15.0	20.0*	45.5	17.5	20.0*	6.0
	7	18.0*	18.0*	20.0*	40.5	18.0*	20.0*	5.5
	8	18.0*	18.0*	20.0*	37.0	18.0*	20.0*	5.0
	9	18.0*	18.0*	20.0*	33.5	18.0*	20.0*	5.0
12	5	14.5	8.5	14.5	51.0	9.0	14.5	10.0
	6	18.0*	12.0	20.5	45.5	13.5	20.5	6.0
	7	18.0*	16.5	24.0*	40.5	18.0*	24.0*	5.5
	8	18.0*	18.0*	24.0*	37.0	18.0*	24.0*	5.0
	9	18.0*	18.0*	24.0*	33.5	18.0*	24.0*	5.0

*Indicates that the max spacing of $2h$, $3h$, $5h$, or 18 in. governs.

per Section 13.5.3 of the Code. The minimum reinforcement ratio (referring to the total slab thickness) is 0.0020 and 0.0018 for yield strengths of 50 and 60 ksi respectively, but not exceeding $5h$ or 18 in. An asterisk, adjacent to the spacing, indicates that the latter requirements govern. The tabulation is separated now as to interior location, having $^3/_4$ in. concrete cover, and to exterior location, having 1½ and 2 in. cover, for 3 to 5 bar sizes and for 6 or larger sizes, respectively, per Section 7.14.1.1.

For one-way slabs the minimum reinforcement ratio (referring to the effective depth of the slab) for flexure is controlled by $200/f_y$ per Section 10.5 of the

Code and the maximum spacing is $3h$ or 18 in. per Section 7.4.3 of the Code. An asterisk indicates that the maximum spacing governs.

For two-way slab systems the temperature reinforcing requirements, as discussed above, or the maximum spacing of $2h$ governs. An asterisk indicates that

Table 3-10. Crack control of beams.

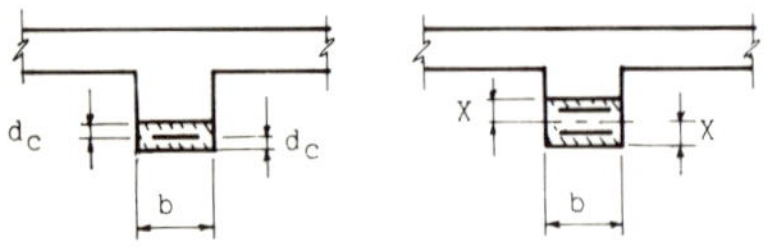

Layers of reinf.	One	Two
#8	$d_c = 2\frac{1}{2}''$	$d_c = 2\frac{1}{2}''$
#18	$d_c = 3''$	$d_c = 3''$

$$z = f_s \sqrt[3]{d_c A}, \quad z \leqslant \begin{cases} 175 \text{ for interior exposure} \\ 145 \text{ for exterior exposure} \end{cases} \qquad f_s = 0.6 \times 60 = 36 \text{ ksi}$$

$A = 2d_c b$ (one layer of reinforcing); $A = 2 \times b$ (two layers of reinforcing)

Values of z for one layer of reinforcement:

Width of beams	$d_c = 2\frac{1}{2}''$					$d_c = 3''$				
	$n = 2$	$n = 3$	$n = 4$	$n = 5$	$n = 6$	$n = 2$	$n = 4$	$n = 6$	$n = 8$	$n = 10$
12″	152	133	120	112	105	171	136	119	108	100
16″	167	146	133	123	116	189	150	131	119	110
18″	174	152	138	128	120	196	156	136	124	115
20″	180	157	143	133	125	203	161	141	128	119
21″	183	160	145	135	127	207	164	143	130	121
24″	191	167	152	141	133	216	171	150	136	126
28″	201	176	160	148	140	227	180	158	143	133
30″	206	180	164	152	143	233	185	161	147	136
36″	219	191	174	161	152	247	196	171	156	145

n = number of bars

Values of z for two layers of reinforcement:

Width of beams	$d_c = 2\frac{1}{2}''$					$d_c = 3''$				
	$n = 2$	$n = 6$	$n = 8$	$n = 10$	$n = 12$	$n = 6$	$n = 8$	$n = 10$	$n = 12$	$n = 14$
12″	135	118	107	99	93	136	124	115	108	103
16″	148	130	118	109	103	150	136	126	119	113
18″	154	135	122	114	107	156	142	131	124	117
20″	160	140	127	118	111	161	147	136	128	122
21″	162	142	129	120	113	164	149	138	130	124
24″	170	148	135	125	118	171	156	145	136	129
28″	179	156	142	132	124	180	164	152	143	136
30″	183	160	145	135	127	185	168	156	147	139
36″	194	170	154	143	135	196	178	166	156	148

the $2h$ spacing is critical. The limitations for crack control, per Section 10.6, are shown in separate columns. The crack control figure does not govern for interior locations; the requirements, however, exceed the maximum spacing for two-way slab systems of 6 inch slabs or thicker at exterior locations, especially when 2 in. of cover are applied for size 6 and larger bars.

The minimum number of bars in concrete beams for crack control are shown in Table 3-10.

Equation (10-2) provides a simplified method for crack control in concrete beams, where the steel yield strength exceeds 40 ksi. Corresponding values of z are printed for one and two layers of reinforcement, and for different number of bars (n) and beam widths (b). The concrete cover to the center of bars located closest to the beam face (d_c) is provided for 2.5 and 3.0 in., which corresponds to #8 and #14 bars, respectively. The value of X represents the center of gravity of bars to the face of the beam, which equals d_c for one layer of reinforcement. At service load 36 ksi of flexural stress is assumed, using the 60 percent allowance by the ACI Code. The quantity of z is limited to 175 kips/in. for interior exposure and to 145 for exterior exposure.

The "Hemmingway" Apartments, Chicago, Illinois. Thirty stories, flat plate, shear wall. Architects: Solomon & Cordwell, Chicago. Structural Engineers: Rogers-Cohen-Barreto-Marchertas, Inc.

4

Deep Flexural Beams

When the ratio of depth-to-span is larger than 0.40 for continuous spans, and larger than 0.80 for single spans, then such deep beams are subject to nonlinear stress distribution even within the elastic range. For very deep walls, supported on columns, a depth-to-span ratio of one is assumed, since this provides for optimum stress distribution.

The theory of deep beams was originally developed by Dischinger; the Portland Cement Association has translated and published some of the Dischinger charts. Since either the original or translation may be difficult to obtain, Fig. 4-1 provides charts for a variety of depth-to-span ratios and these include all necessary data for design. Needless to say, these charts have been in vogue while the Working Stress Design dominated the concrete design. Thus, their accuracy within the Strength Design may be subject to question. It is believed, however, that by properly factoring the loads on deep beams, the results remain on the conservative side. It is to be noted also, that the concentration of tensile forces, to be resisted by reinforcement, is *not* near the extreme fibers and proper placing of such reinforcement is very important.

Figure 4-2 presents the design computations for such a deep beam, which serves as a continuous lintel for a large hyperbolic cooling tower.

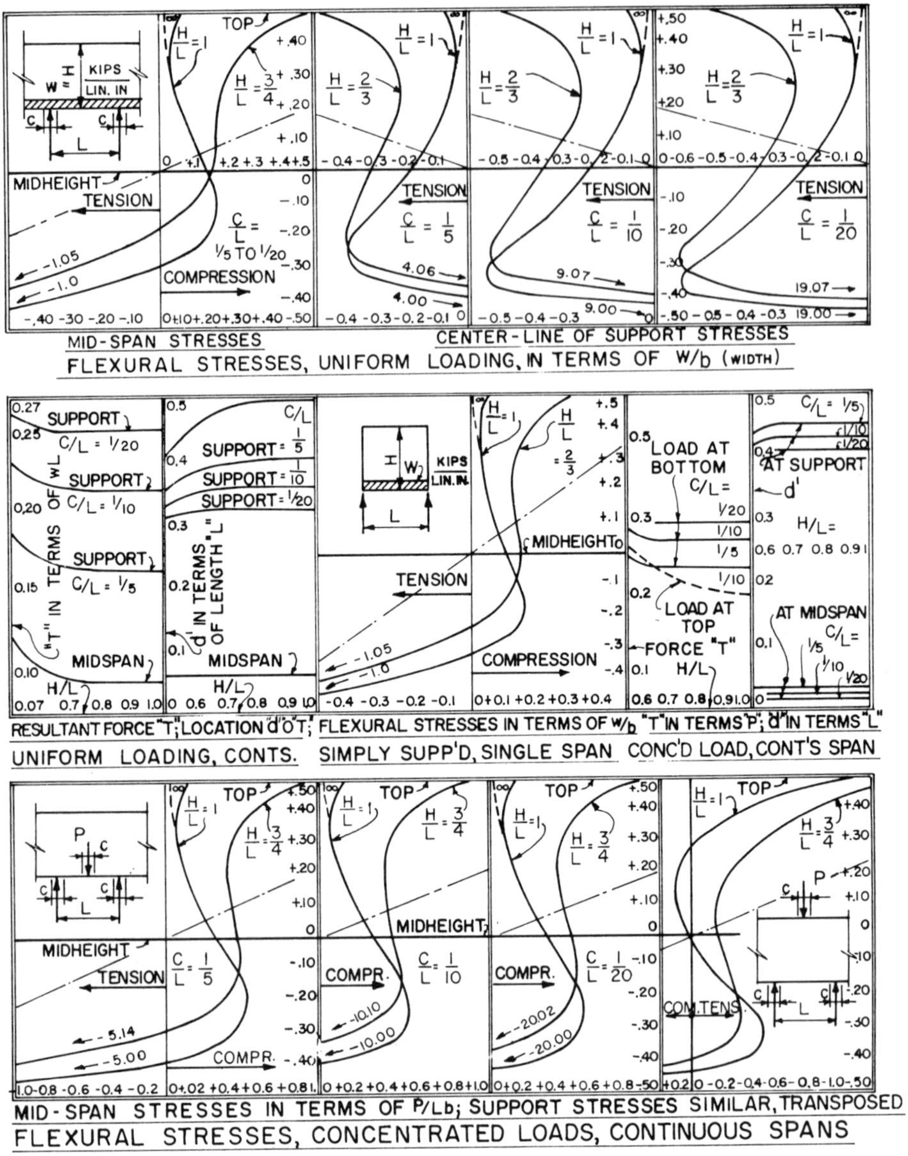

MID-SPAN STRESSES
CENTER-LINE OF SUPPORT STRESSES
FLEXURAL STRESSES, UNIFORM LOADING, IN TERMS OF W/b (WIDTH)
MIDHEIGHT
TENSION
COMPRESSION
TOP
KIPS LIN. IN
RESULTANT FORCE "T"; LOCATION "d'" OF "T";
UNIFORM LOADING, CONTS.
FLEXURAL STRESSES IN TERMS OF W/b
SIMPLY SUPP'D, SINGLE SPAN
"T" IN TERMS "P"; "d'" IN TERMS "L"
CONC'D LOAD, CONT'S SPAN
LOAD AT BOTTOM
LOAD AT TOP
FORCE "T"
AT SUPPORT
AT MIDSPAN
MID-SPAN STRESSES IN TERMS OF P/Lb; SUPPORT STRESSES SIMILAR, TRANSPOSED
FLEXURAL STRESSES, CONCENTRATED LOADS, CONTINUOUS SPANS

Fig. 4-1.

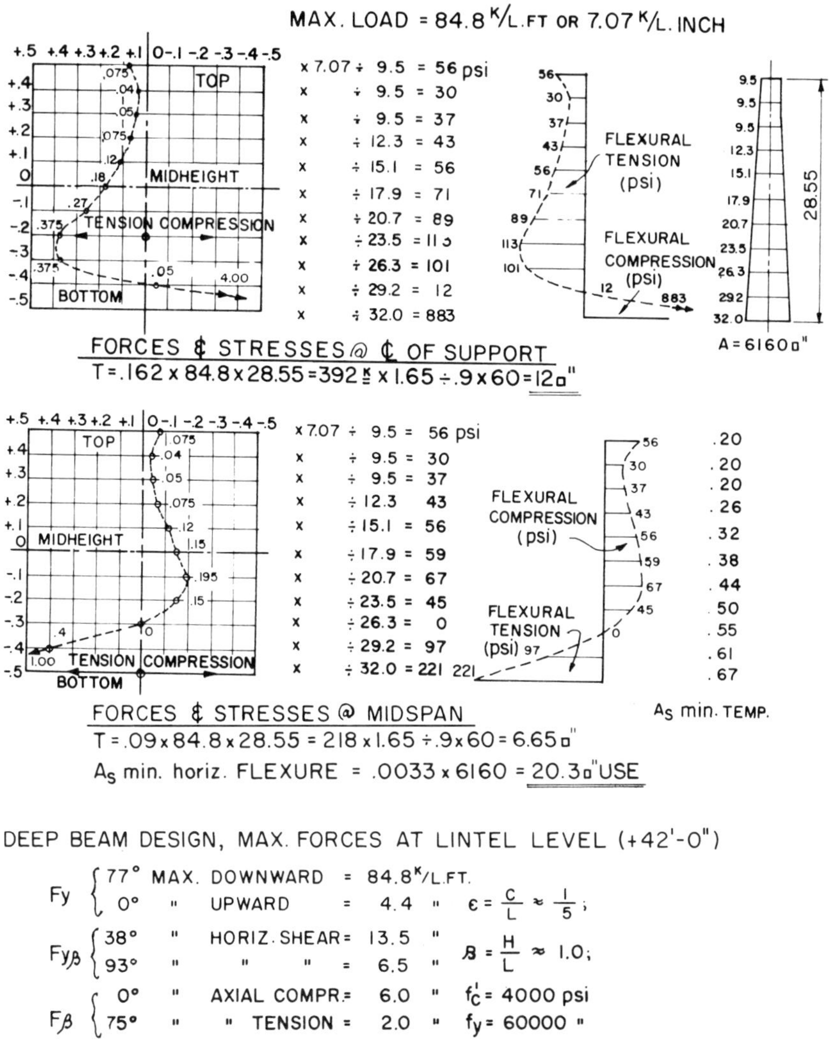
MAX. LOAD = 84.8 K/L.FT OR 7.07 K/L. INCH
+.5 +.4 +.3 +.2 +.1 0 -.1 -.2 -.3 -.4 -.5
TOP
MIDHEIGHT
TENSION COMPRESSION
BOTTOM
.075 .04 .05 .075 .12 .18 .27 .375 .375 .05 4.00
x 7.07 ÷ 9.5 = 56 psi
x ÷ 9.5 = 30
x ÷ 9.5 = 37
x ÷ 12.3 = 43
x ÷ 15.1 = 56
x ÷ 17.9 = 71
x ÷ 20.7 = 89
x ÷ 23.5 = 113
x ÷ 26.3 = 101
x ÷ 29.2 = 12
x ÷ 32.0 = 883
FLEXURAL TENSION (psi)
FLEXURAL COMPRESSION (psi)
9.5 9.5 9.5 12.3 15.1 17.9 20.7 23.5 26.3 29.2 32.0
28.55
A = 6160□"
FORCES ¢ STRESSES @ ¢ OF SUPPORT
T = .162 x 84.8 x 28.55 = 392 K x 1.65 ÷ .9 x 60 = 12□"
+.5 +.4 +.3 +.2 +.1 0 -.1 -.2 -.3 -.4 -.5
TOP
MIDHEIGHT
TENSION COMPRESSION
BOTTOM
.075 .04 .05 .075 .12 .15 .195 .15 0 .4 1.00
x 7.07 ÷ 9.5 = 56 psi
x ÷ 9.5 = 30
x ÷ 9.5 = 37
x ÷ 12.3 43
x ÷ 15.1 = 56
x ÷ 17.9 = 59
x ÷ 20.7 = 67
x ÷ 23.5 = 45
x ÷ 26.3 = 0
x ÷ 29.2 = 97
x ÷ 32.0 = 221
FLEXURAL COMPRESSION (psi)
FLEXURAL TENSION (psi)
.20 .20 .20 .26 .32 .38 .44 .50 .55 .61 .67
As min. TEMP.
FORCES ¢ STRESSES @ MIDSPAN
T = .09 x 84.8 x 28.55 = 218 x 1.65 ÷ .9 x 60 = 6.65□"
As min. horiz. FLEXURE = .0033 x 6160 = 20.3□" USE
DEEP BEAM DESIGN, MAX. FORCES AT LINTEL LEVEL (+42'-0")
Fy 77° MAX. DOWNWARD = 84.8 K/L.FT.
0° " UPWARD = 4.4 "
ε = C/L ≈ 1/5 ;
Fyβ 38° " HORIZ. SHEAR = 13.5 "
93° " " " = 6.5 "
β = H/L ≈ 1.0;
Fβ 0° " AXIAL COMPR. = 6.0 "
75° " " TENSION = 2.0 "
f'c = 4000 psi
fy = 60000 "

Fig. 4-2.

Tennessee Valley Authority. Paradise Power Station–437-ft high hyperbolic cooling tower. Structural Engineers: Paul Rogers & Associates.

5

Shear

Since reinforced concrete is deficient in resisting tension forces, reinforcement is provided wherever such forces occur. Unreinforced concrete develops diagonal cracks near the supports.

Shear, as a measure of diagonal tension, has customarily been resisted partly by the concrete and partly by stirrups and/or bent-up main reinforcing bars. The usefulness of bent-up bars has diminished lately because the Code limits the effectiveness of bent bars to three-fourths of the inclined portion. Strength Design permits shallow members, while the development lengths are increased to cover the augmented forces in the bars. Consequently, it is difficult to utilize bent-up bars efficiently. In seismic areas bent-up bars are undesirable due to possible stress reversals.

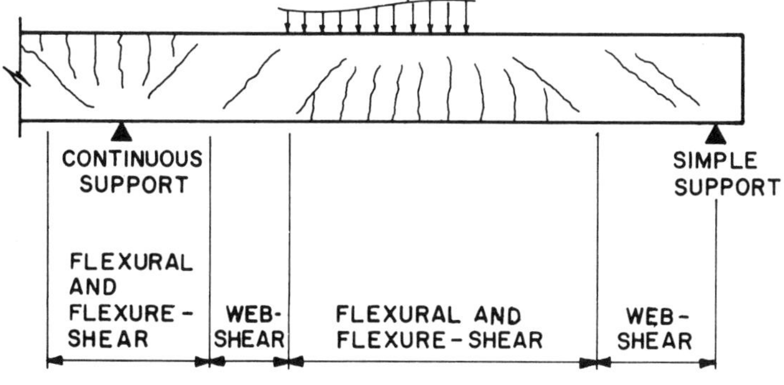

Fig. 5-1.

The ACI 318-71 Code requires, for the first time, that flexural members (with certain exceptions) shall be provided with full-length web reinforcement. Another innovation is the permitted use of welded wire fabric (WWF) for stirrups and column ties.

The minimum area of shear reinforcement should be:

$$A_v = 50\frac{b_w s}{f_y} \quad \text{or} \quad \frac{A_v f_y}{50\, b_w} \leqslant \frac{1}{2}d < 24''$$

where

b_w = web width
s = stirrup spacing
f_y = yield point of reinforced steel

The spacing s is to be reduced to one-half when $(v_u - v_c) > 4\sqrt{f'_c}$. The design yield strength of shear reinforcement must not exceed 60 ksi. The nominal shear stress, v_u, as a measure of diagonal tension, is expressed as:

$$v_u = \frac{V_u}{\phi b_w d}$$

where

V_u = computed shear force,
ϕ = strength reduction factor, 0.85.

For continuous beams, frames, and one-way slabs (but not for simply supported bearing ends) the maximum shear force, V_u, may be computed at a distance d from the face of the support. For slabs in two-way action, the shear stress, v_u, is to be computed at a $\frac{1}{2}d$ distance from the face of support.

The shear stress carried by the concrete must not exceed:

$v_c = 2\sqrt{f'_c}$ for regular weight concrete
$= 1.7\sqrt{f'_c}$ for sand-light weight concrete
$= 1.5\sqrt{f'_c}$ for all-light weight concrete

Tables 3-6 and 3-7 include the shear capacities, v_c, of slabs and beams for a large variety of sizes.

The above values may be increased through the use of the following formula:

$$v_c = 1.9\sqrt{f'_c} + 2500\,\rho_w \frac{V_u d}{M_u} \leqslant \sqrt{3.5}\, f'_c{}^*$$

This refinement requires a preliminary design.

**Important note:* For lightweight concrete substitute $f_{ct}/6.7$ for $\sqrt{f'_c}$.

In design of shear reinforcement, when stirrups perpendicular to the longitudinal reinforcement are used, the required area of such stirrups is to be:

$$A_v = \frac{(v_u - v_c)\, b_w s}{f_y};\ (v_u - v_c) \leq 8\sqrt{f'_c}$$

Sections between the face of support and d shall be designed for shear at d.
The spacing of stirrups is obtained as:

$$s = \frac{A_v f_y}{(v_u - v_c)\, b_w}$$

where

A_v is the cross-sectional area of *two* legs of stirrups of a predetermined diameter

For office use, tables and/or charts of several handbooks may be used. Extreme refinement is not warranted and spacings should be grouped for ease of construction, but not to exceed $\frac{1}{2}d$ (or $\frac{1}{4}d$ as it applies.)

Example 1

See Fig. 5-2; b = 12 in., d = 21.5 in., f'_c = 4 ksi and f_y = 40 ksi.

$$V_D = \frac{18.5}{2} \times 1.5 = 13.875 \times 1.4 \text{ (load factor)} = 19.425\ k = V_{UD}$$

$$V_L = \frac{18.5}{2} \times 3.5 = 32.375 \times 1.7 \text{ (load factor)} = 55.038\ k = V_{UL}$$

$$46.250 \times 1.6 \qquad = 74.463\ k = V_{UT}$$

$$V_{UT_d} = 5.0 \times 1.6 \times 7.46 = 60.05\ k$$

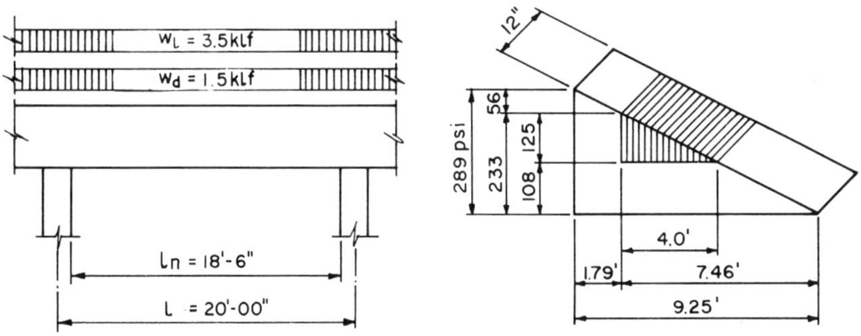

Fig. 5-2.

V_c = (from Table 3-7) = 27.75 k (ϕ included)

V'_u = = 32.30 k

$$v'_u = \frac{V'_u}{b_w d} = \frac{32{,}300}{12 \times 21.5} = 125 \text{ psi} \qquad v_c = 108 \text{ psi}$$

$$v_{u\,max} = \frac{60{,}050}{\phi\, 12 \times 21.5} = 276 \text{ psi} \qquad v_u - v_c = 168 \text{ psi} < 4\sqrt{f'_c}$$

Total shear volume: $\Sigma v'_u = \dfrac{4 \times 12 \times 12 \times 0.125}{2} = 36.0\ k$

Capacity of #3 stirrups (two legs): $A_v f_y \phi = 2 \times 0.11 \times 40 \times 0.85 = 7.5\ k$

No. of stirrups required: $\dfrac{36.0}{7.5} = \approx 5;\ s = \dfrac{A_v f_y}{50\, b_w} = \dfrac{.22 \times 40{,}000}{50 \times 12} = 14.7''$

Maximum spacing of $\frac{1}{2}d$, or $10\frac{1}{2}$ in. governs throughout the beam.

Example 2

Use $30'' \times 54''$ beam; $b = 30''\ (2.5')$; $d = 51''\ (4.25')$; $w_D = 1.7$ klf

$w_L = 1.0$ klf

$w_T = 2.7$ klf

$P_D = 100$ k

$P_L = 50$ k

$P_T = 150$ k

$w_D = 1.4 \times 1.7 = 2.38$ klf

$w_L = 1.7 \times 1.0 = 1.70$ klf

$w_T = (1.51) \times 2.7 = 4.08$ klf

$P_D = 1.4 \times 100 = 140.00$ k

$P_L = 1.7 \times 50 = 85.00$ k

$P_T = (1.5) \times 150 = 225.00$ k

$V_{u\,support} = 225 + 12 \times 4.08 = 274\ k/30 \times 51 = v_{u_1} = 179$ psi

$V_{u_d} = 225 + 7.25 \times 4.08 = 255\ k/30 \times 51 = v_{u_2} = 167$ psi

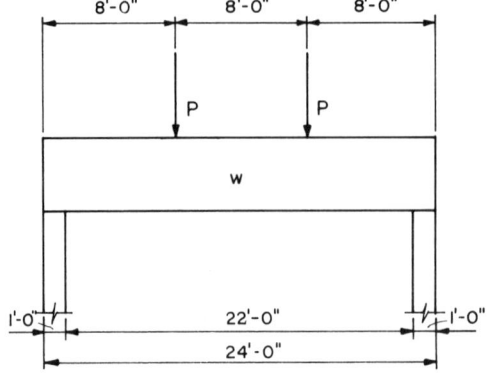

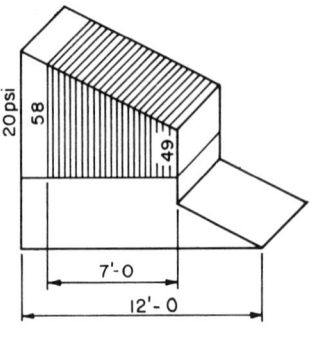

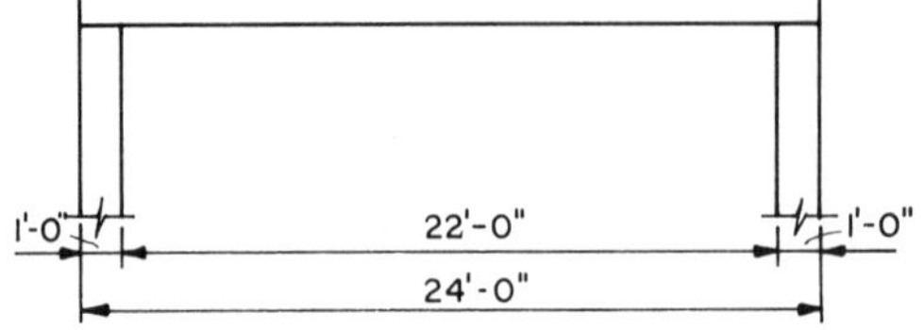

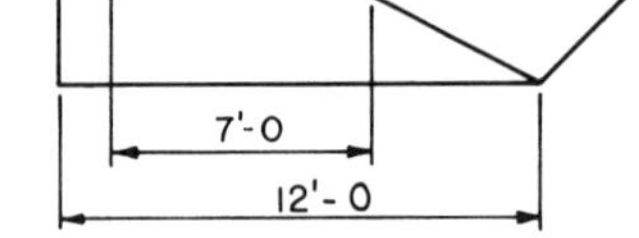

Fig. 5-3.

V_{u_p} = 225 + 4.0 × 4.08 = 241 $k/30$ × 51 = v_{u_3} = 158 psi

M_{u_d} = 269 × 4.25 = 1126 fk $\quad \dfrac{M_{u_d}}{M_{u_c}} = \dfrac{1126}{6095} = 0.184$

M_{u_c} = 2.5 × 2438 = 6095 fk

a_u = 4.37

$$A_{s_{\text{req'd}}} = \frac{1126}{4.37 \times 51} = 5.05 \text{ sq in.}$$

$A_{s_{\min}}$ = 0.0033 × 30 × 51 = 5.05 sq in. Use: 4 - #10(5.08)

$$\frac{M_{u_d}}{V_{u_d}} = \frac{1126}{2.55 \times 4.25} = 1.038\text{; reciprocal value: } \frac{1}{1.038} = 0.963$$

$v_c = [1.9\sqrt{4000} + 2500 \times 0.033 \times 0.963]\ 0.85 = 109$ psi

$v_{u_1} - v_c$ = 179 - 109 = 70 psi

$v_{u_2} - v_c$ = 167 - 109 = 58 psi

$v_{u_3} - v_c$ = 158 - 109 = 49 psi

Total shear volume: $\dfrac{58 + 49}{2} \times 7.0 \times 12 \times 30/1000 = 134.82\ k$

Try #4 stirrups, $A_v = 2 \times 0.20 = 0.40$ sq in.

Capacity of one stirrup: 0.40 × 40 × 0.85 = 13.6 k; no.: $\dfrac{134.82}{13.6} \approx 10$

Required spacing: 7 × 12/9 = 9.33″; use s = 9″ (midspan: s = 24″)

Min $A_v = 50 \times b_w \times s/f_y = 50 \times 30 \times 9/40{,}000 = 0.337 < 0.40$ (O.K.)

Example 3. Deep beam

$f'_c = 4$ ksi; $f_y = 60$ ksi; $w_u = 1.4 \times 84.8 = 118.8$ klf

$l_n = 0.80 \times 28.55 = 22.8$ ft

Critical section for shear:

$l_{n_{cr}} = 0.15 \times 22.8 = 3.4$ ft (ACI-11-9.3)

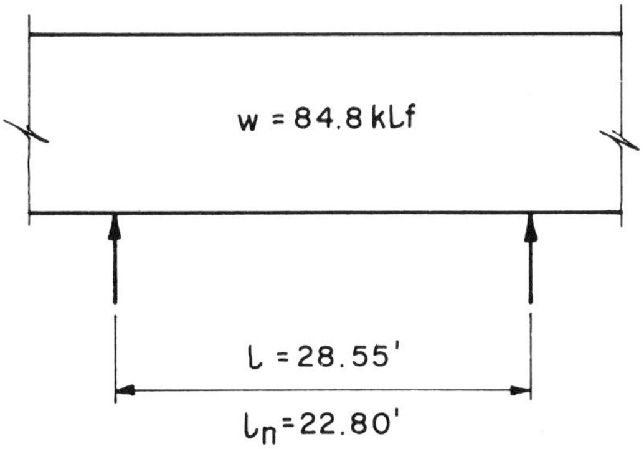

Fig. 5-4.

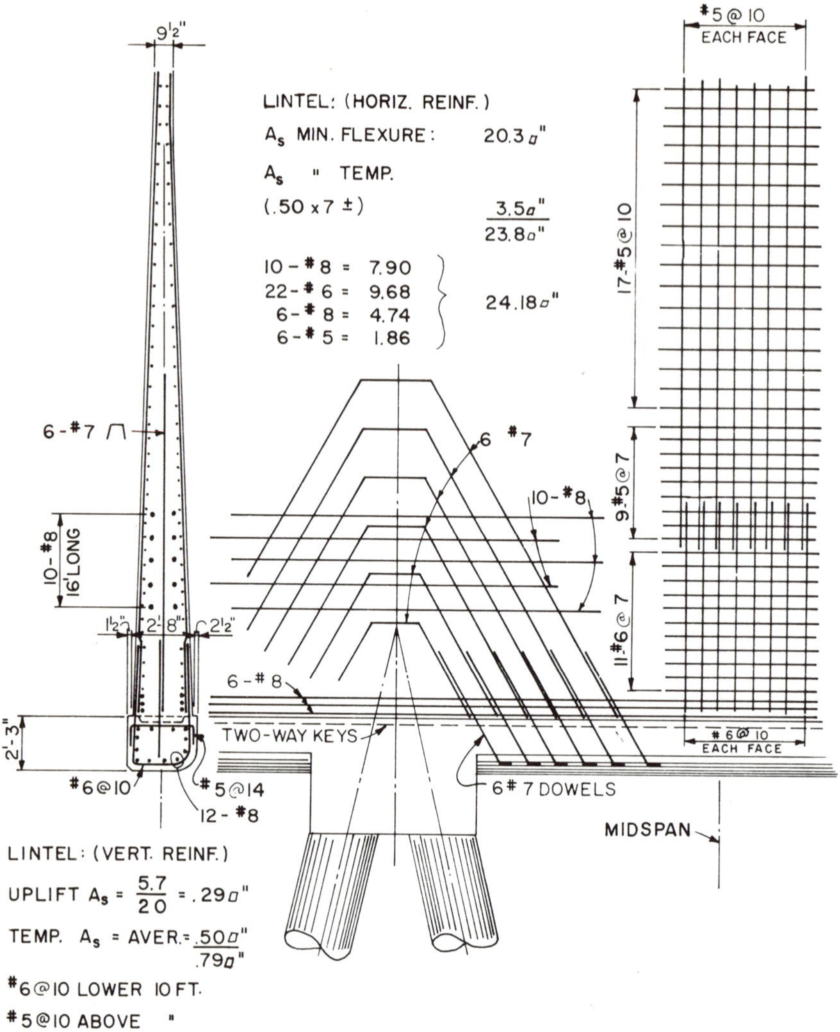

Fig. 5-5 Reinforcing details.

$$V_u = 118.8\,\frac{(22.8 - 2 \times 3.4)}{2} = 950\text{ k}; M_{u_{cr}} = \frac{1354 + 950}{2} \times 3.4 = 3916\text{ fk}$$

$$\frac{M_u}{V_u d} = \frac{3916}{950 \times 25} = 0.164; \frac{V_u d}{M_u} = \frac{1}{0.164} = 6.10$$

$$v_c = (3.5 - 2.5 \times 0.164)(1.9\sqrt{4000} + 2500 \times 0.0033 \times 6.10)$$

(ACI-11-22)

$$v_c = 2.5\,(120 + 50.32) = 426 \text{ psi}$$
$$v_{c_{max}} = 6\sqrt{4000} = 380 \text{ psi}$$
$$v_u = \frac{950{,}000}{19 \times 25 \times 12 \times 0.85} = 196 \text{ psi (O.K.)}$$

Minimum reinforcement:

$$A_{s_{vert.}} = 0.0015 \times b \times s$$
$$A_{s_{horiz.}} = 0.0025 \times b \times s$$

s = assumed to be 10″; b = varies from $9\frac{1}{2}$″ to 32″

In the example considerably more than minimum vertical reinforcement was used in order to resist vertical uplift forces.

Diagonal shear reinforcement, although not required by the Code, is always employed by the author for deep beams.

For members subjected to axial compression, shear, and bending:

$$v_c = 1.9\sqrt{f'_c} + \frac{2500\,\rho_w\,V_u d}{M_u - N_u \dfrac{(4h - d)}{8}}$$

or alternatively:

$$v_c = 2\,(1 + 0.0005\,N_u/A_g)\sqrt{f_c}$$

however, v_c shall not exceed

$$v_c = 3.5\sqrt{f'_c}\sqrt{1 + 0.002\,N_u/A_g)}$$

For members subjected to significant axial tension, shear, and bending, the web reinforcement is to carry the total shear, unless a more detailed analysis is made using:

$$v_c = 2\,(1 + 0.002\,N_u/A_g)\sqrt{f'_c}$$

The ACI 318-71 Code includes, for the first time, special provisions for the design of brackets and corbels.

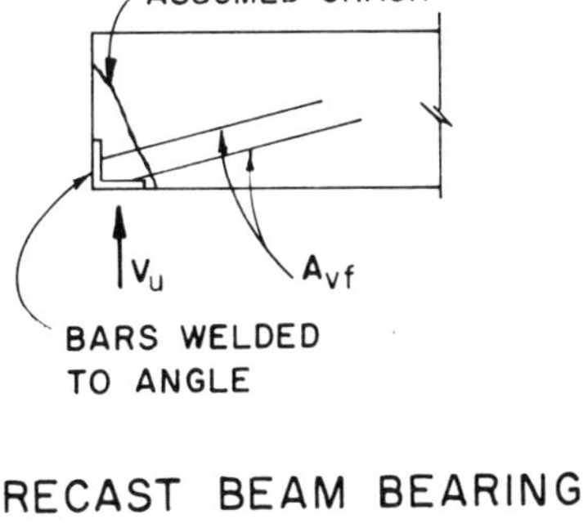

Fig. 5-6.

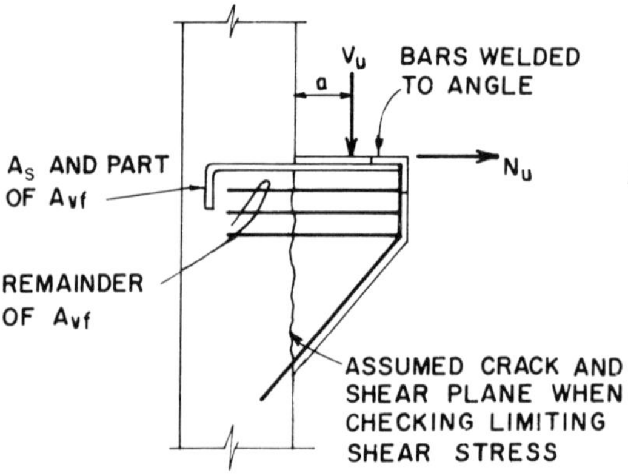

b) CORBEL

Fig. 5-7.

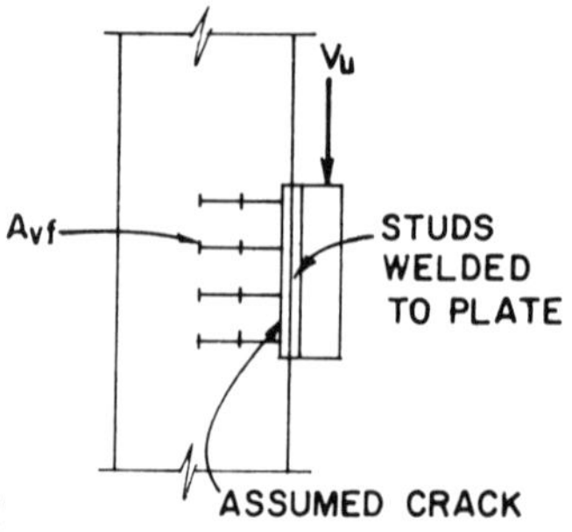

c) COLUMN FACE PLATE

Fig. 5-8.

For design of these the reader should follow the formulas and expressions of the Code and its commentary.

Shear Friction

This is an entirely new concept in the ACI Code and it pertains to an assumed cracked portion of a member which, however, may still be safe due to the holding power of special reinforcement across the crack.

The shear stress, v_u, should not exceed $0.2\,f'_c$ nor 800 psi. The required area of reinforcement perpendicular to the crack must be computed by the following formula:

$$A_{v_f} = \frac{V_u}{\phi f_y \mu}$$

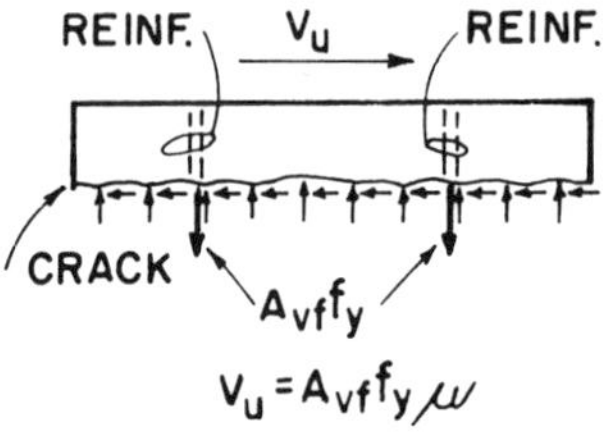

Fig. 5-9. Stresses and forces at crack surface.

where

f_y = design yield strength not exceeding 60,000 psi
= coefficient of friction to be 1.4 for concrete cast monolytically; 1 for concrete placed against hardened concrete, and 0.7 for concrete placed against rolled structural steel

The direct tension across the assumed crack must be resisted by additional reinforcement. Successful application of shear friction depends on the proper location of the assumed crack.

Shear Heads

Metallic shear heads have been employed by experienced engineers for a long time. Their use was implicitly permitted in the previous ACI Codes. The 1971 issue of the Code now officially recognizes the use of metallic shear heads, and practical specifications are given for their usage.

Shear provisions for slabs and footings will be dealt with in the proper chapters.

Example of Bracket

Design shear, $V_u = 1.4D + 1.7L = 103.5$ k

Allowable bearing stress $= 0.85\,\phi f'_c = 0.85 \times 0.7 \times 5 = 2.97$ ksi

Required bearing plate area: $\dfrac{103.5}{2.97} = 35$ sq in.; use 4″ × 10″ plate

Assume shear span of twice the distance between the end of beam and the face of the column, plus one-half of the bearing plate width:

$$a = 2 + \tfrac{1}{2}(4) = 4 \text{ in.}$$

Assume shear span-to-depth ratio: $a/d = 0.3$ ($d = 13.5''$):

$$v_u = \frac{103.5}{(0.85)(14)(13.5)} = 0.645 \text{ ksi;}$$

$$v_u = \text{(by Section 11.14.3, ACI-318-71)}$$
$$= 6.5\,((1 - 0.5)(0.3))\,(1 + 64\,\rho_v)\sqrt{5000}$$
$$= 390\,(1 + \rho_v) \text{ or } \rho_v = \left(\frac{645}{390} - 1\right)\frac{1}{64} = 0.0102$$

Check: $0.20 f'_c/f_y = 0.20 \times 5000/40{,}000 = 0.025 > 0.0102$ (O.K.)

The "Dunes" Hotel, Las Vegas, Nevada. Twenty-two stories, filler block-flat plates. Architects: Milton Schwartz & Associates. Structural Engineers: Paul Rogers & Associates.

6

Torsion

The ACI 318-71 Code requires, for the first time, that torsion effect shall be included in the design of reinforced concrete members, if the torsion stress, v_{t_u} is greater than 1.5 $\sqrt{f'_c}$. Otherwise, the reduction of the shear and flexural strengths is so small that this may be neglected.

The effect of torsion is similar to that of shear, and the resistance of the member to torsion is made up of two parts: the concrete compression zone and the web reinforcement. The contribution of the concrete compression zone to transverse shear strength is approximately equal to the shear at diagonal tension

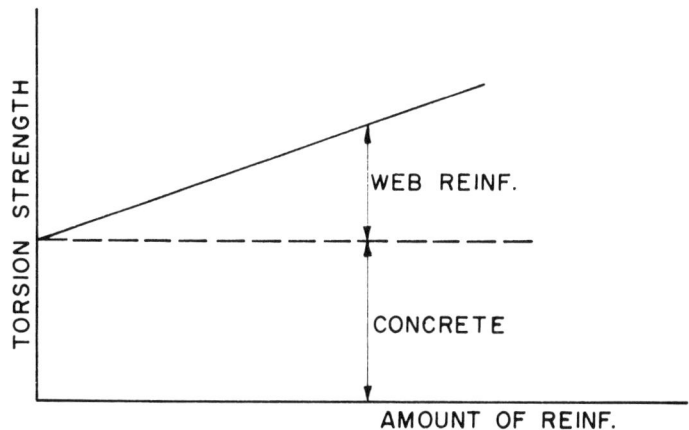

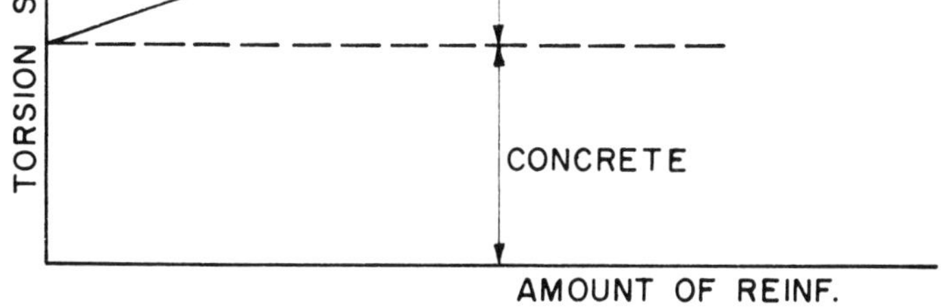

Fig. 6-1.

cracking. Thus, the Code specifies that the concrete could absorb a shear stress of $2\sqrt{f'_c}$ if torsion is not present.

In the case of pure torsion, the contribution of the concrete compression zone to the ultimate torsional strength is equal to about 40 percent of the cracking torque of a beam without web reinforcement. Since the nominal torsion stress at diagonal tension cracking is about $6\sqrt{f'_c}$, the Code specifies that the concrete can absorb a torsion stress of $2.4\sqrt{f'_c}$ if torsion acts only.

When both shear and torsion act simultaneously, they interact with each other. The shear and torsion which the member can resist simultaneously is less than the shear and the torsion that the member can resist when either shear or torsion acts alone. Figure 6-2 shows the shear and torsion stresses which can be carried by the concrete when a beam with web reinforcement is subject to combined torsion and shear.

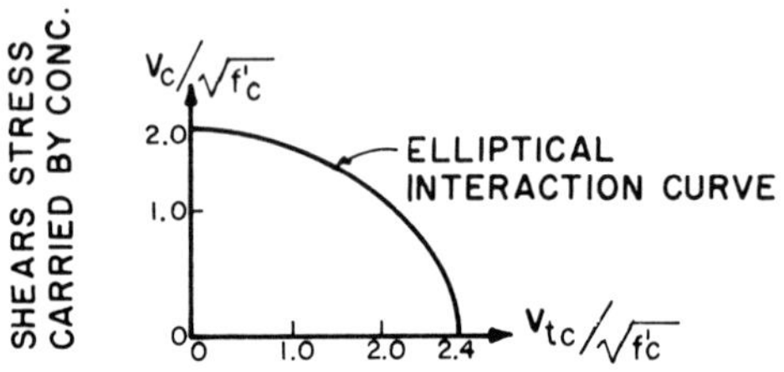

Fig. 6-2.

When the torsion and shear stress v_{tu} and v_u due to design load, using

$$v_{tu} = \frac{3T_u}{\phi \Sigma x^2 y} \qquad \text{(ACI-11-16)}$$

$$v_u = \frac{V_u}{\phi \Sigma b_w d} \qquad \text{(ACI-11-3)}$$

are greater than the allowable torsion and shear stresses, without reinforcement, using

$$v_{tc} = \frac{2.3\sqrt{f'_c}}{\sqrt{1 + (1.2\, v_u/v_{tu})^2}} \qquad \text{(ACI-11-17)}$$

or

$$v_c = \frac{2\sqrt{f'_c}}{\sqrt{1 + (v_{tu}/1.2\, v_u)^2}} \qquad \text{(ACI-11-9)}$$

closed stirrups must be provided.

The stirrups required for torsion are added to those required for shear to produce the total web reinforcement due to combined torsion and shear. It is necessary to provide web reinforcement in each face of the beam, so closed stirrups must be used for torsion. The required area of closed stirrups to carry the torsion stresses, is computed by

$$A_t = \frac{(v_{tu} - v_{tc})\, s \Sigma x^2 y}{3\alpha_t x_1 y_1 (f_y)} \qquad \text{(ACI-11-19)}$$

where $\alpha_t = 0.66 + 0.33\,(y_1/x_1)$, but not more than 1.50.

The stirrups must be spaced not farther apart than one-quarter the sum of the height and depth of the stirrup, or 12 in., whichever is less. The yield strength of the stirrups, f_y, must not be taken as more than 60 ksi.

The required web reinforcement to carry the shear stresses is computed by

$$A_v = \frac{(v_u - v_c)\, b_w s}{f_y} \qquad \text{(ACI-11-13)}$$

and are checked that $(A_v + 2A_t)$ is not less than $50 b_w s/f_y$. Finally, to have the total required web reinforcement

$$\frac{A_v}{s}\,(\text{total}) = \frac{2A_t}{s} + \frac{A_v}{s}$$

The required longitudinal reinforcement to resist torsion, is computed by

$$A_1 = 2A_t \frac{x_1 + y_1}{s} \qquad \text{(ACI-11-20)}$$

or by

$$A_1 = \left[\frac{400\, xs}{f_y} \; \frac{v_{tu}}{v_{tu} + v_u} - 2A_t\right] \frac{x_1 + y_1}{s} \qquad \text{(ACI-11-21)}$$

whichever is greater.

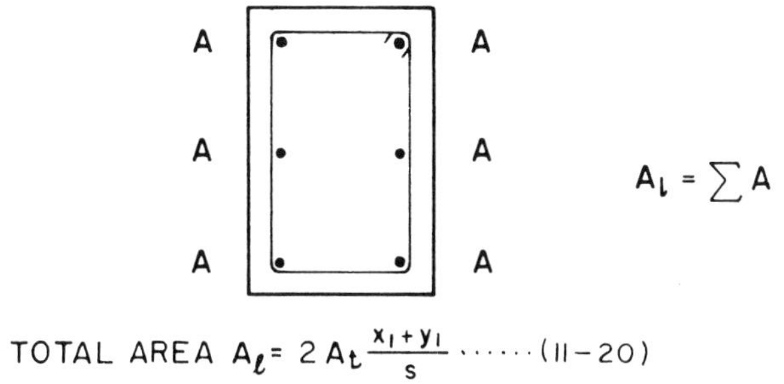

Fig. 6-3.

The value of A_1 computed according to equation (ACI-11-21) need not exceed that obtained by substituting

$$\frac{50\, b_w s}{f_y} \quad \text{for } 2A_t$$

The spacing of longitudinal bars, not less than #3 in size, distributed around the perimeter of the stirrups, shall not exceed 12 in.

Example for Torsion

Given: b = 45 in., b_w = 15 in., h = 36 in., h_f = 8 in., d = 33 in., f'_c = 4,000 psi, f_y = 60,000 psi.

Loads @ d distance from the support: V = 100,000 lb; T = 500,000 lb-in.

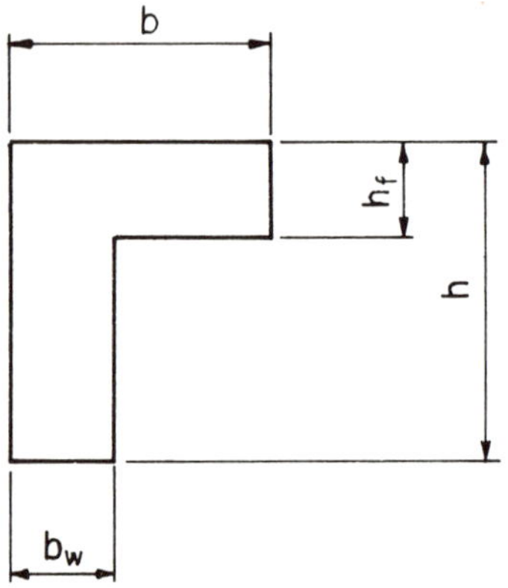

Fig. 6-4.

Design loads, assuming 1.7 load factor: V_u = 170,000 lb; T_u = 850,000 lb-in. Check shear and torsion stresses:

$$v_u = \frac{V_u}{b_w d\phi} = \frac{170{,}000}{0.85(15)33} = 404.00 \text{ psi}$$

$$v_{tu} = \frac{3T_u}{\phi \Sigma x^2 y}$$

x_a = 15 in., x_b = 8 in., y_a = 45 in., y_b = 30 in.,
$\Sigma x^2 y = 15^2(45) + 8^2(30) = 10{,}125 + 1920 = 12{,}045$ in.^{3}

$$v_{tu} = \frac{3(850{,}000)}{0.85(12{,}045)} = 249 \text{ psi}$$

$$v_{tu} > 1.5\sqrt{f'_c}$$

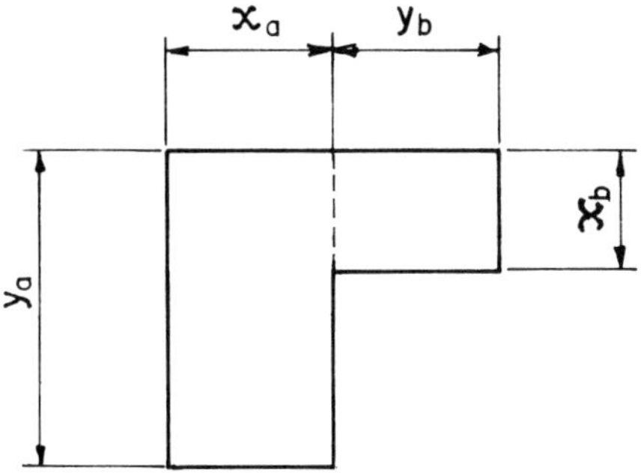

Fig. 6-5.

so torsion should not be neglected. Maximum allowable torsion stress by (ACI-11-18):

$$\frac{12\sqrt{f'_c}}{\sqrt{1+(1.2\,v_u/v_{tu})^2}}$$

$v_u/v_{tu} = 404.0/249.0 = 1.622; (1.2v_u/v_{tu})^2 = 3.79$
$\sqrt{1+3.79} = 2.188$
$\dfrac{12\sqrt{4{,}000}}{2.188} = 346.7 \text{ psi} > 249 \text{ psi (O.K.)}$

Calculate v_c and v_{tc}:

$$v_c = \frac{2\sqrt{f'_c}}{\sqrt{1+(v_{tu}/1.2\,v_u)}} = \frac{2\sqrt{4{,}000}}{\sqrt{1+(249/1.2(404)}} = 112.4 \text{ psi}$$

$$v_u - v_c = 404.00 - 112.40 = 291.60 \text{ psi} < 8\sqrt{f'_c} \text{ (O.K.)}$$

$$v_{t_c} = \frac{2.4\sqrt{f'_c}}{\sqrt{1+(1.2\,v_u/v_{tu})^2}} \qquad \text{(ACI-11-17)}$$

$$v_{tc} = \frac{2.4(63.25)}{2.188} = 69.37 \text{ psi}$$

Required web reinforcement to carry the shear and torsion stresses: Assume #5 stirrups with $1\frac{1}{2}$ in. clear cover:

$$\alpha_t = 0.66 + 0.33\,(y_1/x_1); y_1/x_1 = \frac{36 - 2\,(1.75)}{15 - 2\,(1.75)} = 2.826$$

$$\alpha_t = 0.66 + 0.33\,(2.826) = 1.592; \text{ use } \alpha_t = 1.50 \text{ as max}$$

$$v_{t_u} - v_{tc} = 249.00 - 69.37 = 179.63 \text{ psi}$$

$$A_t = \frac{(v_{tu} - v_{tc})\,s\,\Sigma x^2 y}{3\alpha_t x_1 y_1\,(f_y)} \qquad \text{(ACI-11-19)}$$

$$A_t = \frac{179.63s\,(12{,}045)}{3(1.50)11.50(32.5)60{,}000} = 0.021s$$

$$2A_t = 2(0.021s) = 0.042s$$

$$v_u - v_c = 404.0 - 112.4 = 291.6 \text{ psi}$$

$$v_u - v_c > 4\sqrt{f'_c} = 253 \text{ psi so max } s = \frac{d}{4}$$

$$A_v = \frac{(v_u - v_c)\,b_w s}{f_y} \qquad \text{(ACI-11-13)}$$

$$A_v = \frac{253.00(15)s}{60{,}000} = 0.063s$$

Total required web reinforcement: $2A_t + A_v = 0.042s + 0.063s = 0.105s$
For #5 stirrups,

$$2(0.31) = 0.105s, \; s = \frac{0.62}{0.105} = 5.9 \text{ in.}$$

The Code requires that the spacing of closed stirrups shall not exceed

$$(x_1 + y_1)/4 = \frac{11.5 + 32.5}{4} = 11.0 \text{ in. or } 12 \text{ in.}$$

whichever is smaller. Use #5 closed stirrups @ $5\frac{1}{2}$ in. o.c. Check the longitudinal reinforcement required to resist torsion:

$$A_1 = 2A_t \frac{x_1 + y_1}{s} \qquad \text{(ACI-11-20)}$$

$$A_1 = 0.042s \frac{11.5 + 32.5}{s} = 1.85 \text{ in.}^2\text{, or}$$

$$A_1 = \left[\frac{400xs}{f_y}\left(\frac{v_{tu}}{v_{tu} + v_u}\right) - 2A_t\right]\left(\frac{x_1 + y_1}{s}\right) \qquad \text{(ACI-11-21)}$$

$$A_1 = \left[\frac{400(15)s}{60{,}000}\left(\frac{249}{249 + 404}\right) - 0.042s\right] 44/s$$

$$A_1 = [0.0381 - 0.042]\; 44 < 0, \text{ so use } A_1 = 1.85 \text{ in.}^2$$

The Code requires at least #3 longitudinal bars, spaced not farther apart than 12 in., and at least one bar in each corner of the stirrups.

Assume approximately $1.85/4 = 0.46$ in.^{2} at top, at bottom, and at third-heights on the sides. For A_1 use: 2 - #4 = 0.40 in.^{2} @ third-heights each side. Add $(1.85 - 0.80)/2 = 0.525$ in.^{2} each to top and bottom bars.

REINFORCED CONCRETE BEAM COMPUTATION TABLE,

STRENGTH DESIGN METHOD, ACI 318-71

GIRDER MARK	FLOOR	DESIGN DATA (RESULTS FROM FRAME ANALYSIS)[1]														
		BENDING MOMENTS (FT-KIPS)														
		LEFT (SOUTH, WEST) END						MIDSPAN			RIGHT (NORTH, EAST) END					
		NEGATIVE				POSITIVE[2]		POSITIVE		NEG.[3]	NEGATIVE				POSITIVE[2]	
		D	L	W	E	W	E	D	L	L	D	L	W	E	W	E
(1)	(2)	(3)	(4)	(5)	(6)	(7)	(8)	(9)	(10)	(11)	(12)	(13)	(14)	(15)	(16)	(17)

1) D = DEAD LOAD (IT MAY INCLUDE EFFECT OF SETTLEMENT, CREEP, SHRINKAGE & TEMPERATURE)
L = LIVE LOAD, REDUCED AS PERMITTED BY LOCAL CODE. (IMPACT EFFECTS SHALL BE CONSIDERED)
W = WIND LOAD, AS SPECIFIED.
E = EARTH-QUAKE LOAD, AS SPECIFIED OR ACCORDING TO UBC OR LOCAL CODE.

2) WIND OR EARTH-QUAKE MAY REVERSE END MOMENTS.

3) ALTERNATE LOADING AND/OR UNEVEN SPANS MAY CREATE UPWARD MOMENTS.

REINFORCED CONCRETE BEAM COMPUTATION TABLE, STRENGTH DESIGN METHOD ACI 318-71									
GIRDER MARK	FLOOR	DESIGN DATA (RESULTS FROM FRAME ANALYSIS) [1)]							
		SHEARS, (KIPS) AT "d" DISTANCE FOR CONTINUOUS ENDS, AT FACE OF SUPPORT FOR FREE ENDS.							
		LEFT (S,W) END				RIGHT (N,E) END			
		D	L	W	E	D	L	W	E
(1)	(2)	(18)	(19)	(20)	(21)	(22)	(23)	(24)	(25)

REINFORCED CONCRETE BEAM COMPUTATION TABLE,
STRENGTH DESIGN METHOD, ACI 318-71

GIRDER MARK	FLOOR	ULTIMATE BENDING MOMENTS, (FT-KIPS) [4]																	
		LEFT (SOUTH, WEST) END								MIDSPAN		RIGHT (NORTH, EAST) END							
		NEGATIVE			POSITIVE					POS.	NEG.	NEGATIVE			POSITIVE				
		1.4 (3)D + 1.7 (4)L	0.75 [1.4 (3)D + 1.7 (4)L + 1.7 (5)W]	0.75 [1.4 (3)D + 1.7 (4)L + 1.87 (6)E]	1.7 (4)L − 1.4 (3)D	0.75 [1.7 (13)L + 1.7 (16)W − 1.4 (12)D]	0.75 [1.7 (13)L + 1.87 (17)E − 1.4 (12)D]	1.3 (7)W − 0.9 (3)D	1.43 (8)E − 0.9 (3)D	1.4 (9)D + 1.7 (10)L	1.7 (11)L − 1.4 (10)D	1.4 (12)D + 1.7 (13)L	0.75 [1.4 (12)D + 1.7 (13)L + 1.7 (14)W]	0.75 [1.4 (12)D + 1.7 (13)L + 1.87 (15)E]	1.7 (13)L 1.4 (12)D	0.75 [1.7 (13)L + 1.7 (16)W − 1.4 (12)D]	0.75 [1.7 (13)L + 1.87 (17)E − 1.4 (12)D]	[1.3 (16)W − 0.9 (12)D]	[1.43 (17)E − 0.9 (12)D]
(1)	(2)	(26)	(27)	(28)	(29)	(30)	(31)	(32)	(33)	(34)	(35)	(36)	(37)	(38)	(39)	(40)	(41)	(42)	(43)

4) COMBINATIONS, TO OBTAIN MAX. (OR MIN.) MOMENTS AND SHEARS:

I: 1.4 D + 1.7 L, OR 0.75 (1.4 D + 1.7 L) IF SETTLEMENT, CREEP, SHRINKAGE & TEMPERATURE INCLUDED

II: 0.75 (1.4 D + 1.7 L + 1.7 W)

III: 0.75 (1.4 D + 1.7 L + 1.87 E), FOR CALIFORNIA CHECK THE PROVISIONS OF THE SEAC.

IV: 1.3 W − 0.9 D

V: 1.43 E − 0.9 D

REINFORCED CONCRETE BEAM COMPUTATION TABLE,

STRENGTH DESIGN METHOD, ACI 318-71

GIRDER MARK	FLOOR	ULTIMATE SHEARS (KIPS) 4)						DESIGN DATA							
		AT "d" DIST. FOR CONT'S ENDS, AT FACE OF SUPPORT FOR FREE.						MAX. ABSOLUTE VALUES						f_c' (CRUSHING STRENGTH OF CONC.) 5)	f_y (YIELD POINT OF STEEL) 6)
		LEFT (S,W) END			RIGHT (N,E) END			MOMENTS M_u			SHEARS V_u		T_u		
								LEFT	MIDSPAN	RIGHT	LEFT	RIGHT			
		1.4 (18)D + 1.7 (19)L	0.75 [1.4 (18)D + 1.7 (19)L + 1.7 (20)W]	0.75 [1.4 (18)D + 1.7 (19)L + 1.87 (21)E]	1.4 (22)D + 1.7 (23)L	0.75 [1.4 (22)D + 1.7 (23)L + 1.7 (24)W]	0.75 [1.4 (22)D + 1.7 (23)L + 1.87 (25)E]	(25) TO (33)	(34) TO (35)	(36) TO (43)	(44) TO (46)	(47) TO (49)	ULTIMATE TORSIONAL MOMENT		
(1)	(2)	(44)	(45)	(46)	(47)	(48)	(49)	(50)	(51)	(52)	(53)	(54)	(55)	(56)	(57)

5) IN CALIFORNIA THE MAXIMUM SPECIFIED STRENGTH FOR LIGHTWEIGHT CONCRETE IS LIMITED TO 4000 psi.
FOR SEISMIC FORCES: $f'_{c\ min}$ = 3000 psi

6) $f_{y\ max.}$ = 80,000 psi, EXCEPT FOR SHEAR AND TORSION REINFORCEMENT WHERE $f_{y\ max.}$ = 60,000 psi. - IF SEISMIC GOVERNS $f_{y\ max.}$ = 60,000 psi.

NOTE: FOR 4) SEE PREVIOUS TABLE.

REINFORCED CONCRETE BEAM COMPUTATION TABLE,

STRENGTH DESIGN METHOD, ACI 318-71

GIRDER MARK	FLOOR	DESIGN DATA												
		BEAM DIMENSIONS										CONC. CAPACITY M_{uc}		
		SPAN LENGTH, ℄ TO ℄	" CLEAR "	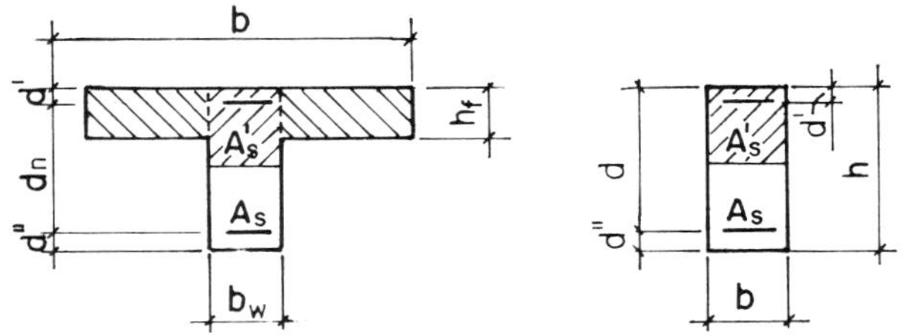								WEB	FLANGE	Σ
		L (FT)	L' (FT)	b_w (IN)	h (IN)	b (IN)	h_f (IN)	d (IN)	d_n (IN)	d' (IN)	d'' (IN)	$K_{umax} \times b_w \times d^2/12$ $K_{umax} \times (60) \times (64)^2/12$	$.85f'_c(b-b_w)\,h_f\,\phi(d-\frac{h_f}{2})/12$ $.0638(56)\times[(62)-(60)]\times(63)[(64)-\frac{(63)}{2}]$	(68) + (69)
(1)	(2)	(58)	(59)	(60)	(61)	(62)	(63)	(64)	(65)	(66)	(67)	(68)	(69)	(70)

REINFORCED CONCRETE BEAM COMPUTATION TABLE,

STRENGTH DESIGN METHOD, ACI 318-71

GIRDER MARK	FLOOR	FLEXURAL REINFORCEMENT												
		CODE REQUIREMENT			LEFT (S,W) END									
				8)	M'	A'_s- COMPR. REINF.			A_s- TENSION REINF.					
							7)		9)	TO BALANCE				
										WEB	FLANGE	A'_s	Σ	
		$\rho_{min} = \frac{200}{f_y} = \frac{200}{(57)}$	$A_{s\,min} = \rho_{min}\, b_w d$ = (71) x (60) x (64)	ρ_{max} OR $(\rho - \rho')_{max.}$ $\Big\} = 0.75\,\rho_b$	$M_{ureq'd} - M_{uconc.}$ = (50) - (70)	REQ'D YES OR NO	$M'/\bar{M}'_{COMPR.}$ (74) / $\bar{M}'_{COMPR.}$	NO. OF BARS SIZE	a_u	$M_{ureq'd}/a_u d$ (50) / (78) x (64)	$.85 f'_c h_f (b - b_w)/f_y$ .85 (56) (63) [(62) - (60)] / (57)	7) $M'/\bar{M}'_{TENS'N}$ (74) / —"—	(79) + (80) + (81)	NO. OF BARS SIZE
(1)	(2)	(71)	(72)	(73)	(74)	(75)	(76)	(77)	(78)	(79)	(80)	(81)	(82)	(83)

7) SEE TABLE
FOR VALUES OF
$M'_{C\,COMPR.}$ & $\bar{M}'_{TENS'N.}$

8) FOR SEISMIC: THE MAXIMUM ρ SHALL NOT EXCEED 0.5 OF ρ_b

9) SEE CHART 3-8 FOR a_u.

REINFORCED CONCRETE BEAM COMPUTATION TABLE,

STRENGTH DESIGN METHOD, ACI 318-71

GIRDER MARK	FLOOR	FLEXURAL REINFORCEMENT									
		MIDSPAN									
		M'	A's COMPR. REINF.			A_s TENSION REINF.					
							TO BALANCE				
				7)		9)	WEB	FLANGE	A's 7)	Σ	
		$M_{ureq'd} - M_{u\,conc.}$ = (51) − (70)	REQ'D YES OR NO	$M'/\bar{M}'_{COMPR.}$ (84)/$\bar{M}'_{COMPR.}$	NO OF BARS SIZE	a_u	$M_{ureq'd}/a_u \times d$ (51)/(88) × (64)	$.85 \times f'_c \times h_f (b-b_w)/f_y$ .85(56)×(63)[(62)-(60)]/(57)	$M'/\bar{M}'_{TENS'N}$ (84)/ —"—	(89)+(90)+(91)	NO OF BARS SIZE
(1)	(2)	(84)	(85)	(86)	(87)	(88)	(89)	(90)	(91)	(92)	(93)

7) SEE TABLE FOR VALUES OF $M'_{C\,COMPR.}$ & $\bar{M}'_{TENS'N}$

REINFORCED CONCRETE BEAM COMPUTATION TABLE,
STRENGTH DESIGN METHOD, ACI 318-71

GIRDER MARK	FLOOR	FLEXURAL REINFORCEMENT — RIGHT (S,W) END										
		M'	A's COMPR. REINF.			As TENSION REINF.						
							TO BALANCE					
							WEB	FLANGE	A's	Σ		
		$M_{ureq'd} - M_{uconc.}$ = (52) - (70)	REQ'D YES OR NO	7) $M'/\bar{M}'_{COMPR.}$ (94) / $M'_{COMPR.}$	NO OF BARS SIZE	9) a_u	$M_{ureq'd}/a_u x d$ (52) / (98) x (64)	$.85 \times f'_c \times h_f (b - b_w)/f_y$ .85 (56) x (63) [(62) - (60)] / (57)	7) $M'/\bar{M}'_{TENS'N}$ (94) / —l—	(99) + (100) + (101)	NO OF BARS SIZE	
(1)	(2)	(94)	(95)	(96)	(97)	(98)	(99)	(100)	(101)	(102)	(103)	

7) SEE TABLE
FOR VALUES OF
$M'_{C\ COMPR.}$ & $\bar{M}'_{TENS'N}$

REINFORCED CONCRETE BEAM COMPUTATION TABLE,

STRENGTH DESIGN METHOD, ACI 318 - 71

GIRDER MARK	FLOOR	AT LEFT END										
		V_u	v_u	v_{tu}	ρ_w	LIMITS FOR v_c						TORSION SHEAR CHECK
										LIGHT WT CONCRETE		
										"SAND" LIGHT WT	"ALL" LIGHT WT	
		10) MAX. OF (44) - (45) - (46)	$v_u = V_u/\phi\, b_w d$ $v_u =$ (104) $/0.85 \times$ (60) $\times$ (64)	$v_{tu} = 3\,T_u/\phi \Sigma x^2 y$ $v_{tu} = 3 \times$ (55) $/0.85 \times \Sigma x^2 y$	$\rho_w = \frac{A_s}{b_w d} = \frac{(82)}{(60) \times (64)}$	$v_c = 1.5\sqrt{f_c'} = 1.5\sqrt{(56)}$	$v_c = 2\sqrt{f_c'} = 2\sqrt{(56)}$	$v_c = 3.5\sqrt{f_c'} = 3.5\sqrt{(56)}$	$v_c = 1.9\sqrt{f_c'} + 2500\,\rho_w \frac{V_u d}{M_u}$ $v_c = 1.9\sqrt{(56)} + 2500 \times (107) \times \frac{(104)(64)}{(50)}$ BUT v_c SHALL NOT BE GREATER THAN (110) AND $\frac{V_u d}{M_u} < 1.0$	$v_c = 1.7\sqrt{f_c'}$ $v_c = 1.7\sqrt{(56)}$	$v_c = 1.5\sqrt{f_c'}$ OR USE $f_{ct}/6.7$ FOR $\sqrt{f_c'}$	IF $v_{tu} > 1.5\sqrt{f_c'}$ (106) > (108) CHECK TORSION SHEAR BY CODE
(1)	(2)	(104)	(105)	(106)	(107)	(108)	(109)	(110)	(111)	(112)	(113)	(114)

10) IF SEISMIC GOVERNS CHECK V_u FROM DESIGN GRAVITY LOADS ON THE MEMBER AND FROM THE MOMENT CAPACITIES OF PLASTIC HINGES AT THE ENDS OF THE MEMBER PRODUCED BY LATERAL DISPLACEMENT. IN CALIFORNIA $V_u \geq \frac{M_u^A + M_u^B}{l} + 1.4\,V_{D+L}$, WHERE M_u^A AND M_u^B ARE ULTIMATE MOMENT CAPACITIES OF OPPOSITE SENSE AT EACH END OF THE MEMBER AND V_{D+L} IS THE SIMPLE SPAN SHEAR.

REINFORCED CONCRETE BEAM COMPUTATION TABLE, STRENGTH DESIGN METHOD, ACI 318-71											
GIRDER MARK	FLOOR	SHEAR REINFORCEMENT									
		AT LEFT END									
		@ "d" DISTANCE FROM FACE OF SUPPORT									
		v_c	v_u'	SIZE	A_v	SPACING OF STIRRUPS 11)					
						REQUIRED					FURNISHED
		SELECTED VALUE FROM (108) - (113)	$v_u' = v_u - v_c$ $v_u' =$ (105) - (115) $v_u' < 8\sqrt{f_c'}$	MIN. # 3	AREA OF 2 - LEGS STIRRUPS	$S = \frac{A_v \quad f_y}{(v_u - v_c')\ b_w}$ $S = \frac{(118) \times (57)}{(116) \times (60)}$	$S_{max.} = 0.5\,d$ $S_{max.} = 0.5 \times$ (64)	IF $v_u - v_c > 4\sqrt{f_c'}$ $S_{max.} = 0.25\,d$ $S_{max.} = 0.25 \times$ (64)	$S_{max.} = \frac{A_v f_y}{50\,b_w}$ $S_{max.} = \frac{(118) \times (57)}{50 \times (60)}$	$S_{max.} = 24''$	SPACING
(1)	(2)	(115)	(116)	(117)	(118)	(119)	(120)	(121)	(122)	(123)	(124)

11) IF SEISMIC GOVERNS, WITHIN A DISTANCE EQUAL TO FOUR TIMES THE EFFECTIVE DEPTH "d" FROM THE END OF THE MEMBER, THE SPACING SHALL NOT BE GREATER THAN $\frac{A_v d}{0.15 A_s'}$ OR $\frac{A_v d}{0.15 A_s}$ OR $\frac{d}{4}$. –

IN CALIFORNIA USE MAXIMUM SPACING OF NOT OVER d/4, 8BAR DIAMETERS, 24 STIRRUP TIE DIAMETERS, OR 12" WHICH EVER THE LEAST AT EACH END OF ALL FLEXURAL MEMBERS. THE FIRST STIRRUP TIE SHALL BE LOCATED NOT MORE THAN 2" FROM THE COLUMN FACE, AND THE LAST, A DISTANCE OF AT LEAST TWICE THE DEPTH OF THE MEMBER.

REINFORCED CONCRETE BEAM COMPUTATION TABLE,

STRENGTH DESIGN METHOD, ACI 318 - 71

GIRDER MARK	FLOOR	SHEAR REINFORCEMENT AT LEFT END WITHIN DISTANCE "S_1" 12)						
		S_1	Q	$v'_u \times b_w$ / $B \times Q$	SIZE	Av	SPACING OF STIRRUPS REQUIRED	SPACING OF STIRRUPS FURN'D.
		$S_1 = \frac{v_u'(l'-2d)}{2 v_u}$	SEE TABLE BELOW	B = 1.0	MIN # 3	AREA OF 2 - LEGS STIRRUPS	FOR SPACING READ FROM ACI-SPEC. PUBL. NO. 17 (SHEARS) BUT NOT MORE THAN (119), (120), (121), (122), (123), (130)	SPACING
		$S_1 = \frac{(116) \times (59) - 2\,(64)}{2\,(105)}$		$\frac{(116) \times (60)}{1.0 \times (126)}$				
(1)	(2)	(125)	(126)	(127)	(128)	(129)	(130)	(131)

12) IN CALIFORNIA STIRRUPS SHALL BE SPACED AT NO MORE THAN d/2 THROUGHOUT THE LENGTH OF THE MEMBER.—

VALUES OF Q FOR STIRRUPS ⌐ OR ⊔

	fy psi		
BARS	40,000	50,000	60,000
# 3	7,480	9,350	11,220
# 4	13,600	17,000	20,400
# 5	21,080	26,350	31,620
# 6	29,920	37,400	44,880

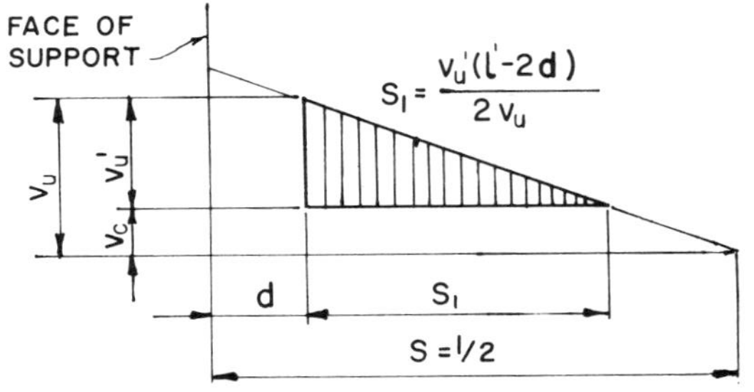

REINFORCED CONCRETE BEAM COMPUTATION TABLE,

STRENGTH DESIGN METHOD, ACI 318-71

GIRDER MARK	FLOOR	SHEAR REINFORCEMENT AT RIGHT END										
		V_u	v_u	v_{tu}	p_w	LIMITS FOR v_c						TORSION SHEAR CHECK
										LIGHT WT CONCRETE "SAND." LIGHT WT	LIGHT WT CONCRETE "ALL" LIGHT WT	
		MAX. OF (47) (48) (49) 10)	$v_u = V_u / \phi b_w d$ $v_u =$ (132) / 0.85 × (60) × (64)	SAME AS (106)	$p_w = \frac{A_s}{b_w d} = \frac{(102)}{(60) \times (64)}$	SAME AS (108)	SAME AS (109)	SAME AS (110)	$v_c = 1.9\sqrt{f_c'} + 2500 p_w \frac{V_u d}{M_u}$ $v_c = 1.9\sqrt{(56)} + 2500 (135) \frac{(132)(64)}{(52)}$ BUT v_c SHALL NOT BE GREATER THAN (138) AND $\frac{V_u d}{M_u}$	SAME AS (112)	SAME AS (113)	IF $v_{tu} > 1.5\sqrt{f_c'}$ (134) > (136) CHECK TORSION SHEAR BY CODE
(1)	(2)	(132)	(133)	(134)	(135)	(136)	(137)	(138)	(139)	(140)	(141)	(142)

REINFORCED CONCRETE BEAM COMPUTATION TABLE,

STRENGTH DESIGN METHOD, ACI 318 – 71

		SHEAR REINFORCEMENT									
		AT RIGHT END									
		@ "d" DISTANCE FROM FACE OF SUPPORT									
		v_c	v_u'	SIZE	A_v	SPACING OF STIRRUPS 11)					
						REQUIRED					FURNISHED
GIRDER MARK	FLOOR	SELECTED VALUE FROM (136) – (141)	$v_u' = v_u - v_c$ $v_u' =$ (133) – (143) $v_u' = < 8\sqrt{f_c'}$	MIN. # 3	AREA OF 2-LEGS STIRRUPS	$S = \frac{A_v \; f_y}{(v_u - v_c')\, b_w}$ $S = \frac{(146) \times (57)}{(144) \times (60)}$	SAME AS (120)	SAME AS (121)	$S_{max.} = \frac{A_v \, f_y}{50\, b_w}$ $S_{max.} = \frac{(146) \times (57)}{50 \times (60)}$	$S_{max.} = 24''$	SPACING
(1)	(2)	(143)	(144)	(145)	(146)	(147)	(148)	(149)	(150)	(151)	(152)

REINFORCED CONCRETE BEAM COMPUTATION TABLE,

STRENGTH DESIGN METHOD, ACI 318-71

GIRDER MARK	FLOOR	SHEAR REINFORCEMENT AT RIGHT END WITHIN DISTANCE "S_1" 12)						
		S_1	Q	$\frac{v_u' \times b_w}{B \times Q}$		A	SPACING OF STIRRUPS REQUIRED	
		$S_1 = \frac{v_u'(l' - 2d)}{2 v_u}$ $S_1 = \frac{(144) \times (59) - 2 (64)}{2 (133)}$	SEE TABLE BELOW	B = 1.0 $\frac{(144) \times (60)}{1.0 \times (154)}$	MIN. # 3	AREA OF 2 - LEGS STIRRUPS	FOR SPACING READ FROM ACI-SPEC. PUBL. NO. 17 (SHEAR 5) BUT NOT MORE THAN (147), (148), (149), (150), (151)	SPACING
(1)	(2)	(153)	(154)	(155)	(156)	(157)	(158)	(159)

FOR "S_1" AND "Q" SEE (125) AND (127)

REINFORCED CONCRETE BEAM COMPUTATION TABLE, STRENGTH DESIGN METHOD, ACI 318-71								
GIRDER MARK	FLOOR	DEVELOPMENT OF LONGITUDINAL REINFORCEMENT						
		BASIC DEVELOPMENT LENGTH IN TENSION						
		#	d_b	A_b	# 11 OR SMALLER BARS	# 14 BARS	# 18 BARS	DEFORMED WIRE
		BAR SIZE	NOMINAL DIAMETER OF BAR OR WIRE	AREA OF AN INDIVIDUAL BAR	$0.04\ A_b f_y/\sqrt{f_c'}$ $\frac{0.04 \times (162) \times (57)}{\sqrt{(56)}}$ BUT NOT LESS THAN: $0.0004\ d_b f_y =$ $0.0004 \times (161) \times (57)$	$0.085\ f_y/\sqrt{f_c'}$ $\frac{0.085 \times (57)}{\sqrt{(56)}}$	$0.11\ f_y/\sqrt{f_c'}$ $\frac{0.11 \times (57)}{\sqrt{(56)}}$	$0.03\ d_b f_y/\sqrt{f_c'}$ $\frac{0.03 \times (161) \times (57)}{\sqrt{(56)}}$
(1)	(2)	(160)	(161)	(162)	(163)	(164)	(165)	(166)

REINFORCED CONCRETE BEAM COMPUTATION TABLE,

STRENGTH DESIGN METHOD, ACI 318-71

GIRDER MARK	FLOOR	DEVELOPMENT OF LONGITUDINAL REINFORCEMENT [13)] — DEVELOPMENT LENGTH IN TENSION (l_d) — MODIFICATION FACTORS										
		TOP REINF.	BARS WITH $f_y > 60$	ALL LIGHT WEIGHT CONC.	SAND LIGHT WEIGHT CONC.	REINFORCED SPACED LEAST @ 6" o.c. ₵ LEAST 3" FROM SIDE FACE OF BEAM		BARS ENCLOSED WITHIN SPIRAL 1/4"φ - 4" PITCH	REQ'D DEVEL. LENGTH	EQUIVALENT EMBEDMENT LENGTH OF STANDARD HOOKS - l_e		TOTAL DEVEL. LENGTH
		1.4	$2 - \frac{60{,}000}{f_y}$	BASIC DEVELOPMENT TO BE MULTIPLIED WITH 1.33	BASIC DEVELOPMENT TO BE MULTIPLIED WITH 1.18	0.8	$\frac{A_s \text{ REQ'D.}}{A_s \text{ PROVIDED}}$	0.75	PRODUCT OF BASIC DEVELOP. LENGTH ₵ SELECTED MODIFICATION F.	$f_h = \xi\sqrt{f_c'}$ FOR ξ SEE TABLE BELOW	USE f_h FOR f_y AND l_e FOR l_d	(174) + (176)
(1)	(2)	(167)	(168)	(169)	(170)	(171)	(172)	(173)	(174)	(175)	(176)	(177)

13) IN CALIFORNIA: LENGTH OF ANCHORAGE SHALL BE DETERMINED BY PROVISIONS OF SEAC. —

ξ VALUES

BAR SIZE	fy = 60 ksi TOP BARS	fy = 60 ksi OTHER BARS	fy = 40 ksi ALL BARS
#3 TO #5	540	540	360
#6	450	540	360
#7 TO #9	360	540	360
#10	360	480	360
#11	360	420	360
#14	330	330	330
#18	220	220	220

REINFORCED CONCRETE BEAM COMPUTATION TABLE,

STRENGTH DESIGN METHOD, ACI 318-71

GIRDER MARK	FLOOR	DEVELOPMENT OF LONGITUDINAL REINFORC'MT.				CRACK CONTROL ($f_y > 40$ ksi)				
		DEVELOPMENT LENGTH IN COMPR. (l_d)				f_s	d_c (in)	A (sq in)	S_b (in)	Z
		l_d (in)	REDUCTION FACTORS		TOTAL DEV. LENGTH					
		$0.02\, f_y d_b/\sqrt{f_c}$ $\dfrac{0.02 \times (57) \times (161)}{\sqrt{(56)}}$ BUT NOT LESS THAN $0.0003\, f_y d_b$ OR 8"	$\dfrac{\text{REQ'D. } A_s'}{\text{PROVIDED } A_s'}$	IF REINFORCEMENT ENCLOSED BY SPIRALS. (MIN. 1/4" ϕ ₵ MIN. 4" PITCH) 0.25	(178) × (179) OR (180)	$f_s = 0.60\, f_y$ $f_s = 0.60 \times$ (57)	THICKNESS OF CONCRETE COVER MEASURED FROM THE EXTREME TENSION FIBER TO CENTER OF THE BAR	EFFECTIVE TENSION AREA OF CONCRETE SURROUNDING THE BAR = $2 d_c S_b$	REINFORCING BAR SPACING	$Z = f_s \sqrt[3]{d_c A} < \begin{matrix} 175 \text{ (FOR INT. EXP.)} \\ 145 \text{ (FOR EXT. EXP.)} \end{matrix}$
(1)	(2)	(178)	(179)	(180)	(181)	(182)	(183)	(184)	(185)	(186)

The "Constellation," Chicago, Illinois. Twenty-seven stories, filler block and flat plate. Architects: Milton Schwartz & Associates. Structural Engineers: Paul Rogers & Associates.

7

Columns and Axially Loaded Members

The design of members subjected to flexure and axial loads is covered in a unified manner in the new ACI 318-71 Code. Thus, there are no separate provisions for columns and axially loaded members.

The 1963 ACI Code contained several complex formulas for computing the capacity of column cross sections. In the new Code these formulas have been replaced by the provisions of Section 10.2.1, whereby the strength design of members for flexure and axial loads are based on the assumptions given in the Code, and on satisfaction of the applicable conditions on equilibrium and compatibility of strains. The assumptions referred to are essentially the same as those specified in the old 1963 Code.

In accordance with the provisions of the new Code, the strains in steel and concrete are linearly distributed and the maximum usable compression strain in the extreme fiber is limited to 0.003. Once the strains have been assumed, it is possible to compute the stresses in the reinforcement and the compression zone. Having obtained the stresses, by computation of forces, summing of internal forces parallel to the axis of the column, and the moment of the internal forces parallel to the axis of the column, and the moments of the internal forces about the centroid, it is possible to compute P_u and M_u. This method of computation is illustrated on two numerical examples as follows:

Figures 7-3 and 7-4, as interaction diagrams, are the results of these computations, showing the resulting loads, P_u, and moments, M_u.

The detailed computation examples are too complicated for daily design use. For this reason the following Tables 7-1 and 7-2 are prepared, providing the ultimate design capacities of different spiral and tied columns with varying

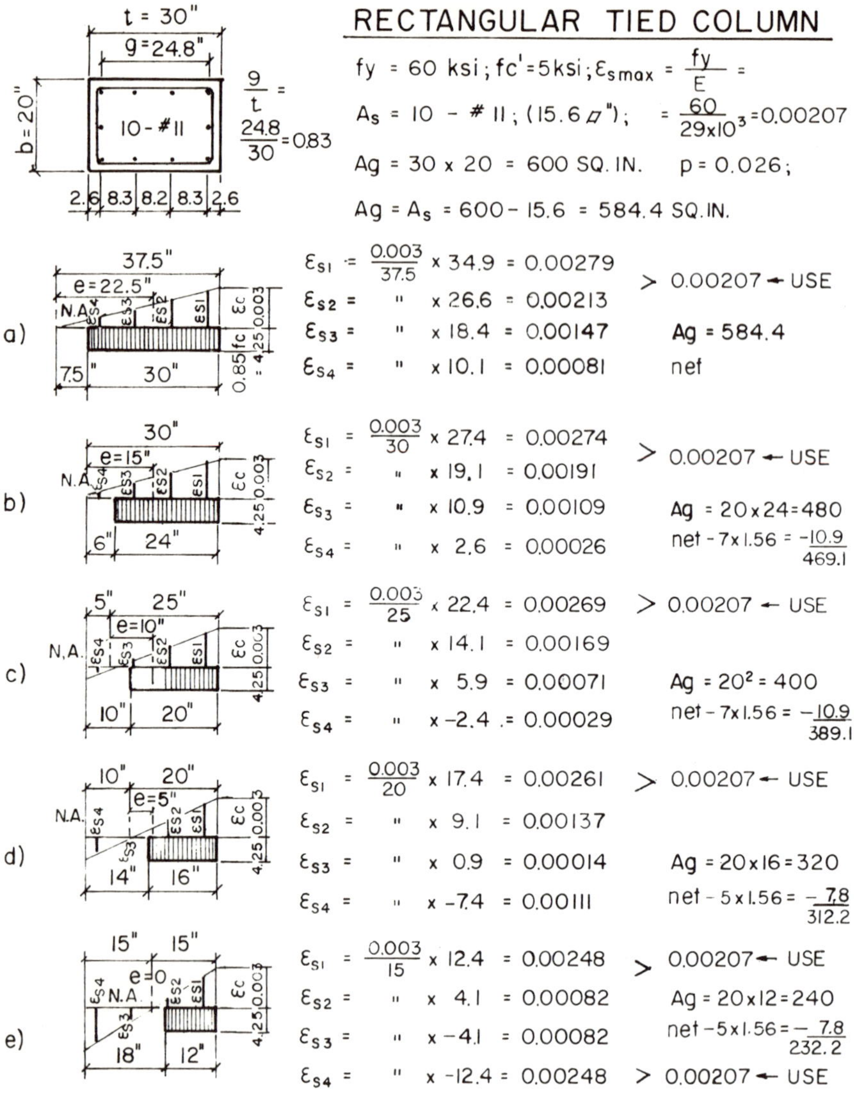

Fig. 7-1.

Capacity:

a) Conc.:	$\phi \times A_{g_{net}} \times 0.85 \times f_c'$	=		$P_u(K) \times Arm('') = M_u(''K)$
	$0.70 \times 0.85 \times 5 \times A_{g(net)}$	=	$2.98 \times 584.4 = 1740 \times$	$0 = 0''K$

(continued)

Steel: $\phi \times A_{s(\text{bar})} \times N_{o(\text{bars})} \times E_s \times \epsilon_s =$

$\underbrace{0.70 \times 1.56 \times 29 \times 10^3}_{31.65 \times 10^3}$

		31.65×10^3	× 3	× 0.00207 =	197 ×	12.4 =	2420
		31.65 ×	× 2	× 0.00207 =	131 ×	4.1 =	537
		31.65 ×	× 2	× 0.00147 =	93 ×	−4.1 =	−382
		31.65 ×	× 3	× 0.00081 =	77 ×	−12.4 =	−943
		$e/t = 0.73/30 = 0.024$			2238 ×	0.73 =	1632″K
					P_u ×	(ecc.):	136′K
b)	Conc.:		2.98 ×	469.1 =	1400 ×	3.0 =	4200
	Steel:	31.65 ×	3 ×	2.07 =	197 ×	12.4 =	2420
		31.65 ×	2 ×	1.91 =	121 ×	4.1 =	496
		31.65 ×	2 ×	1.09 =	69 ×	−4.1 =	−283
		31.65 ×	3 ×	0.26 =	25 ×	−12.4 =	−306
		$e/t = 3.6/30 = 0.12$			1812 ×	3.6 =	6527″K
					P_u ×	(ecc.):	545′K
c)	Conc.:		2.98 ×	389.1 =	1160 ×	5.0 =	5800
	Steel:	31.65 ×	3 ×	2.07 =	197 ×	12.4 =	2420
		31.65 ×	2 ×	1.69 =	107 ×	4.1 =	439
		31.65 ×	2 ×	0.71 =	45 ×	−4.1 =	−185
		31.65 ×	−3 ×	0.29 =	−28 ×	−12.4 =	342
		$e/t = 5.95/30 = 0.197$			1481 ×	5.95 =	8816″K
					P_u ×	(ecc.):	735′K
d)	Conc.:		2.98 ×	312.2 =	932 ×	7.0 =	6520
	Steel:	31.65 ×	3 ×	2.07 =	197 ×	12.4 =	2420
		31.65 ×	2 ×	1.37 =	87 ×	4.1 =	356
		31.65 ×	2 ×	0.14 =	9 ×	−4.1 =	−36
		31.65 ×	−3 ×	1.11 =	−105 ×	−12.4 =	1308
		$e/t = 9.45/30 = 0.314$			1120 ×	9.45 =	10568″K
					P_u ×	(ecc.):	870′K
e)	Conc.:		2.98 ×	232.2 =	692 ×	9.0 =	6230
	Steel:	31.65 ×	3 ×	2.07 =	197 ×	12.4 =	2420
		31.65 ×	2 ×	0.82 =	51 ×	4.1 =	208
		31.65 ×	−2 ×	0.82 =	−51 ×	−4.1 =	208
		31.65 ×	−3 ×	2.07	−197 ×	−12.4 =	2420
		$e/t = 16.6/30 = 0.553$			692 ×	16.6 =	11486″K
					P_u ×	(ecc.):	956′K

eccentricities. Before using the Tables the slenderness must be checked to determine if the column is a short or long member. If it is a short one proceed with no further adjustment to P_u; one may select the P_u in accordance with the actual eccentricity. However, if it is a long member or a slender column, the effects of slenderness should be checked and an adjusted eccentricity will determine the value of P_u.

To compute the adjusted eccentricities Table 7-3 Moment Magnification Factors (δ) for square long columns was prepared, with varying concrete

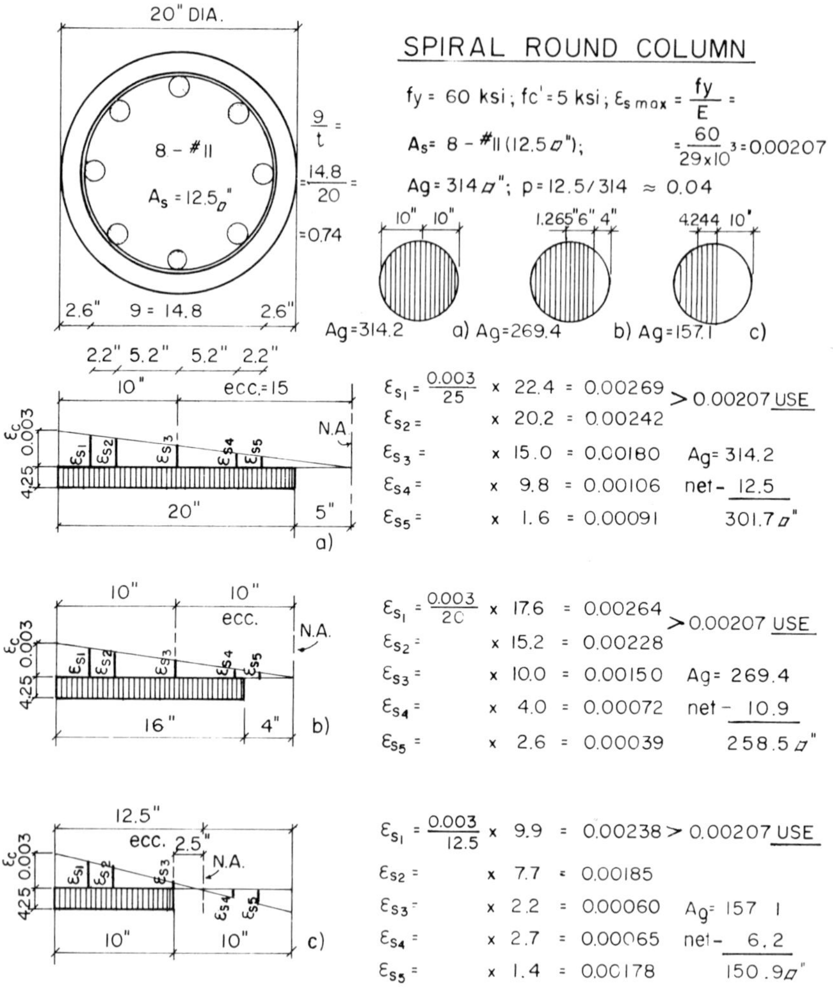

Fig. 7-2.

Capacity:

a) Conc.: $\phi \times A_{g_{\text{net}}} \times 0.85 \times f_c' \; = \qquad P_u(K) \times Arm('') = M_u(''K)$

$0.75 \times 0.85 \times 5 \times A_{g(\text{net})} = 3.19 \times 301.7 = 962 \times 0 = 0''K$

(continued)

Steel:	$\phi \times A_{s(\text{bar})} \times N_{o(\text{bars})} \times E_s \times \epsilon_s$ =					
	$0.75 \times 1.56 \times 29 \times 10^3$	× 1	× 0.00207 =	70 ×	7.2 =	504
	33.93×10^3	× 2	× 0.00207 =	140 ×	5.2 =	728
	33.93 ×	× 2	× 0.00180 =	122 ×	0 =	0
	33.93 ×	× 2	× 0.00106 =	72 ×	−5.2 =	−374
	33.93 ×	× 1	× 0.00091 =	31 ×	−7.2 =	−222
	$e/t = 0.455/20 = 0.023$			1397 ×	0.455 =	636″K
				P_u ×	(ecc.):	53′K
b) Conc.:		× 3.19 ×	258.5 =	825 ×	1.265 =	1043
Steel:	33.93 ×	× 1 ×	2.07 =	70 ×	7.2 =	504
	33.93 ×	× 2 ×	2.07 =	140 ×	5.2 =	728
	33.93 ×	× 2 ×	1.50 =	102 ×	0 =	0
	33.93 ×	× 2 ×	0.72 =	49 ×	−5.2 =	−254
	33.93 ×	× 1 ×	0.39 =	13 ×	−7.2 =	−95
	$e/t = 1.61/20 = 0.081$			1199 ×	1.61 =	1926″K
				P_u ×	(ecc.):	161′K
c) Conc.:		= 3.19 ×	150.9 =	481 ×	4.244 =	2043
Steel:	33.93 ×	× 1 ×	2.07 =	70 ×	7.2 =	504
	33.93 ×	× 2 ×	1.85 =	126 ×	5.2 =	653
	33.93 ×	× 2 ×	0.60 =	41 ×	0 =	0
	33.93 ×	× −2 ×	0.65 =	−44 ×	−5.2 =	229
	33.93 ×	× −1 ×	1.78 =	−60 ×	−7.2 =	435
	$e/t = 6.3/20 = 0.315$			614 ×	6.3 =	3864″K
				P_u ×	(ecc.):	322′K

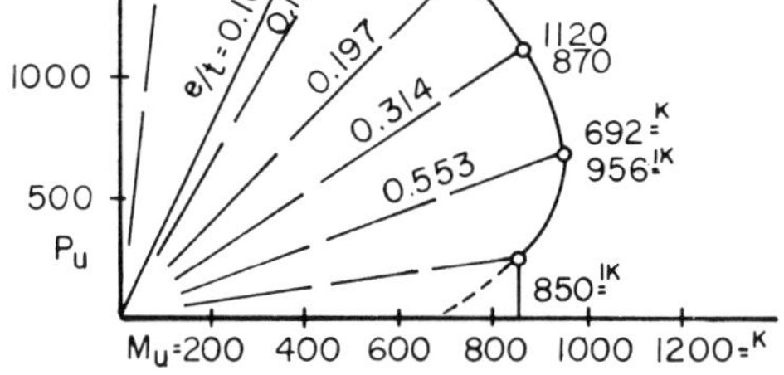

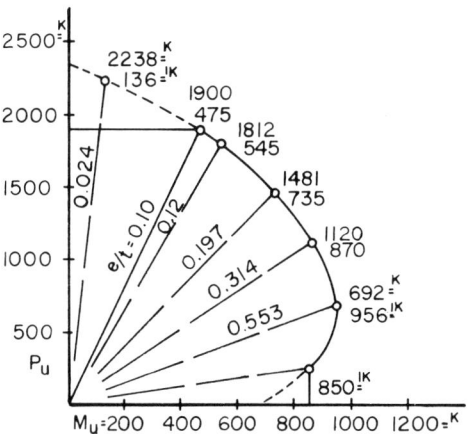

Fig. 7-3 Tied column.

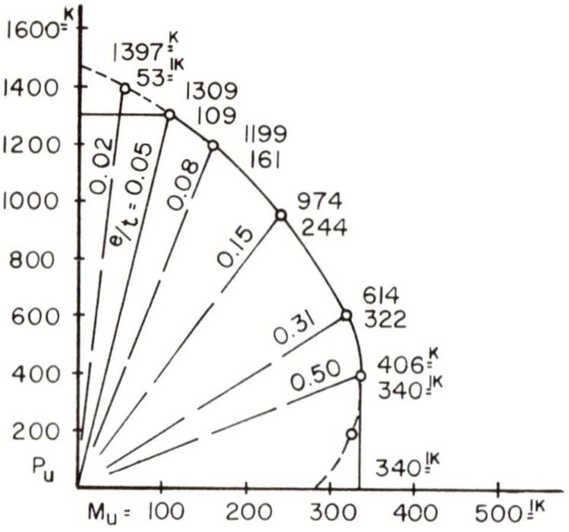

Fig. 7-4 Spiral column.

strengths, column sizes, ratios of maximum design dead load moment to maximum design total load moment (β_d), products of unsupported length of columns, effective length factors (kl_u), and values of C_m. Equations (10-5), (10-6), and (10-8) of Section 10.11.5 of the new Code were utilized, assuming a range of different values of axial design load (P_u).

Equations $kl_u/r = 34 - 12\, M_1/M_2$ for braced frames and $kl_u/r = 22$ for unbraced frames are used to determine if slenderness can be neglected in accordance with the provisions of Section 10.11.4 of the Code.

Provisions for considering the effect of slenderness in compression members are completely rewritten in the new Code. The moment magnifier method outlined in Section 10.11 of the Code is similar to the one used in the *AISC Manual* and is a function of the

1) Ratio of axial design load to critical load of column (P_u/P_c)
2) Ratio of moments at the ends of the column (M_1/M_2)
3) Deflected shape of the column.

The slenderness of the column is expressed in terms of kl_u/r, where k is the effective length factor; l_u is the unsupported length of column; r is the radius of gyration of the cross section of the column. (See Figs. 7-5 and 7-6 for interpretation of l_u and r. The effective length (kl_u) depends on two factors, the degree of end restraint acting on the column, and whether sidesway is prevented or not prevented (See Figs. 7-7, 7-8 and 7-9).

For the case of no sidesway and no restraint at the ends, the effective length kl_u is, of course, equal to the unsupported length (l_u), so $k = 1.0$.

When partial moment restraint is provided at the ends of the column and sidesway is prevented, the actual effective length kl_u, varies between one-half to one times the unsupported length, so k will be 0.5 to 1.0.

TIED COL.

TIED COL.

TABLE 7-1 , COLUMN CAPACITY. STEEL AT 2 FACES. ACI-1963, EQ. 19-1 ,19-3, 19-5 AND 19-10 PAGE 1

f_y = 60.0 Ø = 0.7 ECCENTRICITY = 0.1 X COLUMN SIZE

ρ	12 144	14 196	16 256	18 324	20 400	22 484	24 576	26 676	28 784	30 900	32 1024	34 1156	36 1296	COL. SIZE COL. AREA
1	290	400	530	680	840	1030	1230	1450	1690	1940	2220	2510	2820	f'_c (KSI) = 4
2	330	460	610	780	980	1190	1420	1680	1950	2250	2560	2900	3260	
3	380	520	700	890	1110	1350	1620	1890	2210	2540	2900	3280	3690	
4	420	590	780	1000	1240	1510	1810	2120	2470	2850	3250	3670	4130	
5	460	640	850	1090	1360	1660	1990	2310	2690	3110	3550	4020	4520	
6	500	700	940	1200	1500	1820	2180	2530	2950	3410	3890	4410	4960	
7	510	730	980	1270	1590	1940	2330	2720	3180	3670	4200	4760	5360	
8	550	790	1060	1370	1720	2100	2520	2940	3440	3970	4540	5140	5790	
1	350	480	640	820	1020	1240	1490	1750	2040	2350	2680	3040	3410	f'_c (KSI) = 5
2	390	550	720	930	1150	1410	1680	1980	2310	2660	3030	3430	3850	
3	440	610	810	1030	1290	1570	1870	2200	2560	2950	3370	3810	4280	
4	480	670	890	1140	1420	1730	2070	2430	2820	3250	3710	4200	4720	
5	520	720	960	1240	1540	1880	2250	2610	3040	3510	4010	4540	5110	
6	560	790	1050	1340	1670	2040	2440	2830	3300	3810	4350	4930	5540	
7	570	810	1090	1410	1760	2150	2590	3020	3530	4070	4650	5280	5940	
8	610	870	1170	1510	1890	2310	2770	3240	3780	4370	4990	5660	6370	

Table 7-1. Square Tied Columns.
The vertical ultimate load capacities of short columns are tabulated including the effect of ϕ, for varying column sizes, eccentricity ratios of 0.1 to 2.0, reinforcement ratios of 1% to 8%, and variable concrete strengths. The assumed yield strength of reinforcement is 60 ksi. The quantities of d' are 2½", 3" and 3½" for reinforcement ratios of 1–3, 4–6 and 7–8%, respectively. The location of reinforcement is concentrated at two faces only.

TABLE 7-1, COLUMN CAPACITY. STEEL AT 2 FACES. ACI-1963, EQ. 19-1 ,19-3, 19-5 AND 19-10 PAGE 2

f_y = 60.0 Ø = 0.7 ECCENTRICITY = 0.3 X COLUMN SIZE

ρ	12 144	14 196	16 256	18 324	20 400	22 484	24 576	26 676	28 784	30 900	32 1024	34 1156	36 1296	COL. SIZE COL. AREA
1	180	260	350	450	570	700	840	990	1160	1340	1530	1740	1950	f'_c (KSI) = 4
2	210	300	410	530	660	810	970	1150	1350	1560	1780	2020	2270	
3	240	340	460	600	750	920	1110	1300	1520	1760	2020	2290	2580	
4	270	390	520	670	850	1040	1250	1460	1710	1980	2270	2570	2900	
5	290	420	570	740	930	1140	1370	1570	1640	2140	2450	2790	3150	
6	320	460	620	810	1020	1250	1510	1730	2030	2350	2700	3070	3460	
7	310	460	630	830	1060	1310	1580	1840	2160	2510	2890	3290	3710	
8	330	490	680	900	1140	1420	1710	1990	2340	2720	3130	3560	4020	
1	220	310	420	550	690	840	1010	1200	1400	1620	1850	2100	2360	f'_c (KSI) = 5
2	250	360	480	620	780	960	1150	1360	1590	1840	2100	2380	2680	
3	280	400	540	690	870	1070	1290	1510	1760	2040	2330	2650	2980	
4	310	440	590	770	970	1180	1420	1670	1950	2260	2580	2930	3300	
5	330	470	640	830	1040	1280	1540	1770	2080	2410	2760	3140	3540	
6	360	510	690	900	1140	1390	1680	1930	2260	2620	3010	3420	3860	
7	340	510	700	920	1170	1450	1750	2030	2390	2780	3190	3630	4100	
8	370	540	750	990	1260	1550	1880	2190	2570	2990	3430	3910	4410	

TIED COL.

TIED COL.

TABLE 7-1, COLUMN CAPACITY. STEEL AT 2 FACES. ACI-1963, EQ. 19-1 ,19-3, 19-5 AND 19-10 PAGE 3

f_y = 60.0 Ø = 0.7 ECCENTRICITY = 0.5 X COLUMN SIZE

ρ	12 144	14 196	16 256	18 324	20 400	22 484	24 576	26 676	28 784	30 900	32 1024	34 1156	36 1296	COL. SIZE COL. AREA
1	110	150	210	270	340	420	510	600	700	810	930	1050	1190	f'_c (KSI) = 4
2	160	220	300	400	500	610	740	880	1030	1190	1370	1550	1750	
3	180	260	350	450	570	700	850	990	1160	1350	1550	1760	1980	
4	200	290	390	510	640	790	950	1120	1310	1520	1740	1980	2230	
5	210	310	420	550	700	870	1050	1190	1400	1630	1870	2140	2410	
6	230	340	470	610	770	950	1150	1310	1540	1790	2060	2350	2660	
7	220	330	470	620	790	990	1200	1390	1640	1910	2200	2510	2840	
8	240	360	500	670	860	1070	1300	1510	1780	2070	2390	2720	3080	
1	120	170	230	300	380	470	560	670	780	910	1040	1180	1330	f'_c (KSI) = 5
2	180	260	360	470	590	720	870	1040	1210	1400	1610	1830	2060	
3	210	300	400	520	660	810	980	1150	1350	1560	1790	2030	2290	
4	230	330	450	580	730	900	1090	1270	1490	1730	1980	2250	2540	
5	240	350	480	620	790	970	1180	1340	1580	1830	2110	2400	2710	
6	260	380	520	680	860	1060	1280	1460	1720	2000	2300	2620	2960	
7	240	370	520	680	880	1090	1320	1530	1810	2110	2430	2770	3140	
8	260	400	550	740	940	1170	1420	1650	1950	2270	2610	2980	3380	

TIED COL.

TIED COL.

TABLE 7-1, COLUMN CAPACITY. STEEL AT 2 FACES. ACI-1963, EQ. 19-1 ,19-3, 19-5 AND 19-10 PAGE 4

f_y = 60.0 f'_c = 0.7 ECCENTRICITY = 1.0 X COLUMN SIZE

ρ	12	14	16	18	20	22	24	26	28	30	32	34	36	COL. SIZE
	144	196	256	324	400	484	576	676	784	900	1024	1156	1296	COL. AREA
1	30	50	60	90	110	140	170	200	240	280	320	360	410	f'_c(KSI) = 4
2	60	90	120	160	200	250	310	370	430	500	580	660	740	
3	80	120	170	220	280	350	430	500	590	690	790	900	1020	
4	100	150	210	280	360	440	540	630	740	860	990	1130	1270	
5	120	190	260	340	440	540	660	740	880	1020	1180	1350	1530	
6	140	210	290	380	480	600	720	820	970	1130	1300	1490	1680	
7	130	200	280	380	490	610	750	860	1020	1190	1380	1580	1790	
8	140	210	300	410	530	660	810	930	1110	1300	1500	1720	1950	
1	30	50	70	90	110	140	170	210	240	280	320	370	420	f'_c (KSI) = 5
2	60	90	120	160	210	260	320	380	450	520	600	680	770	
3	90	130	180	230	300	370	450	520	620	720	820	940	1060	
4	110	160	220	290	370	460	560	660	770	900	1040	1180	1340	
5	120	180	260	340	430	540	650	740	870	1020	1180	1350	1530	
6	160	230	320	420	530	660	800	910	1080	1250	1450	1650	1870	
7	140	220	310	420	540	670	820	950	1130	1320	1520	1740	1980	
8	150	240	330	450	580	720	890	1020	1210	1420	1640	1880	2130	

TIED COL.

TIED COL.

TABLE 7-1, COLUMN CAPACITY. STEEL AT 2 FACES. ACI-1963, EQ. 19-1 ,19-3, 19-5 AND 19-10 PAGE 5

f_y = 60.0 Ø = 0.7 ECCENTRICITY = 2.0 X COLUMN SIZE

ρ	12	14	16	18	20	22	24	26	28	30	32	34	36	COL. SIZE
	144	196	256	324	400	484	576	676	784	900	1024	1156	1296	COL. AREA
1	10	20	20	30	40	50	60	70	90	100	120	130	150	f'_c (KSI) = 4
2	20	30	50	60	80	100	120	150	170	200	230	260	300	
3	30	50	70	90	120	150	180	210	250	290	340	380	440	
4	40	70	90	120	160	200	240	280	330	380	440	510	570	
5	50	80	110	150	190	230	290	320	380	450	520	590	670	
6	60	90	130	170	220	280	340	380	450	530	610	700	800	
7	50	90	130	180	230	290	360	420	500	580	680	780	890	
8	60	100	150	200	260	330	410	470	560	660	760	880	1000	
1	10	20	20	30	40	50	60	70	90	100	120	130	150	f'_c (KSI) = 5
2	20	30	50	60	80	100	120	150	170	200	230	270	300	
3	30	50	70	90	120	150	180	210	250	300	340	390	440	
4	40	70	90	120	160	200	240	280	330	390	450	510	580	
5	50	80	110	150	190	240	290	320	390	450	520	600	680	
6	60	90	130	180	230	280	340	390	460	540	620	710	810	
7	60	90	130	180	240	300	370	420	510	590	690	790	900	
8	60	100	150	200	270	340	420	480	570	670	780	900	1020	

TIED COL.

TIED COL.

TABLE 7-2 CIRCULAR COLUMN CAPACITY PAGE 1

fy = 60.0 ∅ = 0.75 ECCENTRICITY = 0.05 X COLUMN DIA.

ρ	12 113	14 154	16 201	18 254	20 314	22 380	24 452	26 531	28 616	30 707	32 804	34 908	36 1018	COL. DIA. COL. AREA
1	260	360	480	610	760	920	1100	1300	1510	1740	1980	2240	2520	f'_c (KSI) = 4
2	300	420	550	710	880	1070	1280	1500	1750	2010	2290	2590	2910	
3	340	480	630	800	1000	1210	1450	1700	1980	2280	2600	2940	3300	
4	380	530	700	900	1110	1350	1620	1900	2220	2550	2910	3290	3690	
5	420	590	780	990	1230	1500	1780	2100	2440	2810	3200	3620	4070	
6	460	640	850	1090	1350	1640	1950	2300	2670	3080	3510	3970	4460	
7	500	690	920	1170	1460	1780	2090	2470	2880	3310	3780	4280	4810	
8	540	750	990	1270	1580	1920	2260	2670	3110	3580	4090	4630	5200	
1	320	440	580	740	920	1120	1340	1580	1830	2110	2400	2720	3050	f'_c (KSI) = 5
2	360	500	660	840	1040	1260	1510	1780	2070	2380	2710	3070	3440	
3	400	550	730	930	1160	1410	1680	1980	2300	2650	3020	3410	3830	
4	440	610	800	1030	1270	1550	1850	2180	2540	2920	3330	3760	4220	
5	480	660	880	1120	1390	1690	2010	2370	2760	3160	3620	4100	4600	
6	520	720	950	1220	1510	1840	2180	2570	2990	3450	3930	4440	4990	
7	550	770	1020	1300	1620	1970	2320	2740	3190	3680	4200	4750	5340	
8	590	820	1090	1400	1740	2110	2490	2940	3420	3940	4500	5090	5720	

Table 7-2. Circular Column Capacity.
Vertical ultimate load capacity of short columns are tabulated including the effect of ϕ, for varying column diameters, eccentricity ratios of 0.05 to 2.0, reinforcement ratios of 1% to 8%, and for varying conc. strengths.

SPIRAL COL.

SPIRAL COL.

TABLE 7-2 CIRCULAR COLUMN CAPACITY PAGE 2

f_y = 60.0 Ø = 0.75 ECCENTRICITY = 0.30 x COLUMN DIA.

ρ	12 113	14 154	16 201	18 254	20 314	22 380	24 452	26 531	28 616	30 707	32 804	34 908	36 1018	COL. DIA. COL. AREA
1	120	180	240	310	390	480	580	680	800	920	1060	1200	1350	f'_c (KSI) = 4
2	140	210	280	360	450	560	670	800	930	1080	1230	1400	1580	
3	160	230	320	410	520	640	770	910	1060	1230	1410	1600	1800	
4	180	260	360	460	580	720	860	1020	1200	1380	1580	1800	2030	
5	200	290	400	510	650	790	950	1120	1310	1520	1740	1980	2230	
6	220	320	430	560	710	870	1040	1230	1440	1670	1920	2180	2450	
7	230	340	460	600	760	940	1090	1290	1520	1770	2030	2310	2610	
8	250	370	500	650	830	1020	1170	1400	1650	1910	2200	2500	2830	
1	150	210	290	370	470	580	700	830	970	1120	1280	1450	1640	f'_c (KSI) = 5
2	170	240	330	430	540	660	790	940	1100	1270	1460	1650	1860	
3	190	270	370	480	600	740	890	1050	1230	1420	1630	1850	2080	
4	210	300	410	530	660	820	980	1170	1360	1580	1810	2050	2310	
5	230	330	450	580	730	900	1060	1260	1480	1710	1960	2230	2510	
6	250	360	490	630	790	970	1160	1380	1610	1860	2140	2430	2730	
7	260	380	510	670	840	1040	1200	1430	1680	1950	2240	2550	2880	
8	280	400	550	720	910	1120	1290	1540	1810	2100	2410	2740	3100	

SPIRAL COL.

SPIRAL COL.

TABLE 7-2 CIRCULAR COLUMN CAPACITY PAGE 3

f_y = 60.0 ϕ = 0.75 ECCENTRICITY = 0.50 X COLUMN DIA.

ρ	12	14	16	18	20	22	24	26	28	30	32	34	36	COL. DIA.
	113	154	201	254	314	380	452	531	616	707	804	908	1018	COL. AREA
1	60	80	120	150	190	240	290	340	400	460	530	600	680	f'_c (KSI) = 4
2	90	130	170	230	290	350	430	510	590	690	790	900	1010	
3	120	170	220	290	360	440	540	640	750	860	990	1130	1270	
4	130	190	260	330	420	520	630	750	870	1010	1160	1320	1490	
5	140	210	280	370	470	580	690	820	960	1110	1280	1450	1640	
6	160	230	310	410	520	640	760	900	1060	1230	1410	1600	1810	
7	160	240	330	430	550	680	780	940	1110	1290	1480	1690	1910	
8	180	260	360	470	600	740	850	1020	1200	1390	1600	1830	2070	
1	60	90	130	170	210	260	310	370	440	500	580	660	740	f'_c (KSI) = 5
2	100	140	190	250	320	390	470	560	650	760	870	990	1110	
3	130	180	240	320	400	490	590	700	820	950	1090	1240	1400	
4	150	210	290	380	480	590	720	850	1000	1120	1280	1460	1640	
5	160	230	320	420	530	650	770	920	1080	1250	1440	1630	1840	
6	180	260	350	450	570	710	840	1000	1180	1360	1570	1780	2010	
7	180	270	370	480	610	750	870	1040	1220	1420	1630	1860	2110	
8	200	290	390	520	660	810	930	1110	1310	1530	1760	2010	2270	

SPIRAL COL.

SPIRAL COL.

TABLE 7-2 CIRCULAR COLUMN CAPACITY PAGE 4

f_y = 60.0 Ø = 0.75 ECCENTRICITY = 1.00 X COLUMN DIA.

P	12 113	14 154	16 201	18 254	20 314	22 380	24 452	26 531	28 616	30 707	32 804	34 908	36 1018	COL. DIA. COL. AREA
1	10	20	30	40	50	70	80	100	110	130	150	180	200	f'_c (KSI) = 4
2	30	40	60	80	100	130	150	190	220	250	290	330	380	
3	40	60	90	110	150	180	220	270	310	370	420	480	540	
4	50	80	110	150	190	240	290	340	400	470	540	620	700	
5	70	100	140	180	230	290	340	410	480	560	640	730	830	
6	80	110	160	210	270	330	390	470	560	650	740	850	960	
7	90	140	170	230	290	370	420	500	600	700	810	920	1040	
8	100	150	210	260	330	410	470	560	660	780	900	1020	1160	
1	10	20	30	40	50	70	80	100	120	130	160	180	200	f'_c (KSI) = 5
2	30	40	60	80	100	130	160	190	220	260	300	340	390	
3	40	60	90	120	150	190	230	270	320	380	430	490	560	
4	50	80	110	150	200	240	300	350	420	480	560	640	720	
5	70	100	140	190	240	300	350	420	500	580	660	760	860	
6	80	120	160	220	280	340	410	490	580	670	780	880	1000	
7	100	130	180	240	310	380	430	520	620	730	840	960	1090	
8	110	170	200	270	340	430	490	580	690	810	940	1070	1220	

SPIRAL COL.

SPIRAL COL.

TABLE 7-2 CIRCULAR COLUMN CAPACITY

PAGE 5

f_y = 60.0 ϕ = 0.75 ECCENTRICITY = 2.00 X COLUMN DIA.

ρ	12 113	14 154	16 201	18 254	20 314	22 380	24 452	26 531	28 616	30 707	32 804	34 908	36 1018	COL. DIA. COL. AREA
1	0	10	10	10	20	20	30	30	40	50	60	60	70	f'_c (KSI) = 4
2	10	10	20	30	40	50	60	70	80	100	110	130	150	
3	10	20	30	40	60	70	90	110	130	150	170	190	220	
4	20	30	40	60	80	100	120	140	170	200	230	260	290	
5	30	40	60	80	100	120	140	170	200	240	280	320	360	
6	30	50	70	90	120	140	170	210	240	290	330	380	430	
7	30	50	70	100	130	160	180	220	270	310	360	410	470	
8	50	60	90	110	150	190	210	250	300	350	410	470	530	
1	0	10	10	10	20	20	30	30	40	50	60	60	70	f'_c (KSI) = 5
2	10	20	20	30	40	50	60	70	80	100	110	130	150	
3	10	20	30	50	60	70	90	110	130	150	170	200	220	
4	20	30	40	60	80	100	120	140	170	200	230	260	290	
5	30	40	60	80	100	120	140	170	210	240	280	320	360	
6	30	50	70	90	120	150	170	210	250	290	330	380	430	
7	30	50	80	100	130	160	190	220	270	310	360	420	480	
8	40	60	90	120	150	190	210	260	300	360	410	480	540	

SPIRAL COL.

SPIRAL COL.

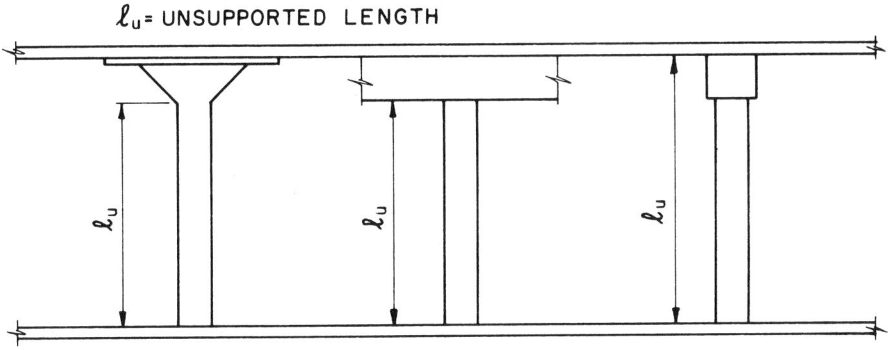

Fig. 7-5.

The Code (10.11.3) suggests that the conservative value of $k = 1.0$ can always be used and, if a more detailed analysis is made, lower values can be substituted. In the extremely important case where sidesway is not prevented by some kind of bracing, the effective length of the column, kl_u, is always greater than the unsupported length, l_u, and it is mandatory in this case that accurate determination of k be made. The Commentary of the Code suggests that the Jackson-Moreland alignment charts (Fig. 7-10) can be utilized to provide such effective length factors.

Summarizing: The effective length factor, k, for braced frames shall be 1.0, unless analysis shows it to be less. For unbraced frames, k shall be greater than 1.0. It "shall be determined with due consideration of cracking and reinforcement on relative stiffness." Furthermore, "a value of k less than 1.2 for columns not braced against sidesway, normally would not be realistic."

To classify a compression member, in a story, as braced against sidesway, the bracing elements (shear walls, shear trusses, etc.) should have a total stiffness,

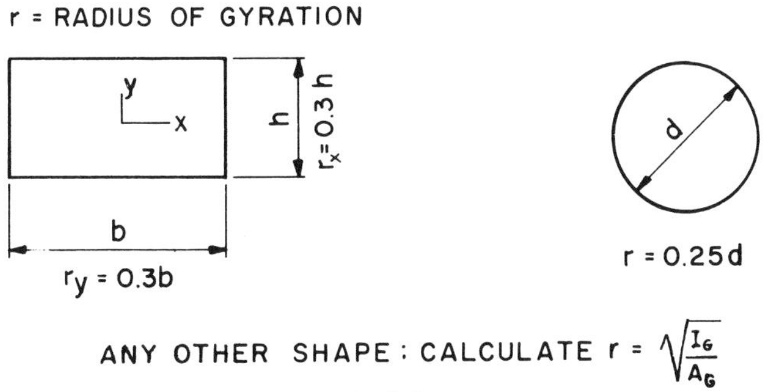

Fig. 7-6.

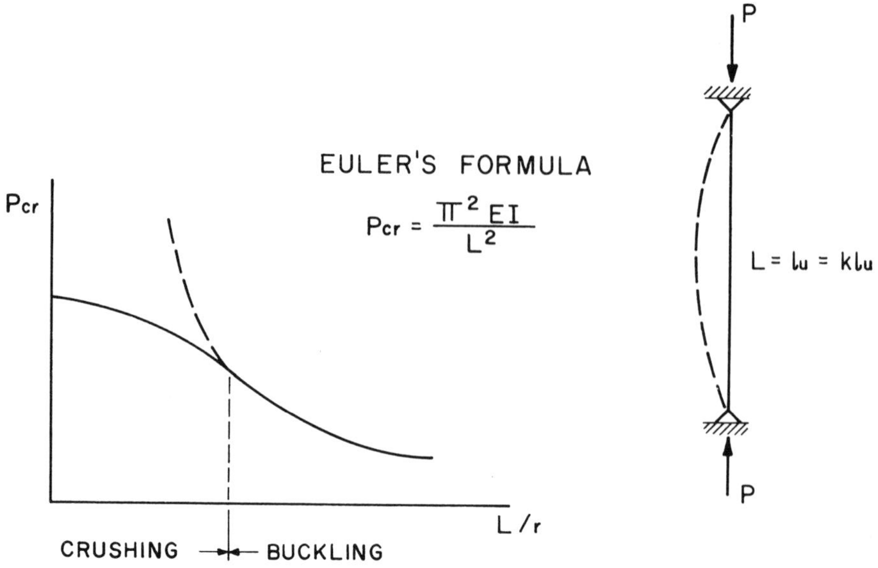

Fig. 7-7.

resisting lateral movement of the story, of at least six times the sum of the stiffness of all the columns resisting lateral movement in that story.

Two equations are used to determine the magnification factor. Equation (10-6),

$$P_c = \frac{\pi^2 EI}{(kl_u)^2}$$

is used to determine the critical load, which is similar to the Euler-type buckling load.

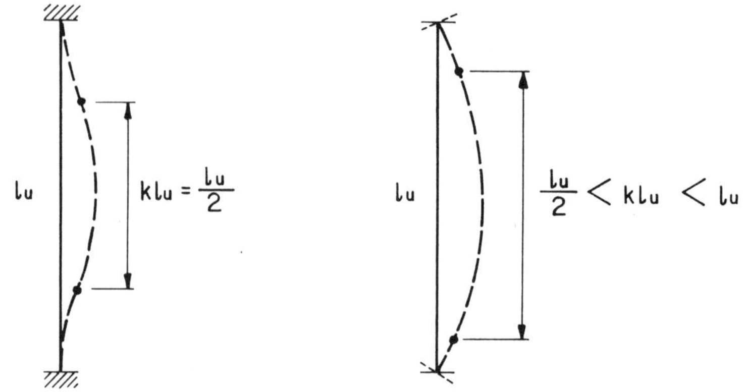

Fig. 7-8.

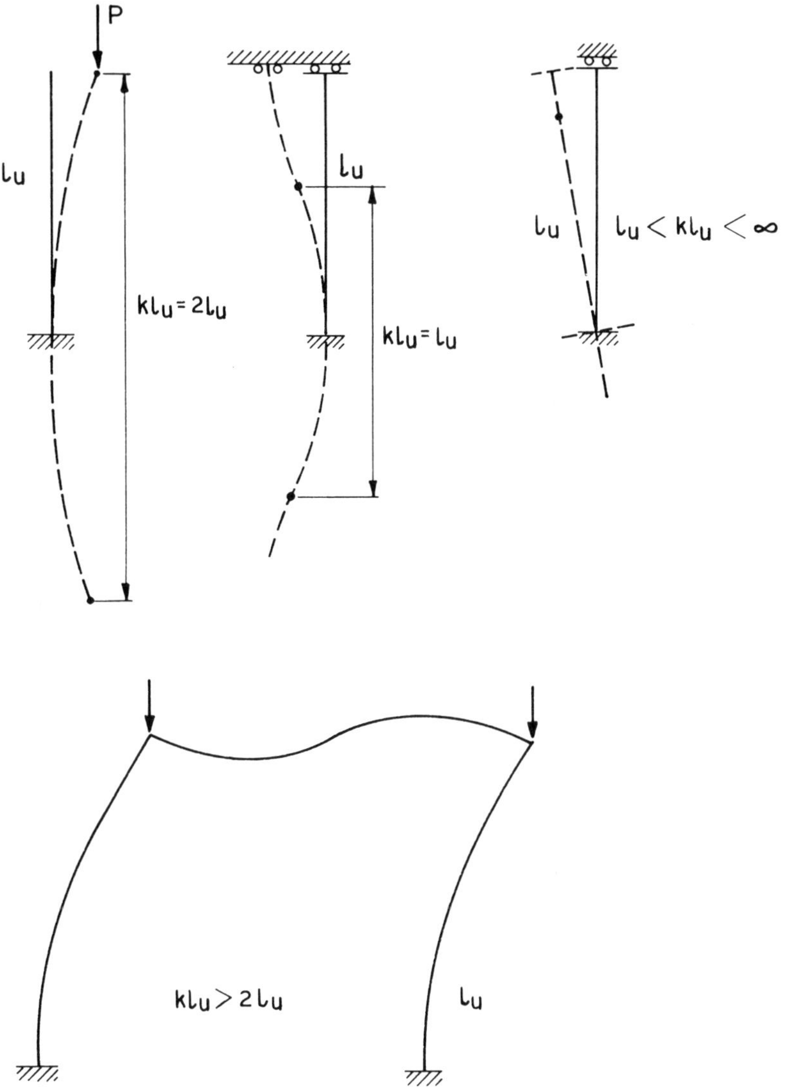

Fig. 7-9.

Equation (10-5),

$$\delta = \frac{C_m}{1 - \dfrac{P_u}{\phi P_c}}$$

is used to determine the magnifier as a function of the equivalent moment coefficient, C_m, and the ratio of the critical load.

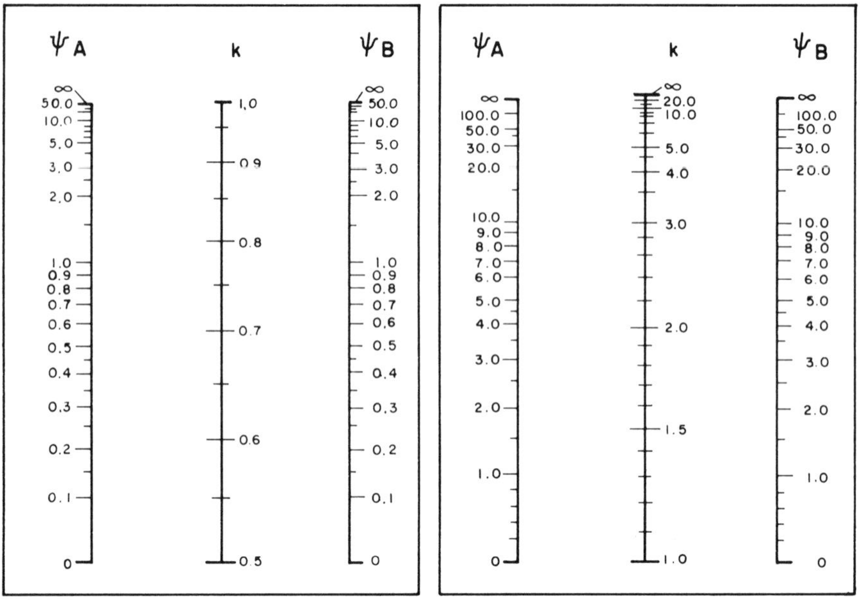

(a) BRACED FRAMES

(b) UNBRACED FRAMES

ψ = RATIO OF $\Sigma(EI/l_c)$ OF COMPRESSION MEMBERS TO $\Sigma(EI/l)$ OF FLEXURAL MEMBERS IN A PLANE AT ONE END OF A COMPRESSION MEMBER.

k = EFFECTIVE LENGTH FACTOR

Fig. 7-10.

The Code specifies that in lieu of a more precise analysis the value of EI may be given by either equation (10-7) or (10-8).

The equivalent moment factor, C_m, is given for braced frames by equation (10-9):

$$C_m = 0.6 + 0.4 \frac{M_1}{M_2}$$

but not less than 0.4. Whenever the frame is not braced against sway, the C_m factor must be taken as 1.0.

For all compression members with kl_u/r greater than 100, the moment magnifier method may not be utilized, and an analysis as defined in Section 10.10.1 needs to be made.

For bending about both axes of a column, moments about each axis shall be magnified by the appropriate δ. Since kl_u and EI may be different in the two

TABLE 7-3 MOMENT MAGNIFICATION FACTORS δ FOR LONG COLUMNS. ACI 10.11.5 PAGE 1

f'_c = (KSI) = 4.00 Ø =0.70

SIZE	Cm	klu	βd	100	200	300	400	500	600	700	800	900	1000	1100	1200	1300	1400	1500	1600	1700	1800	1900	2000	2100
				P(U) KIP																				
16	0.6	16	0.3	1.00	1.00	1.00	1.00	1.07	1.27	1.56	2.03	2.89	5.03	****	****	****	****	****	****	****	****	****	****	****
			0.8	1.00	1.00	1.00	1.17	1.53	2.23	4.10	****	****	****	****	****	****	****	****	****	****	****	****	****	****
		19	0.3	1.00	1.00	1.00	1.19	1.58	2.35	4.59	****	****	****	****	****	****	****	****	****	****	****	****	****	****
			0.8	1.00	1.00	1.23	1.92	4.28	****	****	****	****	****	****	****	****	****	****	****	****	****	****	****	****
16	1.0	10	0.3	1.03	1.07	1.11	1.15	1.20	1.26	1.31	1.37	1.44	1.52	1.60	1.70	1.80	1.92	2.06	2.22	2.40	2.62	2.88	3.20	3.60
			0.8	1.05	1.10	1.16	1.23	1.31	1.40	1.50	1.61	1.75	1.90	2.10	2.33	2.62	3.00	3.50	4.20	5.25	7.01	****	****	****
		13	0.3	1.06	1.13	1.21	1.30	1.40	1.53	1.68	1.86	2.09	2.38	2.77	3.30	4.09	5.37	7.82	****	****	****	****	****	****
			0.8	1.08	1.19	1.31	1.47	1.67	1.93	2.29	2.80	3.63	5.13	8.74	****	****	****	****	****	****	****	****	****	****
		16	0.3	1.09	1.21	1.35	1.54	1.78	2.12	2.60	3.38	4.82	8.39	****	****	****	****	****	****	****	****	****	****	****
			0.8	1.13	1.32	1.57	1.95	2.56	3.72	6.83	****	****	****	****	****	****	****	****	****	****	****	****	****	****
		19	0.3	1.14	1.33	1.59	1.98	2.63	3.92	7.66	****	****	****	****	****	****	****	****	****	****	****	****	****	****
			0.8	1.20	1.52	2.06	3.20	7.13	****	****	****	****	****	****	****	****	****	****	****	****	****	****	****	****
20	0.6	16	0.3	1.00	1.00	1.00	1.00	1.00	1.00	1.00	1.00	1.00	1.00	1.00	1.05	1.12	1.21	1.30	1.41	1.55	1.71	1.90	2.15	2.47
			0.8	1.00	1.00	1.00	1.00	1.00	1.00	1.00	1.00	1.09	1.19	1.33	1.49	1.71	1.99	2.39	2.98	3.97	5.95	****	****	****
		19	0.3	1.00	1.00	1.00	1.00	1.00	1.00	1.00	1.01	1.10	1.22	1.36	1.54	1.77	2.08	2.53	3.22	4.44	7.12	****	****	****
			0.8	1.00	1.00	1.00	1.00	1.00	1.03	1.18	1.37	1.63	2.03	2.66	3.87	7.12	****	****	****	****	****	****	****	****
20	1.0	10	0.3	1.01	1.02	1.04	1.05	1.07	1.09	1.10	1.12	1.14	1.16	1.18	1.20	1.22	1.24	1.26	1.29	1.31	1.33	1.36	1.39	1.42
			0.8	1.01	1.04	1.06	1.08	1.10	1.13	1.15	1.18	1.21	1.24	1.27	1.30	1.33	1.37	1.41	1.45	1.49	1.54	1.58	1.64	1.69
		13	0.3	1.02	1.05	1.07	1.10	1.13	1.16	1.20	1.23	1.27	1.31	1.35	1.40	1.44	1.50	1.55	1.61	1.68	1.75	1.82	1.90	2.00
			0.8	1.03	1.07	1.10	1.15	1.19	1.24	1.30	1.35	1.42	1.49	1.56	1.65	1.75	1.85	1.97	2.11	2.27	2.46	2.67	2.93	3.25
		16	0.3	1.03	1.07	1.12	1.16	1.22	1.27	1.33	1.40	1.48	1.56	1.65	1.76	1.88	2.02	2.17	2.36	2.58	2.85	3.17	3.59	4.12
			0.8	1.05	1.11	1.17	1.24	1.33	1.42	1.53	1.66	1.81	1.99	2.21	2.49	2.85	3.32	3.98	4.98	6.63	9.91	****	****	****
		19	0.3	1.05	1.11	1.18	1.25	1.34	1.43	1.55	1.68	1.84	2.03	2.27	2.56	2.95	3.47	4.22	5.37	7.40	****	****	****	****
			0.8	1.07	1.16	1.26	1.39	1.54	1.73	1.97	2.29	2.73	3.38	4.44	6.46	****	****	****	****	****	****	****	****	****
24	0.6	16	0.3	1.00	1.00	1.00	1.00	1.00	1.00	1.00	1.00	1.00	1.00	1.00	1.00	1.00	1.00	1.00	1.00	1.00	1.00	1.00	1.00	1.00
			0.8	1.00	1.00	1.00	1.00	1.00	1.00	1.00	1.00	1.00	1.00	1.00	1.00	1.00	1.00	1.00	1.00	1.01	1.05	1.10	1.15	1.21
		19	0.3	1.00	1.00	1.00	1.00	1.00	1.00	1.00	1.00	1.00	1.00	1.00	1.00	1.00	1.00	1.00	1.00	1.02	1.07	1.12	1.17	1.23
			0.8	1.00	1.00	1.00	1.00	1.00	1.00	1.00	1.00	1.00	1.00	1.00	1.01	1.07	1.14	1.22	1.31	1.42	1.54	1.69	1.87	2.09
24	1.0	10	0.3	1.00	1.01	1.02	1.02	1.03	1.04	1.04	1.05	1.06	1.07	1.08	1.08	1.09	1.10	1.11	1.12	1.13	1.13	1.14	1.15	1.16
			0.8	1.00	1.01	1.02	1.03	1.04	1.05	1.07	1.08	1.09	1.10	1.11	1.12	1.13	1.15	1.16	1.17	1.19	1.20	1.21	1.23	1.24
		13	0.3	1.01	1.02	1.03	1.04	1.06	1.07	1.08	1.10	1.11	1.12	1.14	1.15	1.17	1.19	1.20	1.22	1.24	1.26	1.27	1.29	1.31
			0.8	1.01	1.03	1.05	1.06	1.08	1.10	1.12	1.14	1.16	1.18	1.21	1.23	1.26	1.28	1.31	1.34	1.37	1.40	1.43	1.46	1.50
		16	0.3	1.01	1.03	1.05	1.07	1.09	1.11	1.13	1.16	1.18	1.21	1.23	1.26	1.29	1.32	1.35	1.38	1.42	1.45	1.49	1.53	1.57
			0.8	1.02	1.05	1.07	1.10	1.13	1.16	1.20	1.23	1.27	1.31	1.36	1.40	1.45	1.50	1.56	1.62	1.69	1.76	1.84	1.92	2.02
		19	0.3	1.02	1.05	1.07	1.10	1.13	1.17	1.20	1.24	1.28	1.32	1.36	1.41	1.46	1.52	1.58	1.64	1.71	1.79	1.87	1.96	2.06
			0.8	1.03	1.07	1.11	1.15	1.20	1.25	1.31	1.37	1.44	1.51	1.59	1.68	1.79	1.90	2.03	2.19	2.36	2.57	2.82	3.11	3.48

Table 7-3. Moment Magnification Factors for Columns.
The moment magnification factors for square columns (δ) are tabulated for varying concrete strengths, square column sizes, ratios of dead load to total load (β_d), products of unsupported lengths of columns and effective length-factors (kl_u), and values of C_m. Equations 10-5, 10-6, 10-8, Sect. 10.11.5 of Code.

TABLE 7-3 MOMENT MAGNIFICATION FACTORS δ FOR LONG COLUMNS. ACI 10.11.5 PAGE 2

f'_c = (KSI) = 4.00 ϕ =0.70

SIZE	C_m	kl_u	β_d	P(U) KIP 400	600	800	1000	1200	1400	1600	1800	2000	2200	2400	2600	2800	3000	3200	3400	3600	3800	4000	4200	4400
28	0.6	16	0.3	1.00	1.00	1.00	1.00	1.00	1.00	1.00	1.00	1.00	1.00	1.00	1.00	1.00	1.00	1.00	1.00	1.00	1.00	1.00	1.00	1.02
			0.8	1.00	1.00	1.00	1.00	1.00	1.00	1.00	1.00	1.00	1.00	1.00	1.00	1.00	1.00	1.02	1.07	1.12	1.18	1.25	1.32	1.40
		19	0.3	1.00	1.00	1.00	1.00	1.00	1.00	1.00	1.00	1.00	1.00	1.00	1.00	1.00	1.00	1.04	1.09	1.14	1.20	1.27	1.35	1.43
			0.8	1.00	1.00	1.00	1.00	1.00	1.00	1.00	1.00	1.00	1.00	1.07	1.14	1.23	1.33	1.45	1.59	1.76	1.97	2.25	2.61	3.10
28	1.0	10	0.3	1.01	1.02	1.03	1.03	1.04	1.05	1.06	1.07	1.07	1.08	1.09	1.10	1.11	1.12	1.13	1.14	1.15	1.16	1.17	1.18	1.19
			0.8	1.02	1.03	1.04	1.05	1.06	1.07	1.08	1.10	1.11	1.12	1.13	1.15	1.16	1.17	1.19	1.20	1.22	1.23	1.25	1.27	1.28
		13	0.3	1.02	1.03	1.05	1.06	1.08	1.09	1.11	1.12	1.14	1.15	1.17	1.19	1.21	1.22	1.24	1.26	1.28	1.30	1.32	1.35	1.37
			0.8	1.03	1.05	1.07	1.09	1.11	1.13	1.15	1.18	1.20	1.23	1.25	1.28	1.31	1.34	1.37	1.41	1.44	1.48	1.52	1.56	1.60
		16	0.3	1.03	1.05	1.08	1.10	1.12	1.15	1.17	1.20	1.23	1.26	1.29	1.32	1.35	1.39	1.42	1.46	1.51	1.55	1.60	1.65	1.70
			0.8	1.05	1.08	1.11	1.14	1.18	1.22	1.26	1.30	1.35	1.40	1.45	1.51	1.57	1.63	1.71	1.79	1.88	1.97	2.08	2.20	2.33
		19	0.3	1.05	1.08	1.11	1.15	1.18	1.22	1.26	1.31	1.36	1.41	1.46	1.52	1.58	1.65	1.73	1.81	1.91	2.01	2.12	2.25	2.39
			0.8	1.07	1.12	1.17	1.22	1.28	1.34	1.41	1.49	1.57	1.67	1.78	1.91	2.05	2.22	2.42	2.65	2.94	3.29	3.75	4.35	5.17
32	0.6	16	0.3	1.00	1.00	1.00	1.00	1.00	1.00	1.00	1.00	1.00	1.00	1.00	1.00	1.00	1.00	1.00	1.00	1.00	1.00	1.00	1.00	1.00
			0.8	1.00	1.00	1.00	1.00	1.00	1.00	1.00	1.00	1.00	1.00	1.00	1.00	1.00	1.00	1.00	1.00	1.00	1.00	1.00	1.00	1.00
		19	0.3	1.00	1.00	1.00	1.00	1.00	1.00	1.00	1.00	1.00	1.00	1.00	1.00	1.00	1.00	1.00	1.00	1.00	1.00	1.00	1.00	1.00
			0.8	1.00	1.00	1.00	1.00	1.00	1.00	1.00	1.00	1.00	1.00	1.00	1.00	1.00	1.00	1.00	1.00	1.00	1.01	1.05	1.09	1.13
32	1.0	10	0.3	1.00	1.01	1.01	1.02	1.02	1.03	1.03	1.04	1.04	1.04	1.05	1.05	1.06	1.06	1.07	1.07	1.08	1.08	1.09	1.09	1.10
			0.8	1.01	1.01	1.02	1.03	1.03	1.04	1.05	1.05	1.06	1.07	1.07	1.08	1.09	1.09	1.10	1.11	1.12	1.12	1.13	1.14	1.15
		13	0.3	1.01	1.02	1.02	1.03	1.04	1.05	1.06	1.06	1.07	1.08	1.09	1.10	1.11	1.12	1.13	1.14	1.15	1.16	1.17	1.18	1.19
			0.8	1.02	1.03	1.04	1.05	1.06	1.07	1.08	1.09	1.11	1.12	1.13	1.15	1.16	1.17	1.19	1.20	1.22	1.23	1.25	1.26	1.28
		16	0.3	1.02	1.03	1.04	1.05	1.07	1.08	1.09	1.10	1.12	1.13	1.15	1.16	1.18	1.19	1.21	1.23	1.24	1.26	1.28	1.30	1.31
			0.8	1.03	1.04	1.06	1.08	1.10	1.11	1.13	1.15	1.17	1.20	1.22	1.24	1.27	1.29	1.32	1.34	1.37	1.40	1.43	1.47	1.50
		19	0.3	1.03	1.04	1.06	1.08	1.10	1.12	1.14	1.16	1.18	1.20	1.22	1.25	1.27	1.30	1.33	1.35	1.38	1.41	1.45	1.48	1.51
			0.8	1.04	1.06	1.09	1.12	1.14	1.17	1.20	1.23	1.27	1.30	1.34	1.38	1.43	1.47	1.52	1.57	1.63	1.69	1.75	1.82	1.89
36	0.6	16	0.3	1.00	1.00	1.00	1.00	1.00	1.00	1.00	1.00	1.00	1.00	1.00	1.00	1.00	1.00	1.00	1.00	1.00	1.00	1.00	1.00	1.00
			0.8	1.00	1.00	1.00	1.00	1.00	1.00	1.00	1.00	1.00	1.00	1.00	1.00	1.00	1.00	1.00	1.00	1.00	1.00	1.00	1.00	1.00
		19	0.3	1.00	1.00	1.00	1.00	1.00	1.00	1.00	1.00	1.00	1.00	1.00	1.00	1.00	1.00	1.00	1.00	1.00	1.00	1.00	1.00	1.00
			0.8	1.00	1.00	1.00	1.00	1.00	1.00	1.00	1.00	1.00	1.00	1.00	1.00	1.00	1.00	1.00	1.00	1.00	1.00	1.00	1.00	1.00
36	1.0	10	0.3	1.00	1.00	1.01	1.01	1.01	1.01	1.02	1.02	1.02	1.03	1.03	1.03	1.03	1.04	1.04	1.04	1.05	1.05	1.05	1.05	1.06
			0.8	1.00	1.01	1.01	1.01	1.02	1.02	1.03	1.03	1.03	1.04	1.04	1.05	1.05	1.05	1.06	1.06	1.07	1.07	1.08	1.08	1.08
		13	0.3	1.00	1.01	1.01	1.02	1.02	1.03	1.03	1.04	1.04	1.05	1.05	1.06	1.06	1.07	1.07	1.08	1.08	1.09	1.09	1.10	1.11
			0.8	1.01	1.01	1.02	1.03	1.03	1.04	1.05	1.05	1.06	1.07	1.08	1.08	1.09	1.10	1.11	1.11	1.12	1.13	1.14	1.15	1.16
		16	0.3	1.01	1.02	1.02	1.03	1.04	1.05	1.05	1.06	1.07	1.08	1.08	1.09	1.10	1.11	1.12	1.13	1.14	1.15	1.15	1.16	1.17
			0.8	1.01	1.02	1.03	1.04	1.06	1.07	1.08	1.09	1.10	1.11	1.12	1.14	1.15	1.16	1.17	1.19	1.20	1.22	1.23	1.24	1.26
		19	0.3	1.01	1.02	1.04	1.05	1.06	1.07	1.08	1.09	1.10	1.11	1.13	1.14	1.15	1.17	1.18	1.19	1.21	1.22	1.24	1.25	1.27
			0.8	1.02	1.04	1.05	1.07	1.08	1.10	1.12	1.13	1.15	1.17	1.19	1.21	1.23	1.25	1.27	1.29	1.31	1.34	1.36	1.39	1.41

δ δ

TABLE 7-3 MOMENT MAGNIFICATION FACTORS δ FOR LONG COLUMNS. ACI 10.11.5 PAGE 3

f'_c = (KSI) = 5.00 Ø =0.70

SIZE	C_m	kl_u	βd	P(U) KIP 100	200	300	400	500	600	700	800	900	1000	1100	1200	1300	1400	1500	1600	1700	1800	1900	2000	2100
16	0.6	16	0.3	1.00	1.00	1.00	1.00	1.00	1.13	1.33	1.62	2.06	2.82	4.49	****	****	****	****	****	****	****	****	****	****
			0.8	1.00	1.00	1.00	1.06	1.31	1.73	2.53	4.71	****	****	****	****	****	****	****	****	****	****	****	****	****
		19	0.3	1.00	1.00	1.00	1.07	1.34	1.79	2.69	5.39	****	****	****	****	****	****	****	****	****	****	****	****	****
			0.8	1.00	1.00	1.11	1.55	2.59	7.78	****	****	****	****	****	****	****	****	****	****	****	****	****	****	****
16	1.0	10	0.3	1.03	1.06	1.10	1.14	1.18	1.22	1.27	1.32	1.38	1.44	1.51	1.58	1.66	1.75	1.85	1.97	2.09	2.24	2.40	2.60	2.82
			0.8	1.04	1.09	1.14	1.20	1.27	1.34	1.42	1.51	1.62	1.74	1.88	2.04	2.24	2.47	2.77	3.14	3.62	4.29	5.25	6.76	9.50
		13	0.3	1.05	1.11	1.18	1.26	1.35	1.45	1.57	1.71	1.87	2.08	2.33	2.66	3.08	3.67	4.54	5.95	8.63	****	****	****	****
			0.8	1.07	1.16	1.27	1.40	1.56	1.76	2.01	2.35	2.84	3.57	4.81	7.36	****	****	****	****	****	****	****	****	****
		16	0.3	1.08	1.18	1.30	1.46	1.64	1.89	2.22	2.70	3.43	4.71	7.49	****	****	****	****	****	****	****	****	****	****
			0.8	1.12	1.27	1.48	1.77	2.19	2.89	4.22	7.85	****	****	****	****	****	****	****	****	****	****	****	****	****
		19	0.3	1.12	1.28	1.49	1.79	2.24	2.99	4.49	8.99	****	****	****	****	****	****	****	****	****	****	****	****	****
			0.8	1.18	1.44	1.85	2.59	4.33	****	****	****	****	****	****	****	****	****	****	****	****	****	****	****	****
20	0.6	16	0.3	1.00	1.00	1.00	1.00	1.00	1.00	1.00	1.00	1.00	1.00	1.00	1.00	1.03	1.09	1.16	1.24	1.32	1.43	1.55	1.69	1.86
			0.8	1.00	1.00	1.00	1.00	1.00	1.00	1.00	1.00	1.00	1.08	1.17	1.29	1.43	1.60	1.81	2.10	2.49	3.06	3.97	5.63	9.72
		19	0.3	1.00	1.00	1.00	1.00	1.00	1.00	1.00	1.00	1.01	1.10	1.20	1.32	1.46	1.65	1.89	2.20	2.64	3.31	4.43	6.67	****
			0.8	1.00	1.00	1.00	1.00	1.00	1.00	1.07	1.20	1.38	1.62	1.95	2.45	3.31	5.08	****	****	****	****	****	****	****
20	1.0	10	0.3	1.01	1.02	1.03	1.05	1.06	1.08	1.09	1.11	1.12	1.14	1.16	1.17	1.19	1.21	1.23	1.25	1.27	1.29	1.31	1.33	1.36
			0.8	1.01	1.03	1.05	1.07	1.09	1.11	1.13	1.16	1.18	1.21	1.23	1.26	1.29	1.32	1.35	1.38	1.42	1.45	1.49	1.53	1.57
		13	0.3	1.02	1.04	1.06	1.09	1.11	1.14	1.17	1.20	1.23	1.27	1.30	1.34	1.38	1.42	1.46	1.51	1.56	1.62	1.67	1.74	1.80
			0.8	1.03	1.06	1.09	1.13	1.17	1.21	1.26	1.30	1.36	1.41	1.48	1.54	1.62	1.70	1.79	1.89	2.00	2.13	2.27	2.43	2.62
		16	0.3	1.03	1.06	1.10	1.14	1.19	1.24	1.29	1.34	1.40	1.47	1.55	1.63	1.72	1.82	1.93	2.06	2.21	2.38	2.58	2.81	3.10
			0.8	1.04	1.09	1.15	1.21	1.28	1.36	1.45	1.55	1.67	1.80	1.96	2.15	2.38	2.67	3.03	3.50	4.15	5.10	6.61	9.39	****
		19	0.3	1.04	1.10	1.15	1.22	1.29	1.37	1.46	1.57	1.69	1.83	2.00	2.20	2.44	2.75	3.15	3.67	4.41	5.52	7.38	****	****
			0.8	1.06	1.14	1.23	1.33	1.45	1.60	1.78	2.01	2.30	2.70	3.25	4.09	5.52	8.48	****	****	****	****	****	****	****
24	0.6	16	0.3	1.00	1.00	1.00	1.00	1.00	1.00	1.00	1.00	1.00	1.00	1.00	1.00	1.00	1.00	1.00	1.00	1.00	1.00	1.00	1.00	1.00
			0.8	1.00	1.00	1.00	1.00	1.00	1.00	1.00	1.00	1.00	1.00	1.00	1.00	1.00	1.00	1.00	1.00	1.00	1.00	1.01	1.05	1.09
		19	0.3	1.00	1.00	1.00	1.00	1.00	1.00	1.00	1.00	1.00	1.00	1.00	1.00	1.00	1.00	1.00	1.00	1.00	1.00	1.02	1.06	1.11
			0.8	1.00	1.00	1.00	1.00	1.00	1.00	1.00	1.00	1.00	1.00	1.00	1.00	1.00	1.04	1.10	1.16	1.24	1.32	1.41	1.52	1.65
24	1.0	10	0.3	1.00	1.01	1.01	1.02	1.03	1.03	1.04	1.05	1.05	1.06	1.07	1.07	1.08	1.09	1.10	1.10	1.11	1.12	1.13	1.13	1.14
			0.8	1.00	1.01	1.02	1.03	1.04	1.05	1.06	1.07	1.08	1.09	1.10	1.11	1.12	1.13	1.14	1.15	1.16	1.17	1.19	1.20	1.21
		13	0.3	1.01	1.02	1.03	1.04	1.05	1.06	1.07	1.08	1.10	1.11	1.12	1.14	1.15	1.16	1.18	1.19	1.21	1.22	1.24	1.25	1.27
			0.8	1.01	1.02	1.04	1.06	1.07	1.09	1.11	1.12	1.14	1.16	1.18	1.20	1.22	1.24	1.27	1.29	1.31	1.34	1.37	1.39	1.42
		16	0.3	1.01	1.03	1.04	1.06	1.08	1.10	1.12	1.14	1.16	1.18	1.20	1.22	1.25	1.27	1.30	1.33	1.35	1.38	1.41	1.45	1.48
			0.8	1.02	1.04	1.06	1.09	1.12	1.14	1.17	1.20	1.24	1.27	1.31	1.34	1.38	1.43	1.47	1.52	1.57	1.63	1.69	1.75	1.82
		19	0.3	1.02	1.04	1.07	1.09	1.12	1.15	1.18	1.21	1.24	1.28	1.31	1.35	1.39	1.44	1.49	1.54	1.59	1.65	1.71	1.78	1.85
			0.8	1.03	1.06	1.10	1.13	1.17	1.22	1.27	1.32	1.37	1.43	1.50	1.57	1.65	1.74	1.83	1.94	2.06	2.20	2.36	2.54	2.76

TABLE 7-3 MOMENT MAGNIFICATION FACTORS δ FOR LONG COLUMNS. ACI 10.11.5 PAGE 4

f'_c = (KSI) = 5.00 ϕ =0.70

SIZE	C_m	kl_u	β_d	P(U) KIP 400	600	800	1000	1200	1400	1600	1800	2000	2200	2400	2600	2800	3000	3200	3400	3600	3800	4000	4200	4400
28	0.6	16	0.3	1.00	1.00	1.00	1.00	1.00	1.00	1.00	1.00	1.00	1.00	1.00	1.00	1.00	1.00	1.00	1.00	1.00	1.00	1.00	1.00	1.00
			0.8	1.00	1.00	1.00	1.00	1.00	1.00	1.00	1.00	1.00	1.00	1.00	1.00	1.00	1.00	1.00	1.00	1.03	1.07	1.12	1.17	1.22
		19	0.3	1.00	1.00	1.00	1.00	1.00	1.00	1.00	1.00	1.00	1.00	1.00	1.00	1.00	1.00	1.00	1.00	1.04	1.09	1.14	1.19	1.25
			0.8	1.00	1.00	1.00	1.00	1.00	1.00	1.00	1.00	1.00	1.00	1.00	1.04	1.10	1.18	1.26	1.35	1.46	1.59	1.74	1.92	2.15
28	1.0	10	0.3	1.01	1.02	1.02	1.03	1.04	1.04	1.05	1.06	1.07	1.07	1.08	1.09	1.10	1.10	1.11	1.12	1.13	1.14	1.15	1.15	1.16
			0.8	1.01	1.02	1.03	1.04	1.05	1.06	1.07	1.08	1.09	1.11	1.12	1.13	1.14	1.15	1.17	1.18	1.19	1.20	1.22	1.23	1.24
		13	0.3	1.02	1.03	1.04	1.05	1.07	1.08	1.09	1.11	1.12	1.13	1.15	1.16	1.18	1.19	1.21	1.23	1.24	1.26	1.28	1.30	1.32
			0.8	1.03	1.04	1.06	1.08	1.10	1.12	1.14	1.16	1.18	1.20	1.22	1.24	1.27	1.29	1.32	1.35	1.38	1.41	1.44	1.47	1.51
		16	0.3	1.03	1.05	1.07	1.09	1.11	1.13	1.15	1.17	1.20	1.22	1.25	1.27	1.30	1.33	1.36	1.39	1.43	1.46	1.50	1.54	1.58
			0.8	1.04	1.07	1.10	1.13	1.16	1.19	1.22	1.26	1.30	1.34	1.38	1.43	1.48	1.53	1.59	1.65	1.72	1.79	1.86	1.95	2.04
		19	0.3	1.04	1.07	1.10	1.13	1.16	1.19	1.23	1.27	1.31	1.35	1.39	1.44	1.49	1.55	1.61	1.67	1.74	1.81	1.90	1.99	2.08
			0.8	1.07	1.10	1.15	1.19	1.24	1.29	1.35	1.41	1.48	1.56	1.64	1.74	1.84	1.96	2.10	2.26	2.44	2.65	2.90	3.21	3.59
32	0.6	16	0.3	1.00	1.00	1.00	1.00	1.00	1.00	1.00	1.00	1.00	1.00	1.00	1.00	1.00	1.00	1.00	1.00	1.00	1.00	1.00	1.00	1.00
			0.8	1.00	1.00	1.00	1.00	1.00	1.00	1.00	1.00	1.00	1.00	1.00	1.00	1.00	1.00	1.00	1.00	1.00	1.00	1.00	1.00	1.00
		19	0.3	1.00	1.00	1.00	1.00	1.00	1.00	1.00	1.00	1.00	1.00	1.00	1.00	1.00	1.00	1.00	1.00	1.00	1.00	1.00	1.00	1.00
			0.8	1.00	1.00	1.00	1.00	1.00	1.00	1.00	1.00	1.00	1.00	1.00	1.00	1.00	1.00	1.00	1.00	1.00	1.00	1.00	1.00	1.03
32	1.0	10	0.3	1.00	1.01	1.01	1.01	1.02	1.02	1.03	1.03	1.04	1.04	1.04	1.05	1.05	1.06	1.06	1.06	1.07	1.07	1.08	1.08	1.09
			0.8	1.01	1.01	1.02	1.02	1.03	1.03	1.04	1.05	1.05	1.06	1.06	1.07	1.08	1.08	1.09	1.09	1.10	1.11	1.11	1.12	1.13
		13	0.3	1.01	1.01	1.02	1.03	1.04	1.04	1.05	1.06	1.06	1.07	1.08	1.09	1.10	1.10	1.11	1.12	1.13	1.14	1.14	1.15	1.16
			0.8	1.01	1.02	1.03	1.04	1.05	1.06	1.07	1.08	1.09	1.10	1.12	1.13	1.14	1.15	1.16	1.18	1.19	1.20	1.21	1.23	1.24
		16	0.3	1.02	1.03	1.04	1.05	1.06	1.07	1.08	1.09	1.10	1.12	1.13	1.14	1.15	1.17	1.18	1.20	1.21	1.23	1.24	1.26	1.27
			0.8	1.02	1.04	1.05	1.07	1.08	1.10	1.12	1.13	1.15	1.17	1.19	1.21	1.23	1.25	1.27	1.30	1.32	1.34	1.37	1.40	1.42
		19	0.3	1.02	1.04	1.05	1.07	1.09	1.10	1.12	1.14	1.16	1.18	1.19	1.22	1.24	1.26	1.28	1.30	1.33	1.35	1.38	1.41	1.43
			0.8	1.03	1.06	1.08	1.10	1.13	1.15	1.18	1.20	1.23	1.26	1.29	1.33	1.36	1.40	1.44	1.48	1.52	1.57	1.62	1.67	1.73
36	0.6	16	0.3	1.00	1.00	1.00	1.00	1.00	1.00	1.00	1.00	1.00	1.00	1.00	1.00	1.00	1.00	1.00	1.00	1.00	1.00	1.00	1.00	1.00
			0.8	1.00	1.00	1.00	1.00	1.00	1.00	1.00	1.00	1.00	1.00	1.00	1.00	1.00	1.00	1.00	1.00	1.00	1.00	1.00	1.00	1.00
		19	0.3	1.00	1.00	1.00	1.00	1.00	1.00	1.00	1.00	1.00	1.00	1.00	1.00	1.00	1.00	1.00	1.00	1.00	1.00	1.00	1.00	1.00
			0.8	1.00	1.00	1.00	1.00	1.00	1.00	1.00	1.00	1.00	1.00	1.00	1.00	1.00	1.00	1.00	1.00	1.00	1.00	1.00	1.00	1.00
36	1.0	10	0.3	1.00	1.00	1.00	1.01	1.01	1.01	1.01	1.02	1.02	1.02	1.02	1.03	1.03	1.03	1.03	1.04	1.04	1.04	1.05	1.05	1.05
			0.8	1.00	1.01	1.01	1.01	1.02	1.02	1.02	1.03	1.03	1.03	1.04	1.04	1.04	1.05	1.05	1.05	1.06	1.06	1.07	1.07	1.07
		13	0.3	1.00	1.01	1.01	1.02	1.02	1.02	1.03	1.03	1.04	1.04	1.05	1.05	1.06	1.06	1.06	1.07	1.07	1.08	1.08	1.09	1.09
			0.8	1.01	1.01	1.02	1.02	1.03	1.04	1.04	1.05	1.05	1.06	1.07	1.07	1.08	1.09	1.09	1.10	1.11	1.11	1.12	1.13	1.14
		16	0.3	1.01	1.01	1.02	1.03	1.03	1.04	1.05	1.05	1.06	1.07	1.07	1.08	1.09	1.10	1.10	1.11	1.12	1.13	1.14	1.14	1.15
			0.8	1.01	1.02	1.03	1.04	1.05	1.06	1.07	1.08	1.09	1.10	1.11	1.12	1.13	1.14	1.15	1.16	1.18	1.19	1.20	1.21	1.23
		19	0.3	1.01	1.02	1.03	1.04	1.05	1.06	1.07	1.08	1.09	1.10	1.11	1.12	1.13	1.14	1.16	1.17	1.18	1.19	1.20	1.22	1.23
			0.8	1.02	1.03	1.05	1.06	1.07	1.09	1.10	1.12	1.13	1.15	1.16	1.18	1.20	1.21	1.23	1.25	1.27	1.29	1.31	1.33	1.35

directions, the two δ's may also be different. Since the design moment,

$$M_c = \delta M_2 \qquad \text{(ACI-10-4)}$$

involves the largest end moment M_2, it will usually be necessary to investigate both ends of a column after magnifying both end moments in both directions.

COLUMN DETAILS

Spiral Column

The minimum number of longitudinal reinforcing bars shall be six and shall be not less than 0.01 nor more than 0.08 times the gross area of the section. The minimum size requirements of the old Code have been eliminated to allow wider utilization of reinforced concrete compression members in low-rise residential or light office buildings, and building systems.

The minimum size of bars to be used in making spirals has been increased to $\frac{3}{8}$ in. for cast-in-place columns. The splicing requirements for spirals have been relaxed to allow a 48 diameter or 12 in. lap or welds. The clear spacing between spirals shall not exceed 3 in. or be less than 1 in.

The elimination of the minimum sizes and a consideration of construction tolerances led to requiring an absolute minimum eccentricity of 1 in. or $0.05h$. For precast members the minimum design eccentricity can be reduced to not less than 0.6 in. provided that the manufacturing and erection tolerances are limited to one-third of the minimum design eccentricity.

Section 10.14 of the new Code deals with bearing stresses on concrete supports which are not laterally reinforced to resist splitting stresses. Bearing stresses must not exceed $0.85\phi f_c'$ with few exceptions.

Tied Column

The minimum number of longitudinal reinforcing bars shall be four and the size of ties is now related to the diameter of the bars tied. Ties must be at least #3 bars for longitudinal bars #10 or smaller and at least #4 in size for #11, #14, #18, and bundled longitudinal bars. Welded wire fabric of equivalent area can be used for ties. The spacing of the ties shall not exceed 16 longitudinal bar diameters, 48 tie bar diameters, or the least dimension of the column. The ties shall be so arranged that every corner and alternate longitudinal bar shall have lateral support provided by the corner of a tie having an included angle of not more than 135 deg, and no bar shall be farther than 6 in. clear on either side from such a laterally supported bar. The first tie above or below a joint must not be farther than half a tie spacing from the floor or footing. In some cases this is reduced to 3 in. Size and amount of longitudinal reinforcing bar requirement is the same as for spiral columns.

Minimum eccentricity requirement is similar to that for spiral columns, but in no case shall the minimum be less than $0.1h$.

The treatment of composite columns is completely different in the new ACI Code. The strength of cross sections is computed in accordance with Section 10.15.2, which refers to the analytical procedure for reinforced concrete columns discussed earlier in this chapter.

Walls can be designed as columns under provisions of Chapter 10 of the new Code, using slenderness provisions in Sections 10.9 and 10.10 or, if the eccentricity is less than one-sixth of the wall thickness, the Empirical Design Procedure given in Section 14.2 may be used.

REINFORCED CONCRETE COLUMN COMPUTATION TABLE,

STRENGTH DESIGN METHOD, ACI 318-71

COLUMN GRID LINES	STORY	DESIGN DATA (RESULTS FROM FRAME ANALYSIS)																			
		D DEAD-LOAD AXIAL IN KIPS MOMENT IN FT.-KIPS					L LIVE - LOAD REDUCED AS PERMITTED, (K) OR (FT-K)					E EARTH-QUAKE LOAD, AS PER CODE, (K) OR (FT-K)					W WIND - LOAD AS PER CODE (K) OR (FT-K)				
		P	MOMENTS				P	MOMENTS				P	MOMENTS				P	MOMENTS			
			N-S		E-W			N-S		E-W			N-S		E-W			N-S		E-W	
			TOP	BOTT.	TOP	BOTT.		TOP	BOTT.	TOP	BOTT		TOP	BOTT	TOP	BOTT		TOP	BOTT.	TOP	BOTT
(1)	(2)	(3)	(4)	(5)	(6)	(7)	(8)	(9)	(10)	(11)	(12)	(13)	(14)	(15)	(16)	(17)	(18)	(19)	(20)	(21)	(22)

REINFORCED CONCRETE COLUMN COMPUTATION TABLE,

STRENGTH DESIGN METHOD, ACI 318 - 71

COLUMN GRID LINES	STORY	COMPUTATION FOR MAXIMUM DESIGN LOADS																								
		0.75(1.4D + 1.7L + 1.87E)					0.75(1.4D + 1.7L + 1.7W)					1.4 x D					1.7 x L					1.4D + 1.7L				
		P	MOMENTS				P	MOMENTS				P	MOMENTS				P	MOMENTS				P	MOMENTS			
			N-S		E-W			N-S		E-W			N-S		E-W			N-S		E-W			N-S		E-W	
			T	B	T	B		T	B	T	B		T	B	T	B		T	B	T	B		T	B	T	B
(1)	(2)	(23)	(24)	(25)	(26)	(27)	(28)	(29)	(30)	(31)	(32)	(33)	(34)	(35)	(36)	(37)	(38)	(39)	(40)	(41)	(42)	(43)	(44)	(45)	(46)	(47)

NOTE: **AS A RULE, ALL COLUMNS** ARE SUBJECT TO MOMENTS IN TWO DIRECTIONS.
FOR EXAMPLE: IF SEISMIC LOAD IS N-S, THEN DESIGN MOMENTS ARE 1.4 (D + L + E) [N-S], AND SIMULTANEOUSLY, 1.4 (D + L) [E-W].
PROPER TABULATION IS MANDATORY.

REINFORCED CONCRETE COLUMN COMPUTATION TABLE,

STRENGTH DESIGN METHOD, ACI 318 - 71

COLUMN GRID LINES	STORY	COMPUTATION FOR MINIMUM DESIGN LOADS.																								
		0.9 x D					1.43 x E					1.30 x W					1.43E - 0.9D					1.30W - 0.9D				
		P	MOMENTS				P	MOMENTS				P	MOMENTS				P	MOMENTS				P	MOMENTS			
			N-S		E-W			N-S		E-W			N-S		E-W			N-S		E-W			N-S		E-W	
			T	B	T	B		T	B	T	B		T	B	T	B		T	B	T	B		T	B	T	B
(1)	(2)	(48)	(49)	(50)	(51)	(52)	(53)	(54)	(55)	(56)	(57)	(58)	(59)	(60)	(61)	(62)	(63)	(64)	(65)	(66)	(67)	(68)	(69)	(70)	(71)	(72)

NOTE: COMPUTATIONS FOR MINIMUM DESIGN LOADS ARE NECESSARY SINCE COLUMNS MAY BE SUBJECT TO UPLIFT AND TWO-DIRECTIONAL SIMULTANEOUS MOMENTS.

REINFORCED CONCRETE COLUMN COMPUTATION TABLE,

STRENGTH DESIGN METHOD, ACI 318-71

COLUMN GRID LINES	STORY	MAX. CONDITIONS					MIN. CONDITIONS					DESIGN DATA DIMENSIONS							
		SELECT VALUES FROM										$r = 0.3t$ OR $0.25D$							
		(24) TO (47)					(63) TO (72)					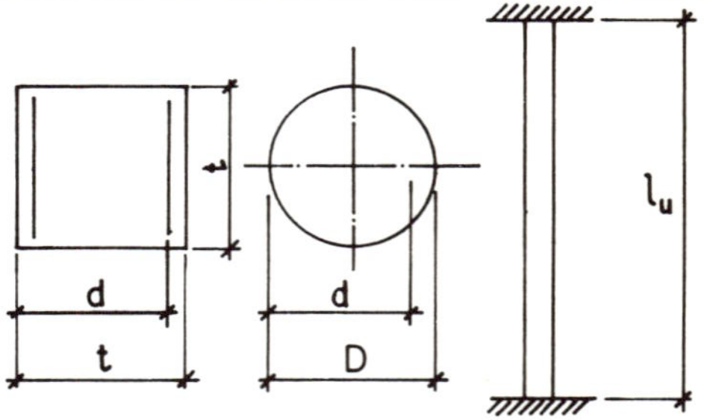							
		P_u (KIP)	M_u (FT.-K)				P_u (KIP)	M_u (FT.-K)											
			N-S		E-W			N-S		E-W									
			T	B	T	B		T	B	T	B	t	D	d	r	A_g	l_u	f'_c	f_y
(1)	(2)	(73)	(74)	(75)	(76)	(77)	(78)	(79)	(80)	(81)	(82)	(83)	(84)	(85)	(86)	(87)	(88)	(89)	(90)

REINFORCED CONCRETE COLUMN COMPUTATION TABLE,
STRENGTH DESIGN METHOD, ACI 318-71

COLUMN GRID LINES	STORY	SLENDERNESS EFFECTS								
		COMPUTATION OF EFFECTIVE LENGTH FACTOR (K)								
		BRACED AGAINST SIDESWAY					NOT BRACED			SELECTED Kl_u
		SLENDERNESS RATIO LIMITS								
		K = 1.0	$34 - 12\frac{M_1}{M_2}$ M_1 & M_2 TO BE SELECTED FROM (74)-(77) OR (79)-(82)	$K\frac{l_u}{r}$ FOR K USE 1.0 OR VALUE BY FIG. 10-3 OF CODE	IF (93) > (92) NEGLECT SLENDERNESS EFFECT	IF (93) > 100 ANALYSIS BY SECTION 10.10.1 OF THE CODE IS REQUIRED	K > 1.0 CHECK IN SEPARATE STEP BY SEC. 10.11.3	$K\frac{l_u}{r}$	IF (97) < 22 NEGLECT SLENDERNESS EFFECT	
(1)	(2)	(91)	(92)	(93)	(94)	(95)	(96)	(97)	(98)	(99)

REINFORCED CONCRETE COLUMN COMPUTATION TABLE,

STRENGTH DESIGN METHOD, ACI 318-71

COLUMN GRID LINES	STORY	COMPUTATION OF MOMENT MAGNIFICATION FACTOR (δ)							
		$\beta_d = \frac{\text{MAX. D.L. MOMENT}}{\text{MAX. TOTAL MOMENT}}$	E_c = MODULUS OF ELASTICITY OF CONCRETE (P.S.I.)	I_g = MOMENT OF INERTIA OF GROSS CONCRETE SECTION	$EI = \frac{E_c I_g / 2.5}{1 + \beta_d}$ $EI = \frac{(101) \times (102) / 2.5}{1 + (100)}$	$P_c = \frac{\pi^2 EI}{(K l_u)^2}$ $P_c = \frac{\pi^2 \times (103)}{(99)^2}$	$\delta = \frac{1.0}{1 - P_u / \phi P_c} \geqslant 1.0$ $\delta = \frac{1.0}{1 - \frac{(73) \text{ OR } (78)}{0.7 \times (104)}}$	IN FRAMES NOT BRACED AGAINST SIDESWAY, THE VALUE OF δ SHALL BE COMPUTED FOR THE ENTIRE STORY ASSUMING ALL COLUMNS TO BE LOADED, IN (105) P_u AND P_c SHALL BE TAKEN AS ΣP_u AND ΣP_c FOR ALL COLUMNS OF STORY	SELECTED δ
(1)	(2)	(100)	(101)	(102)	(103)	(104)	(105)	(106)	(107)

REINFORCED CONCRETE COLUMN COMPUTATION TABLE,
STRENGTH DESIGN METHOD, ACI 318-71

COLUMN GRID LINES	STORY	DESIGN LOADING CONDITIONS													
		MAXIMUM CONDITIONS							MINIMUM CONDITIONS						
			N - S			E - W				N - S			E - W		
		P_u (KIP)	M_u (FT-KIP)	e (IN.)	$\frac{e}{t}$	M_u (FT-KIP)	e (IN.)	$\frac{e}{t}$	P_u (KIP)	M_u (FT-KIP)	e (IN.)	$\frac{e}{t}$	M_u (FT-KIP)	e (IN.)	$\frac{e}{t}$
		(73)	(107) × (74) OR (107) × (75)	$\frac{(109) \times 12}{(108)}$	$\frac{(110)}{(83) \text{ OR } (84)}$	(107) × (76) OR (107) × (77)	$\frac{(112) \times 12}{(108)}$	$\frac{(113)}{(83) \text{ OR } (84)}$	(78)	(107) × (79) OR (107) × (80)	$\frac{(116) \times 12}{(115)}$	$\frac{(117)}{(83) \text{ OR } (84)}$	(107) × (81) OR (107) × (82)	$\frac{(119) \times 12}{(115)}$	$\frac{(120)}{(83) \text{ OR } (84)}$
(1)	(2)	(108)	(109)	(110)	(111)	(112)	(113)	(114)	(115)	(116)	(117)	(118)	(119)	(120)	(121)

REINFORCED CONCRETE COLUMN COMPUTATION TABLE, STRENGTH DESIGN METHOD, ACI 318-71												
COLUMN GRID LINES	STORY	COMPUTATION FOR VERTICAL REINFORCEMENT										
		CONCRETE STRENGTH	YIELD STRENGTH OF STEEL		CAPACITY REDUCTION FACTOR		CODE REQUIREMENT					SELECTED REINFORCEMENT (SEE TABLE....)
							MINIMUM AND MAXIMUM REINFORCEMENT		MINIMUM NUMBER OF VERTICAL BARS ARRANGEMENT		MINIMUM SPIRAL REINFORCEMENT $P_s = 0.45\left(\frac{A_g}{A_c}-1\right)\frac{f_c}{f_y}$ max. $f_y = 60$ ksi	
			f_y		ϕ							
		f'_c	REINF.	TIES OR SPIRAL	TIED	SPIRAL	MIN. 1%	MAX. 8%	RECT.	CIRC.	P_s	A_{ST}
(1)	(2)	(122)	(123)	(124)	(125)	(126)	(127)	(128)	(129)	(130)	(131)	(132)

REINFORCED CONCRETE COLUMN COMPUTATION TABLE, STRENGTH DESIGN METHOD, ACI 318-71															
COLUMN GRID LINES	STORY	COMPUTATION FOR TRANSVERSE REINFORCEMENT													
		DESIGN DATA – RESULT FROM FRAME ANALYSIS								COMPUTATION FOR DESIGN LOADS					
		DEAD-LOAD IN KIPS		LIVE-LOAD REDUCED AS PERMITTED IN KIPS		EARTH QUAKE LOAD, PER CODE IN KIPS		WIND-LOAD AS PER CODE IN KIPS		0.75 (1.4D + 1.7L + 1.87E) *		0.75 (1.4D + 1.7L + 1.7W)		1.4D	
		SHEAR								SHEAR					
		N-S	E-W	N-S	E-W	N-S	E-W	N-S	E-W	N-S	E-W	N-S	E-W	N-S	E-W
(1)	(2)	(133)	(134)	(135)	(136)	(137)	(138)	(139)	(140)	(141)	(142)	(143)	(144)	(145)	(146)

* REFER TO SEAC: 1.4 (D + L + E)

REINFORCED CONCRETE COLUMN COMPUTATION TABLE,

STRENGTH DESIGN METHOD, ACI 318 - 71

COLUMN GRID LINES	STORY	COMPUTATION FOR TRANSVERSE REINFORCEMENT											
		COMPUTATION FOR. DESIGN LOADS								LIMITS FOR v_c			
		1.7 L		1.4 D + 1.7 L		V_u *		SELECTED V_u		$v_c = 2\sqrt{f_c'} = 2 \times \sqrt{(120)}$	$v_c = 3.5\sqrt{f_c'} = 3.5 \times \sqrt{(120)}$	$V_c = 2\left(1 + 0.0005\frac{N_u}{A_g}\right)\sqrt{f_c'}$ BUT V_c SHALL NOT EXCEED: a) $V_c = 3.5\sqrt{f_c'}\sqrt{1 + 0.002\frac{N_u}{A_g}}$ b) $V_c = 2.0\sqrt{f_c'}\sqrt{1 + 0.002\frac{N_u}{A_g}}$ a) N_u = AXIAL COMPR. (+) b) N_u = AXIAL TENSION (-) A_g = GROSS AREA OF SECTION, sq. in.	SELECTED V_c
		SHEAR											
		N-S	E-W	N-S	E-W	N-S	E-W	N-S	E-W				
(1)	(2)	(147)	(148)	(149)	(150)	(151)	(152)	(153)	(154)	(155)	(156)	(157)	(158)

* REFER TO SEAC. OR CODE APPENDIX A.

REINFORCED CONCRETE COLUMN COMPUTATION TABLE,

STRENGTH DESIGN METHOD, ACI 318-71

COLUMN GRID LINES	STORY	COMPUTATION FOR TRANSVERSE REINFORCEMENT											
		$v_u = \frac{V_u}{\phi t d} = \frac{(153) \text{ or } (154)}{0.85 \times t \times (85)}$ FOR CIRCULAR SECTION: $t = D$		CODE REQUIREMENT								ONLY FOR SPECIAL DUCTILE FRAMES	
				SPACING OF TIES OR SPIRALS IN THE COLUMN -S-								REFER TO SEAC OR CODE APPENDIX A.	
		v_u		$0.5\,d = 0.5 \times (85)$	FOR SPIRALS {min. 1 in. max. 3 in.	S = 16 LONG. BAR DIAM.	SIZE OF TIES OR SPIRALS min. 3/8" ϕ	A_v = AREA OF SHEAR REINF	S = 48 x TIE BAR DIA.	$S = \frac{A_v f_y}{(v_u - v_c) t}$ FOR CIRCULAR SECTION: $t = D$	SELECTED SPACING	CHECK BEAM - COLUMN CONNECTION. Σ COL. MOMENT STRENGTH SHALL BE GREATER THAN Σ BEAM MOMENT STR.	PROVIDE SPECIAL TRANSV. REINF. (SPIRAL OR HOOPS) FOR CONFINEMENT OF COL. CORE.
		N-S	E-W										
(1)	(2)	(159)	(160)	(161)	(162)	(163)	(164)	(165)	(166)	(167)	(168)	(169)	(170)

625 North Michigan Ave., Chicago, Illinois. Twenty-seven stories, flat plate office. Owner: Romanek & Gollub, Inc. Structural Engineers: Rogers-Cohen-Barreto-Marchertas, Inc.

8

Slab Systems with Square or Rectangular Panels

Under this category are included:

1) Flat slabs with and without capitals and/or drop panels.
2) Flat plates without beams.
3) Two-way slabs with projecting beams at all four sides.

The ACI 318-71 Code had completely altered the previously used systems. The new theories are based on extensive tests performed at the University of Illinois. Undeniably, the new formulas are more complex than the previous ones, and their application is more time-consuming. The new methods now combine flat plates and two-way slabs. It is the author's opinion that for many two-way slabs the new provisions are unduly time-consuming; for this reason, Method 3 from the ACI 318-63 is also included as an acceptable and legitimate system.

Slab systems design, in general, may be performed by the following:

1) Direct Design Method, provided that the prescribed limitations are maintained (Section 13.3), or
2) Equivalent Frame Method (Section 13.4).

The application of these two methods is illustrated by step-by-step development of the complete solution. Several intermediate formulas have been solved by computers, and a limited number of computer outputs are included in this text.

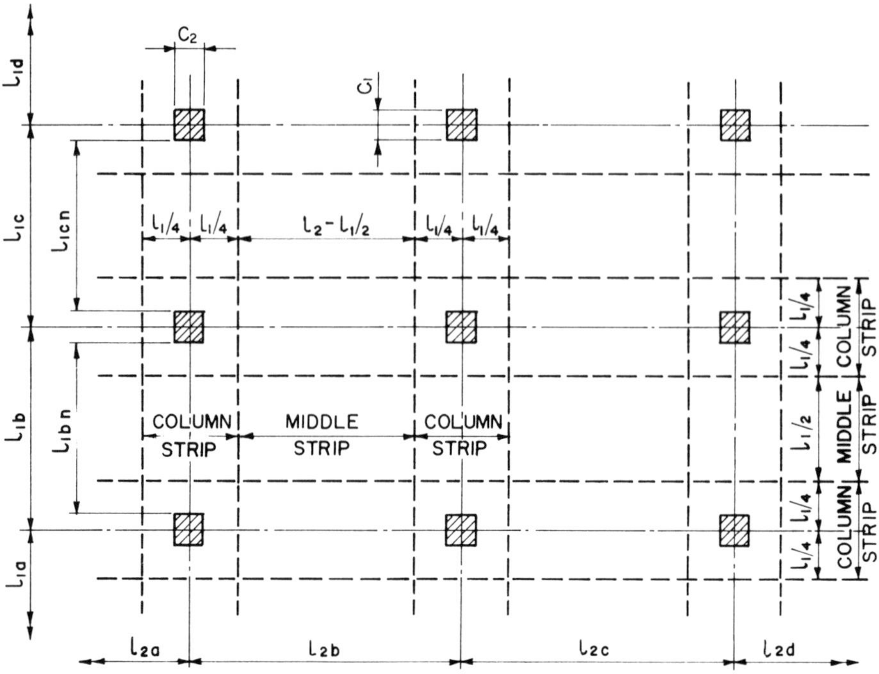

Fig. 8-1.

THE DIRECT-DESIGN METHOD

Limitations:

1) Minimum three continuous spans in each direction.
2) Ratio of long to short spans < 2.0.
3) Adjacent spans to vary not more than one third of the longer span.
4) Columns may be offset 10 percent.
5) Live load $<$ three times dead load.
6) Column strip, on either side, to be 25 percent of the shorter side.
7) For two-way slabs on beams, the relative stiffness of the beams in the two perpendicular directions, $\frac{\alpha_1 l_2^2}{\alpha_2 l_1^2}$ shall not be less than 0.2 or greater than 5.0. Value of α may be taken from Table 8-12.

Step-by-Step Manual Design Procedure for Flat Slabs

In the following tables for flat plate design, simplifications are made for greater flexibility in the expression of design parameters; column sizes are specified as a ratio of the span. The symbols used in the tables are described below.

l_1 = span in the direction moments are computed, c/c of supports
l_2 = span transverse to l_1
γ = ratio of l_1/l_2
c_1 = size of rectangular column in the direction of l_1
m = l_1/c_1 span to column size ratio; the column size is specified as a function of m: $c_1 = l_1/m$.
c_2 = column size transverse to c_1, expressed as a function of the span ratio: $c_1/l_1 = c_2/l_2$; therefore $c_2 = c_1 l_2/l_1$; substitute $c_1 = l_1/m$, and obtain $c_2 = l_2/m$
h = slab thickness
l_c = story height

Monolithic construction is assumed, therefore the modulus of elasticity of concrete for slabs and beams is equal, $E_{cb} = E_{cs}$.

Step 1. Check the requirements for geometry of the structure and live load limitations, according to items 1) to 6) above.

Step 2. Select slab thickness according to ACI formulas (9-6), (9-7), and (9-8). Table 8-2, computer output, furnishes direct solutions for minimum slab thickness.

Step 3. Establish load factors:

$U = 1.4D + 1.7L$ (ACI) or (SEAC, Structural Engineers Assoc.,
$U = 1.5D + 1.8L$ Calif.; in seismic area)

Table 8-1. Minimum α_{min}.

β_α	*Aspect ratio* l_2/l_1	*Relative beam stiffness, a* 0	0.5	1.0	2.0	4.0
2.0	0.5–2.0	0	0	0	0	0
1.0	0.5	0.6	0	0	0	0
	0.8	0.7	0	0	0	0
	1.0	0.7	0.1	0	0	0
	1.25	0.8	0.4	0	0	0
	2.0	1.2	0.5	0.2	0	0
0.5	0.5	1.3	0.3	0	0	0
	0.8	1.5	0.5	0.2	0	0
	1.0	1.6	0.6	0.2	0	0
	1.25	1.9	1.0	0.5	0	0
	2.0	4.9	1.6	0.8	0.3	0
0.33	0.5	1.8	0.5	0.1	0	0
	0.8	2.0	0.9	0.3	0	0
	1.0	2.3	0.9	0.4	0	0
	1.25	2.8	1.5	0.8	0.2	0
	2.0	13.0	2.6	1.2	0.5	0.3

TABLE 8-2. (Minimum slab thickness ratio)

TABLE 8-2 MIN. SLAB THICKNESS RATIO = SPAN/t IN 2-WAY CONSTR'N. ACI CODE, 9.5.5

SLAB CONTINUITY CONFIGURATION (MODE)

MODE =	1	2	3	4	5	6	7	8
NO. OF CONTINUOUS LONG EDGES	2	2	1	2	0	1	1	0
NO. OF CONTINUOUS SHORT EDGES	2	1	2	0	2	1	0	1

SYMBOLS --

α AV) = 0 FOR FLAT SLABS
β = LONG SPAN / SHORT SPAN

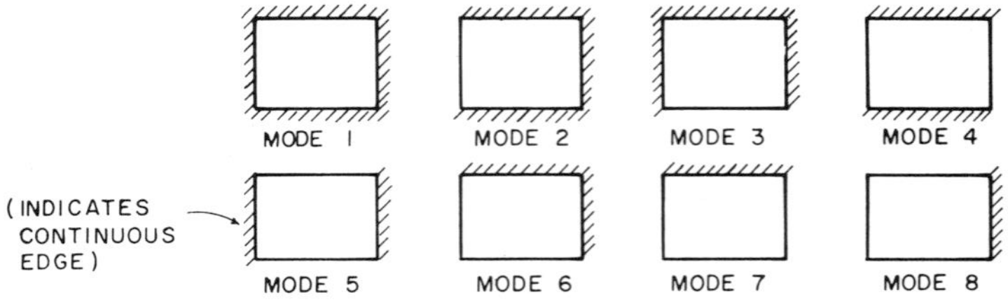

Table 8-2.
GENERAL NOTES: The minimum slab thickness in a two-way construction is controlled by Section 9.5.3.1. This Table (8-2) provides the maximum ratio of span to slab thickness per equations 9-6, 9-7 and 9-8, for varying steel yield stresses, quantities of α and β and slab continuity ratios. The quantity of α is the average value of the flexural stiffness ratios of beams to the slabs for all edges of the panel, and β is the ratio of long to short spans. The tabulation contains data for the eight slab continuity configurations indicated above.

TABLE 8-2. MIN. SLAB THICKNESS RATIO = SPAN/h IN 2-WAY CONSTR'N. ACI CODE, 9.5.5 FY(KSI) = 40.0 PAGE 1

MODE =		1			2			3			4			5			6			7			8		
EQU'N=		9-6	9-7	9-8	9-6	9-7	9-8	9-6	9-7	9-8	9-6	9-7	9-8	9-6	9-7	9-8	9-6	9-7	9-8	9-6	9-7	9-8	9-6	9-7	9-8
α	β																								
0.0	1.0	36	46	36	35	45	36	35	45	36	34	44	36	34	44	36	34	44	36	32	42	36	32	42	36
	1.2	36	48	36	35	47	36	35	46	36	34	45	36	33	45	36	33	45	36	32	44	36	32	43	36
	1.4	36	50	36	35	49	36	34	48	36	34	47	36	33	46	36	33	46	36	32	45	36	31	44	36
	1.6	36	52	36	35	50	36	34	50	36	34	49	36	32	47	36	33	48	36	32	46	36	31	46	36
	1.8	36	54	36	35	52	36	34	51	36	34	51	36	32	48	36	33	49	36	31	48	36	30	47	36
	2.0	36	56	36	35	54	36	34	53	36	34	53	36	31	49	36	32	51	36	31	49	36	30	48	36
0.5	1.0	39	46	36	37	45	36	37	45	36	36	44	36	36	44	36	36	44	36	35	42	36	35	42	36
	1.2	39	48	36	38	47	36	37	46	36	36	45	36	36	45	36	36	45	36	35	44	36	35	43	36
	1.4	40	50	36	38	49	36	38	48	36	37	47	36	36	46	36	36	46	36	35	45	36	35	44	36
	1.6	40	52	36	39	50	36	38	50	36	38	49	36	36	47	36	37	48	36	36	46	36	35	46	36
	1.8	41	54	36	39	52	36	38	51	36	38	51	36	36	48	36	37	49	36	36	48	36	35	47	36
	2.0	41	56	36	40	54	36	39	53	36	39	53	36	36	49	36	37	51	36	36	49	36	35	48	36
1.0	1.0	41	46	36	40	45	36	40	45	36	39	44	36	39	44	36	39	44	36	37	42	36	37	42	36
	1.2	42	48	36	41	47	36	41	46	36	40	45	36	39	45	36	39	45	36	38	44	36	38	43	36
	1.4	43	50	36	42	49	36	41	48	36	41	47	36	40	46	36	40	46	36	39	45	36	38	44	36
	1.6	44	52	36	43	50	36	42	50	36	42	49	36	40	47	36	41	48	36	40	46	36	39	46	36
	1.8	45	54	36	44	52	36	43	51	36	43	51	36	41	48	36	41	49	36	40	48	36	39	47	36
	2.0	46	56	36	45	54	36	44	53	36	44	53	36	41	49	36	42	51	36	41	49	36	40	48	36
2.0	1.0	46	46	36	45	45	36	45	45	36	44	44	36	44	44	36	44	44	36	42	42	36	42	42	36
	1.2	48	48	36	47	47	36	47	46	36	46	45	36	45	45	36	45	45	36	44	44	36	44	43	36
	1.4	50	50	36	49	49	36	48	48	36	48	47	36	47	46	36	47	46	36	46	45	36	45	44	36
	1.6	52	52	36	51	50	36	50	50	36	50	49	36	48	47	36	49	48	36	48	46	36	47	46	36
	1.8	54	54	36	53	52	36	52	51	36	52	51	36	50	48	36	50	49	36	49	48	36	48	47	36
	2.0	56	56	36	55	54	36	54	53	36	54	53	36	51	49	36	52	51	36	51	49	36	50	48	36
5.0	1.0	61	46	36	60	45	36	60	45	36	59	44	36	59	44	36	59	44	36	57	42	36	57	42	36
	1.2	66	48	36	65	47	36	65	46	36	64	45	36	63	45	36	63	45	36	62	44	36	62	43	36
	1.4	71	50	36	70	49	36	69	48	36	69	47	36	68	46	36	68	46	36	67	45	36	66	44	36
	1.6	76	52	36	75	50	36	74	50	36	74	49	36	72	47	36	73	48	36	72	46	36	71	46	36
	1.8	81	54	36	80	52	36	79	51	36	79	51	36	77	48	36	78	49	36	76	48	36	75	47	36
	2.0	86	56	36	85	54	36	84	53	36	84	53	36	81	49	36	82	51	36	81	49	36	80	48	36
10.0	1.0	86	46	36	85	45	36	85	45	36	84	44	36	84	44	36	84	44	36	82	42	36	82	42	36
	1.2	96	48	36	95	47	36	95	46	36	94	45	36	93	45	36	93	45	36	92	44	36	92	43	36
	1.4	106	50	36	105	49	36	104	48	36	104	47	36	103	46	36	103	46	36	102	45	36	101	44	36
	1.6	116	52	36	115	50	36	114	50	36	114	49	36	112	47	36	113	48	36	112	46	36	111	46	36
	1.8	126	54	36	125	52	36	124	51	36	124	51	36	122	48	36	123	49	36	121	48	36	120	47	36
	2.0	136	56	36	135	54	36	134	53	36	134	53	36	131	49	36	132	51	36	131	49	36	130	48	36
20.0	1.0	136	46	36	135	45	36	135	45	36	134	44	36	134	44	36	134	44	36	132	42	36	132	42	36
	1.2	156	48	36	155	47	36	155	46	36	154	45	36	153	45	36	153	45	36	152	44	36	152	43	36
	1.4	176	50	36	175	49	36	174	48	36	174	47	36	173	46	36	173	46	36	172	45	36	171	44	36
	1.6	196	52	36	195	50	36	194	50	36	194	49	36	192	47	36	193	48	36	192	46	36	191	46	36
	1.8	216	54	36	215	52	36	214	51	36	214	51	36	212	48	36	213	49	36	211	48	36	210	47	36
	2.0	236	56	36	235	54	36	234	53	36	234	53	36	231	49	36	232	51	36	231	49	36	230	48	36

ℓ/h

ℓ/h

TABLE 8-2, MIN. SLAB THICKNESS RATIO = SPAN/h IN 2-WAY CONSTR'N. ACI CODE, 9.5.5 FY(KSI) = 60.0 PAGE 2

MODE =		1			2			3			4			5			6			7			8		
EQU'N=		9-6	9-7	9-8	9-6	9-7	9-8	9-6	9-7	9-8	9-6	9-7	9-8	9-6	9-7	9-8	9-6	9-7	9-8	9-6	9-7	9-8	9-6	9-7	9-8
α	β																								
0.0	1.0	33	42	33	32	41	33	32	41	33	30	40	33	30	40	33	30	40	33	29	38	33	29	38	33
	1.2	33	44	33	32	42	33	31	42	33	30	41	33	30	41	33	30	41	33	29	40	33	29	39	33
	1.4	33	45	33	32	44	33	31	44	33	30	43	33	30	42	33	30	42	33	29	41	33	28	40	33
	1.6	33	47	33	32	46	33	31	45	33	30	44	33	29	43	33	30	44	33	29	42	33	28	41	33
	1.8	33	49	33	32	48	33	31	46	33	30	46	33	29	44	33	30	45	33	28	44	33	28	42	33
	2.0	33	51	33	32	49	33	30	48	33	30	48	33	28	45	33	29	46	33	28	45	33	27	43	33
0.5	1.0	35	42	33	34	41	33	34	41	33	33	40	33	33	40	33	33	40	33	32	38	33	32	38	33
	1.2	35	44	33	34	42	33	34	42	33	33	41	33	33	41	33	33	41	33	32	40	33	32	39	33
	1.4	36	45	33	35	44	33	34	44	33	34	43	33	33	42	33	33	42	33	32	41	33	32	40	33
	1.6	36	47	33	35	46	33	35	45	33	34	44	33	33	43	33	33	44	33	32	42	33	32	41	33
	1.8	37	49	33	36	48	33	35	46	33	35	46	33	33	44	33	34	45	33	33	44	33	32	42	33
	2.0	37	51	33	36	49	33	35	48	33	35	48	33	33	45	33	34	46	33	33	45	33	32	43	33
1.0	1.0	37	42	33	36	41	33	36	41	33	35	40	33	35	40	33	35	40	33	34	38	33	34	38	33
	1.2	38	44	33	37	42	33	37	42	33	36	41	33	35	41	33	36	41	33	35	40	33	34	39	33
	1.4	39	45	33	38	44	33	38	44	33	37	43	33	36	42	33	36	42	33	35	41	33	35	40	33
	1.6	40	47	33	39	46	33	38	45	33	38	44	33	36	43	33	37	44	33	36	42	33	35	41	33
	1.8	41	49	33	40	48	33	39	46	33	39	46	33	37	44	33	38	45	33	37	44	33	36	42	33
	2.0	42	51	33	41	49	33	40	48	33	40	48	33	37	45	33	38	46	33	37	45	33	36	43	33
2.0	1.0	42	42	33	41	41	33	41	41	33	40	40	33	40	40	33	40	40	33	38	38	33	38	38	33
	1.2	44	44	33	43	42	33	42	42	33	41	41	33	41	41	33	41	41	33	40	40	33	40	39	33
	1.4	45	45	33	44	44	33	44	44	33	43	43	33	42	42	33	43	42	33	42	41	33	41	40	33
	1.6	47	47	33	46	46	33	45	45	33	45	44	33	44	43	33	44	44	33	43	42	33	43	41	33
	1.8	49	49	33	48	48	33	47	46	33	47	46	33	45	44	33	46	45	33	45	44	33	44	42	33
	2.0	51	51	33	50	49	33	49	48	33	49	48	33	46	45	33	48	46	33	46	45	33	45	43	33
5.0	1.0	55	42	33	54	41	33	54	41	33	53	40	33	53	40	33	53	40	33	52	38	33	52	38	33
	1.2	60	44	33	59	42	33	59	42	33	58	41	33	57	41	33	58	41	33	56	40	33	56	39	33
	1.4	65	45	33	63	44	33	63	44	33	62	43	33	61	42	33	62	42	33	61	41	33	60	40	33
	1.6	69	47	33	68	46	33	67	45	33	67	44	33	65	43	33	66	44	33	65	42	33	64	41	33
	1.8	74	49	33	73	48	33	72	46	33	71	46	33	70	44	33	70	45	33	69	44	33	68	42	33
	2.0	78	51	33	77	49	33	76	48	33	76	48	33	74	45	33	75	46	33	74	45	33	73	43	33
10.0	1.0	78	42	33	77	41	33	77	41	33	76	40	33	76	40	33	76	40	33	75	38	33	75	38	33
	1.2	87	44	33	86	42	33	86	42	33	85	41	33	85	41	33	85	41	33	84	40	33	83	39	33
	1.4	96	45	33	95	44	33	95	44	33	94	43	33	93	42	33	94	42	33	93	41	33	92	40	33
	1.6	105	47	33	104	46	33	104	45	33	103	44	33	102	43	33	103	44	33	101	42	33	101	41	33
	1.8	115	49	33	113	48	33	113	46	33	112	46	33	110	44	33	111	45	33	110	44	33	109	42	33
	2.0	124	51	33	123	49	33	121	48	33	121	48	33	119	45	33	120	46	33	119	45	33	118	43	33
20.0	1.0	124	42	33	123	41	33	123	41	33	121	40	33	121	40	33	121	40	33	120	38	33	120	38	33
	1.2	142	44	33	141	42	33	140	42	33	140	41	33	139	41	33	139	41	33	138	40	33	138	39	33
	1.4	160	45	33	159	44	33	158	44	33	158	43	33	157	42	33	157	42	33	156	41	33	156	40	33
	1.6	178	47	33	177	46	33	176	45	33	176	44	33	175	43	33	175	44	33	174	42	33	173	41	33
	1.8	196	49	33	195	48	33	194	46	33	194	46	33	192	44	33	193	45	33	192	44	33	191	42	33
	2.0	215	51	33	213	49	33	212	48	33	212	48	33	210	45	33	211	46	33	210	45	33	209	43	33

Step 4. Compute total statical moment:

$$M_o = \frac{w l_2 l_n^2}{8} \qquad \text{(ACI-13-2)}$$

where

w = factored dead and live load (reduce live load per panel area)
l_2 = transverse span length, measured center-to-center of supports
l_n = clear span, in direction of moment

Step 5. Compute the value of α_{ec}, based on the properties of the exterior column, which is used for computing the moment distribution coefficients of the positive and negative moments at the end panel.

(1) Compute the flexural stiffnesses of slabs and beams ($K_s + K_b$). The slab-beam system consists of a nonprismatic beam, where the moment of inertia, outside of the column face, is based on the concrete cross-sectional area. At the joint, from centerline to the face of the column, the moment of inertia of the slab-beam structure at that face is divided by $(1 - c_2/l_2)^2$, where l_2 is the span transverse to the direction of moments being considered and c_2 is the column size in the direction of l_2.

(2) Compute the flexural stiffness of the column (K_c), based on infinite moment of inertia at joints.

(3) Compute the torsional stiffness (K_t) of the slab beam attached to the column, transverse to the direction in which moments are being determined.

$$K_t = \sum \frac{9 E_{cs} C}{l_2 (1 - c_2/l_2)^3} \qquad \text{(ACI-13-6)}$$

Where C is computed by dividing the cross section of the slab-beam torsional member into rectangular parts:

$$C = \sum (1 - 0.63\, x/y) x^3 y/3 \qquad \text{(ACI-13-7)}$$

(4) Compute the flexibility ($1/K_{ec}$) of the equivalent column:

$$\frac{1}{K_{ec}} = \frac{1}{K_c} + \frac{1}{K_t} \qquad \text{(ACI-13-5)}$$

Compute the stiffness of the equivalent column (K_{ec}), which is the inverse of the flexibility.

(5) Compute α_{ec}, the ratio of flexural stiffness of the equivalent column to the combined flexural stiffness of the slabs (and the beams, if any) at a joint taken in the direction moments are beind considered:

$$\alpha_{ec} = \frac{K_{ec}}{K_s}$$

TABLE 8-3 VALUE OF K(C) l_c(FT) = 10.0 PAGE 1

h	m	ℓ1 = 15 Y=0.50	ℓ1 = 15 Y=0.75	ℓ1 = 15 Y=1.00	ℓ1 = 15 Y=1.50	ℓ1 = 15 Y=2.00	ℓ1 = 18 Y=0.50	ℓ1 = 18 Y=0.75	ℓ1 = 18 Y=1.00	ℓ1 = 18 Y=1.50	ℓ1 = 18 Y=2.00	ℓ1 = 21 Y=0.50	ℓ1 = 21 Y=0.75	ℓ1 = 21 Y=1.00	ℓ1 = 21 Y=1.50	ℓ1 = 21 Y=2.00
6	16	105.2	70.1	52.6	35.0	26.3	218.3	145.5	109.1	72.7	54.5	404.4	269.6	202.2	134.8	101.1
	18	65.7	43.8	32.8	21.9	16.4	136.2	90.8	68.1	45.4	34.0	252.5	168.3	126.2	84.1	63.1
	20	43.1	28.7	21.5	14.3	10.7	89.4	59.6	44.7	29.8	22.3	165.6	110.4	82.8	55.2	41.4
	24	20.7	13.8	10.3	6.9	5.1	43.1	28.7	21.5	14.3	10.7	79.8	53.2	39.9	26.6	19.9
	28	11.2	7.4	5.6	3.7	2.8	23.2	15.5	11.6	7.7	5.8	43.1	28.7	21.5	14.3	10.7
8	16	108.2	72.1	54.1	36.0	27.0	224.4	149.6	112.2	74.8	56.1	415.8	277.2	207.9	138.6	103.9
	18	67.5	45.0	33.7	22.5	16.8	140.1	93.4	70.0	46.7	35.0	259.6	173.0	129.8	86.5	64.9
	20	44.3	29.5	22.1	14.7	11.0	91.9	61.2	45.9	30.6	22.9	170.3	113.5	85.1	56.7	42.5
	24	21.3	14.2	10.6	7.1	5.3	44.3	29.5	22.1	14.7	11.0	82.1	54.7	41.0	27.3	20.5
	28	11.5	7.6	5.7	3.8	2.8	23.9	15.9	11.9	7.9	5.9	44.3	29.5	22.1	14.7	11.0
10	16	111.9	74.6	55.9	37.3	27.9	232.1	154.7	116.0	77.3	58.0	430.0	286.6	215.0	143.3	107.5
	18	69.8	46.5	34.9	23.2	17.4	144.9	96.6	72.4	48.3	36.2	268.4	178.9	134.2	89.4	67.1
	20	45.8	30.5	22.9	15.2	11.4	95.0	63.3	47.5	31.6	23.7	176.1	117.4	88.0	58.7	44.0
	24	22.1	14.7	11.0	7.3	5.5	45.8	30.5	22.9	15.2	11.4	84.9	56.6	42.4	28.3	21.2
	28	11.9	7.9	5.9	3.9	2.9	24.7	16.4	12.3	8.2	6.1	45.8	30.5	22.9	15.2	11.4
12	16	116.3	77.5	58.1	38.7	29.0	241.2	160.8	120.6	80.4	60.3	446.9	297.9	223.4	148.9	111.7
	18	72.6	48.4	36.3	24.2	18.1	150.6	100.4	75.3	50.2	37.6	279.0	186.0	139.5	93.0	69.7
	20	47.6	31.7	23.8	15.8	11.9	98.8	65.8	49.4	32.9	24.7	183.0	122.0	91.5	61.0	45.7
	24	22.9	15.3	11.4	7.6	5.7	47.6	31.7	23.8	15.8	11.9	88.2	58.8	44.1	29.4	22.0
	28	12.4	8.2	6.2	4.1	3.1	25.7	17.1	12.8	8.5	6.4	47.6	31.7	23.8	15.8	11.9

h	m	ℓ1 = 24 Y=0.50	ℓ1 = 24 Y=0.75	ℓ1 = 24 Y=1.00	ℓ1 = 24 Y=1.50	ℓ1 = 24 Y=2.00	ℓ1 = 28 Y=0.50	ℓ1 = 28 Y=0.75	ℓ1 = 28 Y=1.00	ℓ1 = 28 Y=1.50	ℓ1 = 28 Y=2.00	ℓ1 = 32 Y=0.50	ℓ1 = 32 Y=0.75	ℓ1 = 32 Y=1.00	ℓ1 = 32 Y=1.50	ℓ1 = 32 Y=2.00
10	16	733.5	489.0	366.7	244.5	183.3	1359.0	906.0	679.5	453.0	339.7	2318.4	1545.6	1159.2	772.8	579.6
	18	457.9	305.3	228.9	152.6	114.4	848.4	565.6	424.2	282.8	212.1	1447.4	964.9	723.7	482.4	361.8
	20	300.4	200.3	150.2	100.1	75.1	556.6	371.1	278.3	185.5	139.1	949.6	633.0	474.8	316.5	237.4
	24	144.9	96.6	72.4	48.3	36.2	268.4	178.9	134.2	89.4	67.1	457.9	305.3	228.9	152.6	114.4
	28	78.2	52.1	39.1	26.0	19.5	144.9	96.6	72.4	48.3	36.2	247.2	164.8	123.6	82.4	61.8
12	16	762.5	508.3	381.2	254.1	190.6	1412.6	941.7	706.3	470.8	353.1	2409.9	1606.6	1204.9	803.3	602.4
	18	476.0	317.3	238.0	158.6	119.0	881.9	587.9	440.9	293.9	220.4	1504.5	1003.0	752.2	501.5	376.1
	20	312.3	208.2	156.1	104.1	78.0	578.6	385.7	289.3	192.8	144.6	987.1	658.0	493.5	329.0	246.7
	24	150.6	100.4	75.3	50.2	37.6	279.0	186.0	139.5	93.0	69.7	476.0	317.3	238.0	158.6	119.0
	28	81.3	54.2	40.6	27.1	20.3	150.6	100.4	75.3	50.2	37.6	256.9	171.3	128.4	85.6	64.2
14	16	796.2	530.8	398.1	265.4	199.0	1475.1	983.4	737.5	491.7	368.7	2516.6	1677.7	1258.3	838.8	629.1
	18	497.1	331.4	248.5	165.7	124.2	920.9	613.9	460.4	306.9	230.2	1571.1	1047.4	785.5	523.7	392.7
	20	326.1	217.4	163.0	108.7	81.5	604.2	402.8	302.1	201.4	151.0	1030.8	687.2	515.4	343.6	257.7
	24	157.2	104.8	78.6	52.4	39.3	291.3	194.2	145.6	97.1	72.8	497.1	331.4	248.5	165.7	124.2
	28	84.9	56.6	42.4	28.3	21.2	157.2	104.8	78.6	52.4	39.3	268.3	178.8	134.1	89.4	67.0
16	16	834.7	556.5	417.3	278.2	208.6	1546.5	1031.0	773.2	515.5	386.6	2638.3	1758.8	1319.1	879.4	659.5
	18	521.1	347.4	260.5	173.7	130.2	965.5	643.6	482.7	321.8	241.3	1647.1	1098.0	823.5	549.0	411.7
	20	341.9	227.9	170.9	113.9	85.4	633.4	422.3	316.7	211.1	158.3	1080.6	720.4	540.3	360.2	270.1
	24	164.8	109.9	82.4	54.9	41.2	305.4	203.6	152.7	101.8	76.3	521.1	347.4	260.5	173.7	130.2
	28	89.0	59.3	44.5	29.6	22.2	164.8	109.9	82.4	54.9	41.2	281.3	187.5	140.6	93.7	70.3

Table 8-3.
SYMBOLS: $\gamma = l_1/l_2 - l/\beta$; $m = l_1/c_1$; $c_2 = c_1 \cdot l_2/l_1$; h = slab thickness; l_c = story height. This Table provides design coefficients K_c for two-way slab construction without beams or drop panels, called flat-plates, as per Chapter 13, ACI-318-71. The moment of inertia was computed according to Section 13.4.1.3.

K_c

K_c

TABLE 8-3 VALUE OF K(C) $l_{c(FT)}$ = 12.0 PAGE 2

h	m	ℓ1 = 15 γ=0.50	γ=0.75	γ=1.00	γ=1.50	γ=2.00	ℓ1 = 18 γ=0.50	γ=0.75	γ=1.00	γ=1.50	γ=2.00	ℓ1 = 21 γ=0.50	γ=0.75	γ=1.00	γ=1.50	γ=2.00
6	16	86.7	57.8	43.3	28.9	21.6	179.8	119.8	89.9	59.9	44.9	333.1	222.1	166.5	111.0	83.2
	18	54.1	36.0	27.0	18.0	13.5	112.2	74.8	56.1	37.4	28.0	208.0	138.6	104.0	69.3	52.0
	20	35.5	23.6	17.7	11.8	8.8	73.6	49.1	36.8	24.5	18.4	136.4	90.9	68.2	45.4	34.1
	24	17.1	11.4	8.5	5.7	4.2	35.5	23.6	17.7	11.8	8.8	65.8	43.8	32.9	21.9	16.4
	28	9.2	6.1	4.6	3.0	2.3	19.1	12.7	9.5	6.3	4.7	35.5	23.6	17.7	11.8	8.8
8	16	88.4	58.9	44.2	29.4	22.1	183.4	122.3	91.7	61.1	45.8	339.9	226.6	169.9	113.3	84.9
	18	55.2	36.8	27.6	18.4	13.8	114.5	76.3	57.2	38.1	28.6	212.2	141.4	106.1	70.7	53.0
	20	36.2	24.1	18.1	12.0	9.0	75.1	50.1	37.5	25.0	18.7	139.2	92.8	69.6	46.4	34.8
	24	17.4	11.6	8.7	5.8	4.3	36.2	24.1	18.1	12.0	9.0	67.1	44.7	33.5	22.3	16.7
	28	9.4	6.2	4.7	3.1	2.3	19.5	13.0	9.7	6.5	4.8	36.2	24.1	18.1	12.0	9.0
10	16	90.6	60.4	45.3	30.2	22.6	188.0	125.3	94.0	62.6	47.0	348.3	232.2	174.1	116.1	87.0
	18	56.6	37.7	28.3	18.8	14.1	117.3	78.2	58.6	39.1	29.3	217.4	144.9	108.7	72.4	54.3
	20	37.1	24.7	18.5	12.3	9.2	77.0	51.3	38.5	25.6	19.2	142.6	95.1	71.3	47.5	35.6
	24	17.9	11.9	8.9	5.9	4.4	37.1	24.7	18.5	12.3	9.2	68.8	45.8	34.4	22.9	17.2
	28	9.6	6.4	4.8	3.2	2.4	20.0	13.3	10.0	6.6	5.0	37.1	24.7	18.5	12.3	9.2
12	16	93.2	62.1	46.6	31.0	23.3	193.4	128.9	96.7	64.4	48.3	358.3	238.8	179.1	119.4	89.5
	18	58.2	38.8	29.1	19.4	14.5	120.7	80.5	60.3	40.2	30.1	223.7	149.1	111.8	74.5	55.9
	20	38.2	25.4	19.1	12.7	9.5	79.2	52.8	39.6	26.4	19.8	146.7	97.8	73.3	48.9	36.6
	24	18.4	12.2	9.2	6.1	4.6	38.2	25.4	19.1	12.7	9.5	70.7	47.1	35.3	23.5	17.6
	28	9.9	6.6	4.9	3.3	2.4	20.6	13.7	10.3	6.8	5.1	38.2	25.4	19.1	12.7	9.5

h	m	ℓ1 = 24 γ=0.50	γ=0.75	γ=1.00	γ=1.50	γ=2.00	ℓ1 = 28 γ=0.50	γ=0.75	γ=1.00	γ=1.50	γ=2.00	ℓ1 = 32 γ=0.50	γ=0.75	γ=1.00	γ=1.50	γ=2.00
10	16	594.2	396.1	297.1	198.0	148.5	1100.9	733.9	550.4	366.9	275.2	1878.1	1252.0	939.0	626.0	469.5
	18	370.9	247.3	185.4	123.6	92.7	687.2	458.1	343.6	229.0	171.8	1172.4	781.6	586.2	390.8	293.1
	20	243.4	162.2	121.7	81.1	60.8	450.9	300.6	225.4	150.3	112.7	769.2	512.8	384.6	256.4	192.3
	24	117.3	78.2	58.6	39.1	29.3	217.4	144.9	108.7	72.4	54.3	370.9	247.3	185.4	123.6	92.7
	28	63.3	42.2	31.6	21.1	15.8	117.3	78.2	58.6	39.1	29.3	200.2	133.4	100.1	66.7	50.0
12	16	611.3	407.5	305.6	203.7	152.8	1132.5	755.0	566.2	377.5	283.1	1932.0	1288.0	966.0	644.0	483.0
	18	381.6	254.4	190.8	127.2	95.4	707.0	471.3	353.5	235.6	176.7	1206.1	804.1	603.0	402.0	301.5
	20	250.3	166.9	125.1	83.4	62.5	463.8	309.2	231.9	154.6	115.9	791.3	527.5	395.6	263.7	197.8
	24	120.7	80.5	60.3	40.2	30.1	223.7	149.1	111.8	74.5	55.9	381.6	254.4	190.8	127.2	95.4
	28	65.1	43.4	32.5	21.7	16.2	120.7	80.5	60.3	40.2	30.1	206.0	137.3	103.0	68.6	51.5
14	16	631.1	420.7	315.5	210.3	157.7	1169.2	779.5	584.6	389.7	292.3	1994.7	1329.8	997.3	664.9	498.6
	18	394.0	262.6	197.0	131.3	98.5	729.9	486.6	364.9	243.3	182.4	1245.3	830.2	622.6	415.1	311.3
	20	258.5	172.3	129.2	86.1	64.6	478.9	319.2	239.4	159.6	119.7	817.0	544.6	408.5	272.3	204.2
	24	124.6	83.1	62.3	41.5	31.1	230.9	153.9	115.4	76.9	57.7	394.0	262.6	197.0	131.3	98.5
	28	67.2	44.8	33.6	22.4	16.8	124.6	83.1	62.3	41.5	31.1	212.6	141.7	106.3	70.8	53.1
16	16	653.7	435.8	326.8	217.9	163.4	1211.1	807.4	605.5	403.7	302.7	2066.1	1377.4	1033.0	688.7	516.5
	18	408.1	272.0	204.0	136.0	102.0	756.1	504.0	378.0	252.0	189.0	1289.8	859.9	644.9	429.9	322.4
	20	267.7	178.5	133.8	89.2	66.9	496.0	330.7	248.0	165.3	124.0	846.2	564.1	423.1	282.0	211.5
	24	129.1	86.0	64.5	43.0	32.2	239.2	159.4	119.6	79.7	59.8	408.1	272.0	204.0	136.0	102.0
	28	69.7	46.4	34.8	23.2	17.4	129.1	86.0	64.5	43.0	32.2	220.2	146.8	110.1	73.4	55.0

K_c

K_c

(6) The following tables provide the coefficients described above: Table 8-3, K_c; Table 8-4, K_s; Table 8-5, C; Table 8-6, K_t; Table 8-7, K_{ec}; Table 8-8, α_{ec}.

Step 6. Moments in exterior span, use α_{ec} from step 5 (ACI-13.3.3.3).

Exterior negative design moment: $M_o\left(\dfrac{0.65}{1 + 1/\alpha_{ec}}\right)$ (Table 8-9)

Positive design moment: $M_o\left(0.63 - \dfrac{0.28}{1 + 1/\alpha_{ec}}\right)$ (Table 8-10)

Interior negative design moment: $M_o\left(0.75 - \dfrac{0.10}{1 + 1/\alpha_{ec}}\right)$ (Table 8-11)

Step 7. Moments in interior spans (13.3.3.2).

Negative design moment: $0.65\,M_o$
Positive design moment: $0.35\,M_o$

Step 8. Increase the positive bending moments for the effects of pattern loading. Compute β_a, ratio of dead load to live load per unit area, without load factors: $\beta_a = w'_d/w'_l$. Skip this step when $\beta_a \geqslant 2.0$.

Check all columns, using the procedure described in this step. Compute α_{min}, per Table 8-2, using α, ratio of beam stiffness to the stiffness of a width of slab bounded laterally by the centerline of the panels. For flat slabs $\alpha = 0$. The value of α may be obtained from Table 8-12.

Compute α_c, ratio of column stiffnesses above and below to the sum of slab and beam stiffnesses at the joint in the direction of the moments:

$$\alpha_c = \frac{\sum K_c}{\sum (K_s + K_b)}$$

Values of K_s and K_c may be obtained from Tables 8-3 and 8-4. If the columns satisfy $\alpha_c \geqslant \alpha_{min}$, skip this step.

Compute the coefficient δ_s per equation (13-4):

$$\delta_s = 1 + \frac{2 - \beta_a}{4 + \beta_a}\left(1 - \frac{\alpha_c}{\alpha_{min}}\right) \qquad \text{(ACI-13-4)}$$

Multiply the design moments (computed from steps 6 and 7) by δ_s in the panels supported by those columns.

Step 9. Moments in interior columns:

$$M = \frac{0.08(w_d + 0.5\,w_1)\,l_2\,l_n^2 - w'_d\,l'_2\,(l'_n)^2}{1 + 1/\alpha_{ec}} \qquad \text{(ACI-13-3)}$$

where l'_2 and l'_n refer to the shorter span. The columns above and below the slab shall resist M in proportion to their stiffness.

TABLE 8-4 VALUE OF K(S) Flexural Stiffness of Slab - Moment per Unit Rotation - Chapter 13, ACI-318-71

SYMBOLS

$Y = \ell_1/\ell_2 - 1/\beta$

$m = \ell_1/c_1$

$c_2 = c_1 \cdot \ell_2/\ell_1$

h = Slab thickness

ℓ_c = Story Height

TABLE 8-4 VALUE OF K(S)

h	m	Y =0.50	Y =0.75	Y =1.00	Y =1.50	Y =2.00
6	16	146.5	97.6	73.2	48.8	36.6
	18	145.9	97.3	72.9	48.6	36.4
	20	145.6	97.0	72.8	48.5	36.4
	24	145.1	96.7	72.5	48.3	36.2
	28	144.8	96.5	72.4	48.2	36.2
8	16	347.3	231.5	173.6	115.7	86.8
	18	346.0	230.7	173.0	115.3	86.5
	20	345.1	230.1	172.5	115.0	86.2
	24	344.0	229.3	172.0	114.6	86.0
	28	343.2	228.8	171.6	114.4	85.8
10	16	678.3	452.2	339.1	226.1	169.5
	18	675.9	450.6	337.9	225.3	168.9
	20	674.1	449.4	337.0	224.7	168.5
	24	671.8	447.9	335.9	223.9	167.9
	28	670.4	446.9	335.2	223.4	167.6
12	16	1172.2	781.5	586.1	390.7	293.0
	18	1167.9	778.6	583.9	389.3	291.9
	20	1164.9	776.6	582.4	388.3	291.2
	24	1161.0	774.0	580.5	387.0	290.2
	28	1158.6	772.4	579.3	386.2	289.6
14	16	1861.4	1240.9	930.7	620.4	465.3
	18	1854.7	1236.4	927.3	618.2	463.6
	20	1849.9	1233.2	924.9	616.6	462.4
	24	1843.6	1229.0	921.8	614.5	460.9
	28	1839.8	1226.5	919.9	613.2	459.9

K_s

K_s

TABLE 8-5 TORSIONAL COEFFICIENTS (C), ACI CODE, EQ.(13-7)

h_b	h_s	L SHAPE B=12	L SHAPE B=16	L SHAPE B=20	L SHAPE B=24	L SHAPE B=30	L SHAPE B=36	T SHAPE B=12	T SHAPE B=16	T SHAPE B=20	T SHAPE B=24	T SHAPE B=30	T SHAPE B=36
8	0	1187.8	1870.5	2553.1	3235.8	4259.8	5283.8	1187.8	1870.5	2553.1	3235.8	4259.8	5283.8
	4	1219.4	1902.0	2584.7	3267.4	4291.4	5315.4	1250.9	1933.6	2616.3	3298.9	4322.9	5346.9
	5	1215.8	1898.4	2581.1	3263.8	4287.8	5311.8	1243.8	1926.4	2609.1	3291.8	4315.8	5339.8
	6	1200.4	1883.1	2565.8	3248.4	4272.4	5296.4	1213.1	1895.7	2578.4	3261.1	4285.1	5309.1
	7	1189.9	1872.6	2555.2	3237.9	4261.9	5285.9	1192.0	1874.7	2557.4	3240.0	4264.0	5288.0
10	0	1900.0	3233.3	4566.6	5900.0	7900.0	9900.0	1900.0	3233.3	4566.6	5900.0	7900.0	9900.0
	4	1974.2	3307.5	4640.9	5974.2	7974.2	9974.2	2048.4	3381.8	4715.1	6048.4	8048.4	10048.4
	5	1977.0	3310.4	4643.7	5977.0	7977.0	9977.0	2054.1	3387.5	4720.8	6054.1	8054.1	10054.1
	6	1974.2	3307.5	4640.9	5974.2	7974.2	9974.2	2048.4	3381.8	4715.1	6048.4	8048.4	10048.4
	7	1945.9	3279.3	4612.6	5945.9	7945.9	9945.9	1991.9	3325.3	4658.6	5991.9	7991.9	9991.9
	8	1917.9	3251.3	4584.6	5917.9	7917.9	9917.9	1935.9	3269.2	4602.6	5935.9	7935.9	9935.9
12	0	2557.4	4861.4	7165.4	9469.4	12925.4	16381.4	2557.4	4861.4	7165.4	9469.4	12925.4	16381.4
	4	2674.3	4978.3	7282.3	9586.3	13042.3	16498.3	2791.2	5095.2	7399.2	9703.2	13159.2	16615.2
	5	2717.8	5021.8	7325.8	9629.8	13085.8	16541.8	2878.2	5182.2	7486.2	9790.2	13246.2	16702.2
	6	2717.2	5021.2	7325.2	9629.2	13085.2	16541.2	2877.1	5181.1	7485.1	9789.1	13245.1	16701.1
	7	2717.8	5021.8	7325.8	9629.8	13085.8	16541.8	2878.2	5182.2	7486.2	9790.2	13246.2	16702.2
	8	2674.3	4978.3	7282.3	9586.3	13042.3	16498.3	2791.2	5095.2	7399.2	9703.2	13159.2	16615.2
14	0	3709.4	6567.3	10225.9	13884.6	19372.6	24860.6	3709.4	6567.3	10225.9	13884.6	19372.6	24860.6
	5	3953.1	6811.0	10469.7	14128.3	19616.3	25104.3	4196.9	7054.8	10713.4	14372.1	19860.1	25348.1
	6	4013.2	6871.1	10529.8	14188.4	19676.4	25164.4	4317.1	7174.9	10833.6	14492.3	19980.3	25468.3
	7	4005.5	6863.4	10522.0	14180.7	19668.7	25156.7	4301.6	7159.5	10818.2	14476.8	19964.8	25452.8
	8	4013.2	6871.1	10529.8	14188.4	19676.4	25164.4	4317.1	7174.9	10833.6	14492.3	19980.3	25468.3
	9	3953.1	6811.0	10469.7	14128.3	19616.3	25104.3	4196.9	7054.8	10713.4	14372.1	19860.1	25348.1
16	0	4861.4	8082.7	13544.1	19005.4	27197.4	35389.4	4861.4	8082.7	13544.1	19005.4	27197.4	35389.4
	5	5188.5	8409.8	13871.1	19332.5	27524.5	35716.5	5515.6	8736.9	14198.2	19659.6	27851.6	36043.6
	6	5309.2	8530.6	13991.9	19453.2	27645.2	35837.2	5757.1	8978.4	14439.7	19901.1	28093.1	36285.1
	7	5386.2	8607.5	14068.8	19530.2	27722.2	35914.2	5911.0	9132.3	14593.6	20055.0	28247.0	36439.0
	8	5366.6	8587.9	14049.2	19510.6	27702.6	35894.6	5871.7	9093.1	14554.4	20015.7	28207.7	36399.7
	9	5386.2	8607.5	14068.8	19530.2	27722.2	35914.2	5911.0	9132.3	14593.6	20055.0	28247.0	36439.0
18	0	6013.4	10813.4	16835.0	24611.0	36275.0	47939.0	6013.4	10813.4	16835.0	24611.0	36275.0	47939.0
	5	6423.8	11223.8	17245.4	25021.4	36685.4	48349.4	6834.2	11634.2	17655.8	25431.8	37095.8	48759.8
	6	6605.2	11405.2	17426.8	25202.8	36866.8	48530.8	7197.1	11997.1	18018.7	25794.7	37458.7	49122.7
	7	6766.8	11566.8	17588.4	25364.4	37028.4	48692.4	7520.3	12320.3	18341.9	26117.9	37781.9	49445.9
	8	6859.9	11659.9	17681.5	25457.5	37121.5	48785.5	7706.4	12506.4	18528.0	26304.0	37968.0	49632.0
	9	6822.6	11622.6	17644.2	25420.2	37084.2	48748.2	7631.8	12431.8	18453.4	26229.4	37893.4	49557.4

TABLE 8-5 TORSIONAL COEFFICIENTS (C), ACI CODE, EQ.(13-7) PAGE 2

h_b	h_s	L SHAPE B=12	L SHAPE B=16	L SHAPE B=20	L SHAPE B=24	L SHAPE B=30	L SHAPE B=36	T SHAPE B=12	T SHAPE B=16	T SHAPE B=20	T SHAPE B=24	T SHAPE B=30	T SHAPE B=36
20	0	7165.4	13544.1	19733.3	30400.0	46400.0	62400.0	7165.4	13544.1	19733.3	30400.0	46400.0	62400.0
	5	7659.1	14037.8	20227.0	30893.7	46893.7	62893.7	8152.9	14531.6	20720.8	31387.5	47387.5	63387.5
	6	7901.2	14279.9	20469.1	31135.8	47135.8	63135.8	8637.1	15015.7	21205.0	31871.6	47871.6	63871.6
	7	8147.5	14526.2	20715.4	31382.1	47382.1	63382.1	9129.6	15508.3	21697.5	32364.2	48364.2	64364.2
	8	8353.2	14731.9	20921.1	31587.8	47587.8	63587.8	9541.1	15919.7	22109.0	32775.6	48775.6	64775.6
	9	8460.6	14839.2	21028.5	31695.1	47695.1	63695.1	9755.8	16134.4	22323.7	32990.3	48990.3	64990.3
22	0	8317.4	16274.7	25066.6	35990.2	57286.2	78582.2	8317.4	16274.7	25066.6	35990.2	57286.2	78582.2
	6	9197.2	17154.6	25946.5	36870.0	58166.0	79462.0	10077.1	18034.4	26826.3	37749.9	59045.9	80341.9
	7	9528.2	17485.5	26277.4	37201.0	58497.0	79793.0	10739.0	18696.3	27488.2	38411.8	59707.8	81003.8
	8	9846.6	17803.9	26595.8	37519.4	58815.4	80111.4	11375.7	19333.1	28125.0	39048.5	60344.5	81640.5
	9	10098.6	18055.9	26847.8	37771.4	59067.4	80363.4	11879.8	19837.1	28629.0	39552.6	60848.6	82144.6
	10	10217.4	18174.7	26966.6	37890.2	59186.2	80482.2	12117.4	20074.7	28866.6	39790.2	61086.2	82382.2
24	0	9469.4	19005.4	30400.0	40919.0	68567.0	96215.0	9469.4	19005.4	30400.0	40919.0	68567.0	96215.0
	6	10493.2	20029.2	31423.8	41942.8	69590.8	97238.8	11517.1	21053.1	32447.6	42966.7	70614.7	98262.7
	7	10908.8	20444.8	31839.4	42358.4	70006.4	97654.4	12348.3	21884.3	33278.9	43797.9	71445.9	99093.9
	8	11339.9	20875.9	32270.5	42789.5	70437.5	98085.5	13210.4	22746.4	34141.0	44660.0	72308.0	99956.0
	9	11736.6	21272.6	32667.1	43186.2	70834.2	98482.2	14003.8	23539.8	34934.3	45453.4	73101.4	100749.4
	10	12036.1	21572.1	32966.6	43485.7	71133.7	98781.7	14602.7	24138.7	35533.3	46052.3	73700.3	101348.3
28	0	11773.4	24466.7	41066.6	59351.0	90442.2	134346.2	11773.4	24466.7	41066.6	59351.0	90442.2	134346.2
	6	13085.2	25778.6	42378.5	60662.8	91754.0	135658.0	14397.1	27090.4	43690.3	61974.7	93065.9	136969.9
	7	13670.2	26363.5	42963.4	61247.8	92339.0	136243.0	15567.0	28260.3	44860.2	63144.6	94235.8	138139.8
	8	14326.6	27019.9	43619.8	61904.2	92995.4	136899.4	16879.7	29573.1	46173.0	64457.3	95548.5	139452.5
	9	15012.6	27705.9	44305.8	62590.2	93681.4	137585.4	18251.8	30945.1	47545.0	65829.4	96920.6	140824.6
	10	15673.4	28366.7	44966.6	63251.0	94342.2	138246.2	19573.4	32266.7	48866.6	67151.0	98242.2	142146.2
32	0	14077.4	29928.1	51733.3	77783.0	117900.0	173015.0	14077.4	29928.1	51733.3	77783.0	117900.0	173015.0
	6	15533.2	31383.9	53189.1	79233.8	119355.8	174470.8	16989.1	32839.7	54645.0	80694.7	120811.6	175926.7
	7	16431.5	32282.2	54087.4	80137.1	120254.1	175369.1	18785.6	34636.3	56441.5	82491.2	122608.2	177723.2
	8	17313.2	33163.9	54969.1	81013.8	121135.8	176250.8	20549.1	36399.7	58205.0	84254.7	124371.6	179486.7
	9	18288.6	34139.2	55944.5	81994.2	122111.1	177226.2	22499.8	38350.4	60155.7	86205.4	126322.3	181437.4
	10	19310.7	35161.4	56966.6	83015.3	123133.3	178248.3	24544.1	40394.7	62200.0	88249.7	128366.6	183481.7
36	0	16381.4	35389.4	62400.0	96215.0	153900.0	207152.6	16381.4	35389.4	62400.0	96215.0	153900.0	207152.6
	7	19078.5	38086.5	65097.1	98912.1	156597.1	209849.7	21775.6	40783.6	67794.2	101609.2	159294.2	212546.8
	8	20299.9	39307.9	66318.5	100133.5	157818.5	211071.1	24218.4	43226.4	70237.0	104052.0	161737.0	214989.6
	9	21564.6	40572.6	67583.1	101398.2	159083.1	212335.8	26747.8	45755.8	72766.3	106581.4	164266.3	217519.0
	10	22948.1	41956.1	68966.6	102781.7	160466.6	213719.3	29514.7	48522.7	75533.3	109348.3	167033.3	220285.9
	12	25850.8	44858.8	71869.4	105684.4	163369.4	216622.0	35320.3	54328.3	81338.8	115153.9	172838.8	226091.5

TABLE 8-6 VALUE OF K(t) PAGE 1

h	m	L1 = 15 γ=0.50	γ=0.75	γ=1.00	γ=1.50	γ=2.00	L1 = 18 γ=0.50	γ=0.75	γ=1.00	γ=1.50	γ=2.00	L1 = 21 γ=0.50	γ=0.75	γ=1.00	γ=1.50	γ=2.00
6	16	16.3	24.4	32.6	48.9	65.2	17.6	26.5	35.3	53.0	70.7	18.6	28.0	37.3	56.0	74.7
	18	13.2	19.9	26.5	39.8	53.1	14.6	21.9	29.2	43.9	58.5	15.5	23.3	31.1	46.7	62.3
	20	10.9	16.4	21.9	32.8	43.8	12.2	18.4	24.5	36.8	49.1	13.2	19.8	26.4	39.6	52.9
	24	7.6	11.4	15.2	22.8	30.4	8.8	13.3	17.7	26.6	35.5	9.8	14.7	19.6	29.4	39.2
	28	5.3	7.9	10.6	15.9	21.2	6.5	9.8	13.1	19.7	26.3	7.4	11.2	14.9	22.4	29.9
8	16	32.1	48.2	64.3	96.4	128.6	36.5	54.7	73.0	109.5	146.0	39.6	59.4	79.2	118.8	158.4
	18	25.1	37.6	50.2	75.3	100.4	29.3	44.0	58.7	88.1	117.5	32.4	48.6	64.8	97.2	129.6
	20	19.7	29.5	39.4	59.1	78.8	23.8	35.8	47.7	71.6	95.5	26.8	40.3	53.7	80.6	107.4
	24	13.0	19.6	26.1	39.2	52.3	15.9	23.9	31.9	47.9	63.9	18.9	28.3	37.8	56.7	75.6
	28	9.7	14.6	19.5	29.2	39.0	11.1	16.7	22.3	33.4	44.6	13.4	20.1	26.9	40.3	53.8
10	16	50.0	75.0	100.1	150.1	200.2	60.6	91.0	121.3	182.0	242.7	68.2	102.4	136.5	204.8	273.0
	18	36.6	54.9	73.2	109.8	146.4	46.9	70.4	93.9	140.9	187.9	54.4	81.6	108.8	163.2	217.6
	20	30.6	46.0	61.3	92.0	122.7	36.4	54.6	72.8	109.3	145.7	43.7	65.6	87.4	131.2	174.9
	24	21.0	31.6	42.1	63.2	84.2	24.9	37.3	49.8	74.7	99.6	28.4	42.6	56.8	85.2	113.6
	28	14.6	22.0	29.3	44.0	58.7	18.2	27.4	36.5	54.8	73.1	20.9	31.4	41.9	62.8	83.8
12	16	70.7	106.1	141.4	212.2	282.9	86.5	129.7	173.0	259.5	346.0	102.2	153.3	204.4	306.7	408.9
	18	56.3	84.5	112.7	169.1	225.5	63.2	94.8	126.4	189.7	252.9	78.6	117.9	157.2	235.8	314.5
	20	44.8	67.2	89.7	134.5	179.4	53.0	79.5	106.0	159.0	212.0	60.4	90.6	120.9	181.3	241.8
	24	29.0	43.5	58.1	87.1	116.2	36.4	54.6	72.8	109.2	145.6	42.1	63.2	84.3	126.4	168.6
	28	19.6	29.4	39.2	58.8	78.5	25.3	38.0	50.7	76.1	101.5	30.6	45.9	61.2	91.9	122.5

h	m	L1 = 24 γ=0.50	γ=0.75	γ=1.00	γ=1.50	γ=2.00	L1 = 28 γ=0.50	γ=0.75	γ=1.00	γ=1.50	γ=2.00	L1 = 32 γ=0.50	γ=0.75	γ=1.00	γ=1.50	γ=2.00
10	16	73.9	110.9	147.9	221.8	295.8	79.6	119.4	159.2	238.9	318.5	83.9	125.8	167.8	251.7	335.6
	18	59.9	89.9	119.9	179.9	239.8	65.5	98.3	131.0	196.6	262.1	69.7	104.5	139.4	209.1	278.8
	20	49.2	73.8	98.4	147.6	196.8	54.6	82.0	109.3	164.0	218.6	58.7	88.1	117.5	176.3	235.0
	24	33.7	50.5	67.4	101.1	134.9	39.0	58.5	78.1	117.1	156.2	43.0	64.5	86.1	129.1	172.2
	28	23.1	34.7	46.3	69.4	92.6	28.3	42.5	56.7	85.1	113.5	32.3	48.4	64.6	96.9	129.2
12	16	114.0	171.0	228.0	342.0	456.1	125.8	188.7	251.6	377.4	503.3	134.6	202.0	269.3	404.0	538.7
	18	90.1	135.2	180.3	270.5	360.6	101.7	152.5	203.4	305.1	406.8	110.3	165.5	220.7	331.0	441.4
	20	71.8	107.7	143.6	215.4	287.2	83.1	124.7	166.2	249.4	332.5	91.6	137.4	183.2	274.9	366.5
	24	45.4	68.1	90.8	136.2	181.6	56.4	84.6	112.8	169.3	225.7	64.7	97.0	129.4	194.1	258.9
	28	34.8	52.3	69.7	104.6	139.5	38.1	57.2	76.3	114.5	152.7	46.3	69.4	92.6	138.9	185.3
14	16	159.2	238.8	318.4	477.6	636.9	181.0	271.6	362.1	543.2	724.3	197.4	296.2	394.9	592.4	789.8
	18	121.8	182.7	243.6	365.4	487.2	143.1	214.7	286.3	429.5	572.7	159.2	238.8	318.4	477.6	636.8
	20	93.0	139.5	186.0	279.0	372.0	114.0	171.0	228.0	342.0	456.0	129.7	194.6	259.5	389.3	519.0
	24	65.8	98.7	131.7	197.5	263.4	72.0	108.1	144.1	216.2	288.3	87.4	131.1	174.8	262.3	349.7
	28	47.5	71.3	95.0	142.6	190.1	55.4	83.1	110.8	166.2	221.6	60.2	90.3	120.4	180.6	240.9
16	16	205.0	307.5	410.1	615.1	820.2	242.3	363.5	484.6	727.0	969.3	270.2	405.4	540.5	810.8	1081.1
	18	149.9	224.8	299.8	449.7	599.6	186.3	279.5	372.7	559.1	745.5	213.7	320.5	427.4	641.1	854.9
	20	125.6	188.5	251.3	377.0	502.6	143.3	214.9	286.6	429.9	573.2	170.1	255.2	340.3	510.5	680.7
	24	86.3	129.4	172.6	258.9	345.2	99.9	149.9	199.8	299.8	399.7	107.6	161.4	215.2	322.8	430.4
	28	60.1	90.2	120.3	180.5	240.7	72.6	108.9	145.2	217.8	290.4	82.7	124.0	165.4	248.1	330.8

K_t

Table 8-6.
SYMBOLS: $\gamma = 1/2 - 1/\beta$; $m = 1/1$; $c_2 = c_1 \cdot 2/1$; h = Slab thickness; c = Story Height. This Table furnishes the Torsional Stiffness of the Transverse Beam, (K_t), as per equations 13-6 and 13-7 of the ACI-318-71 Code. $E_{cs} = 1.0$ was used for computing the tabulated quantities.

K_t

h	m	γ=0.50	γ=0.75	γ=1.00	γ=1.50	γ=2.00	γ=0.50	γ=0.75	γ=1.00	γ=1.50	γ=2.00	γ=0.50	γ=0.75	γ=1.00	γ=1.50	γ=?.00
6	16	24.9	28.8	29.1	25.8	21.9	30.4	38.8	42.9	43.1	39.3	34.1	46.3	54.5	61.1	60.3
	18	18.9	20.8	20.3	17.1	14.2	24.0	29.6	31.4	29.9	26.3	27.7	36.6	41.7	44.3	41.9
	20	14.5	15.3	14.4	11.7	9.6	19.2	22.7	23.4	21.2	18.2	22.8	29.1	32.2	32.5	29.7
	24	8.7	8.6	7.7	6.0	4.7	12.5	13.8	13.4	11.3	9.3	15.7	18.9	19.8	18.3	15.9
	28	5.4	5.0	4.4	3.3	2.6	8.4	8.6	8.0	6.4	5.2	11.1	12.6	12.5	10.8	9.1
8	16	40.3	41.2	38.0	30.3	24.4	55.0	63.2	63.4	55.7	47.0	66.5	83.1	89.9	87.5	78.2
	18	28.8	28.1	25.2	19.5	15.5	41.3	45.3	43.8	36.9	30.4	51.8	62.2	64.8	59.8	51.9
	20	20.8	19.7	17.3	13.1	10.3	31.4	33.0	31.0	25.2	20.5	40.8	47.1	47.5	41.9	35.5
	24	11.7	10.4	8.8	6.5	5.0	18.5	18.2	16.4	12.8	10.2	25.8	27.8	26.6	22.0	18.0
	28	7.2	6.0	5.0	3.6	2.7	11.5	10.8	9.4	7.1	5.6	16.7	17.0	15.7	12.4	10.0
10	16	52.8	49.8	43.7	33.1	26.1	79.6	83.6	78.5	63.8	51.8	103.6	119.4	120.2	106.1	89.8
	18	35.7	32.7	28.2	21.0	16.4	57.0	57.3	52.2	41.2	33.0	77.4	85.3	83.0	70.2	58.1
	20	26.2	22.9	19.3	14.1	10.9	41.2	40.1	35.8	27.6	21.9	58.4	61.9	58.5	47.9	39.1
	24	14.5	11.9	9.7	6.9	5.3	23.8	21.6	18.6	13.8	10.8	34.0	34.0	30.9	24.2	19.4
	28	8.4	6.7	5.4	3.8	2.9	14.7	12.6	10.5	7.6	5.9	21.8	20.5	18.0	13.6	10.7
12	16	63.8	56.8	48.2	35.5	27.6	100.7	99.3	89.4	69.6	55.4	140.2	151.1	144.5	119.8	98.3
	18	44.1	37.6	31.2	22.5	17.4	68.7	65.6	58.0	44.3	35.0	100.5	104.0	96.6	77.6	62.7
	20	31.1	25.7	21.0	15.0	11.5	51.1	46.5	40.0	29.8	23.3	72.8	72.9	66.4	52.2	41.8
	24	16.4	13.0	10.4	7.3	5.6	28.8	24.6	20.4	14.8	11.4	43.1	40.1	34.9	26.3	20.7
	28	9.4	7.2	5.7	3.9	3.0	17.0	13.9	11.4	8.1	6.2	26.8	23.6	19.9	14.6	11.3

		ℓ1 = 24					ℓ1 = 28					ℓ1 = 32				
h	m	γ=0.50	γ=0.75	γ=1.00	γ=1.50	γ=2.00	γ=0.50	γ=0.75	γ=1.00	γ=1.50	γ=2.00	γ=0.50	γ=0.75	γ=1.00	γ=1.50	γ=2.00
10	16	123.0	152.6	163.7	157.6	139.9	142.5	189.0	216.8	232.5	221.5	156.4	216.4	260.2	304.8	311.0
	18	95.0	113.2	117.1	107.1	92.4	113.5	145.8	162.0	164.4	151.0	127.1	171.8	201.2	224.0	219.4
	20	74.1	84.9	85.2	74.7	63.0	91.3	113.7	122.4	118.5	105.5	104.5	137.9	157.2	166.8	157.7
	24	46.0	49.4	47.1	38.9	31.9	60.5	70.8	72.1	64.7	55.2	72.4	90.7	98.2	95.9	85.9
	28	29.0	29.7	27.4	21.9	17.6	40.7	45.2	44.2	37.6	31.2	51.2	61.0	63.1	57.8	49.8
12	16	175.5	204.4	207.6	185.3	157.6	213.6	269.4	293.8	290.0	261.4	242.2	322.8	372.2	402.8	386.4
	18	130.7	146.0	143.3	122.6	102.1	165.2	200.8	211.6	198.4	173.4	192.4	248.9	278.1	285.3	263.7
	20	98.3	105.8	101.1	83.8	68.7	129.1	151.4	154.7	139.0	118.8	154.5	193.9	210.3	205.8	184.6
	24	56.6	57.8	53.2	42.3	34.1	80.3	88.6	86.2	72.9	60.4	101.7	120.4	124.0	112.6	96.7
	28	37.5	35.7	31.4	23.9	18.9	50.6	53.5	50.4	41.1	33.5	68.1	76.7	75.8	65.4	54.7
14	16	227.4	251.4	244.9	207.7	172.1	290.7	349.9	365.4	338.5	293.9	341.3	437.8	485.2	491.1	449.9
	18	163.4	173.7	164.5	135.0	110.2	218.4	252.7	255.2	226.1	191.7	264.7	328.0	351.7	338.2	300.2
	20	118.4	122.2	113.3	90.9	73.4	165.5	184.9	181.7	155.6	129.5	207.3	248.5	258.6	238.3	206.4
	24	71.6	68.5	60.5	46.2	36.5	96.4	102.3	96.7	79.3	64.6	129.3	146.4	145.3	125.9	105.5
	28	44.8	40.5	34.7	25.7	20.1	65.0	64.2	58.0	45.2	36.1	83.1	89.8	86.1	71.6	58.8
16	16	275.0	292.1	276.6	226.9	185.1	369.0	426.3	430.1	380.5	322.3	448.6	555.0	594.1	570.2	505.4
	18	190.3	196.0	181.6	145.5	117.5	268.9	299.2	293.0	249.9	207.7	339.3	404.8	419.4	384.4	331.8
	20	144.8	142.0	127.5	99.0	78.7	197.3	213.0	204.0	169.5	139.1	258.8	298.8	301.2	266.2	225.4
	24	84.3	77.1	66.5	49.6	38.9	120.8	121.2	110.5	87.0	69.7	152.3	167.3	162.3	136.8	113.1
	28	51.1	44.6	37.5	27.4	21.2	77.2	73.0	64.2	48.8	38.4	104.1	106.8	98.6	78.8	63.5

Table 8-7.
SYMBOLS: $\gamma = l_1/l_2 = 1/\beta$; $m = l_1/c_1$; $c_2 = c_1 \cdot l_2/l_1$; h = slab thickness; l_c = story height. This Table furnishes the flexural Stiffness of an equivalent column, K_{ec}, as per Equation 13-5, where the sum of the Torsional Stiffness of both members joining to the column, perpendicular to the direction under consideration. The quantities in the Table are not applicable for edge conditions, where the edge is parallel to the direction of design.

TABLE 8-7 VALUE OF K(EC) NO COLUMNS ABOVE l_c (FT) = 12.0 PAGE 2

h	m	ℓ_1 = 15					ℓ_1 = 18					ℓ_1 = 21				
		γ=0.50	γ=0.75	γ=1.00	γ=1.50	γ=2.00	γ=0.50	γ=0.75	γ=1.00	γ=1.50	γ=2.00	γ=0.50	γ=0.75	γ=1.00	γ=1.50	γ=2.00
6	16	23.7	26.5	26.0	22.3	18.5	29.5	36.7	39.6	38.3	34.1	33.5	44.7	51.5	55.7	53.4
	18	17.8	18.9	17.9	14.7	12.0	23.2	27.6	28.6	26.2	22.6	27.1	34.9	38.9	39.8	36.7
	20	13.5	13.7	12.6	10.0	8.0	18.4	21.0	21.0	18.4	15.5	22.1	27.6	29.8	28.9	25.7
	24	8.0	7.6	6.6	5.0	4.0	11.8	12.5	11.8	9.6	7.8	15.1	17.6	17.9	15.9	13.6
	28	4.9	4.4	3.7	2.8	2.1	7.8	7.7	7.0	5.5	4.3	10.5	11.5	11.1	9.3	7.7
8	16	37.2	36.6	32.9	25.5	20.3	52.2	57.7	56.3	47.8	39.6	64.2	77.9	82.0	76.7	67.0
	18	26.3	24.7	21.6	16.4	12.9	38.8	40.9	38.5	31.3	25.5	49.6	57.6	58.3	51.8	44.0
	20	18.8	17.1	14.7	10.9	8.5	29.2	29.4	26.9	21.3	17.1	38.7	43.1	42.2	36.0	29.9
	24	10.4	8.9	7.4	5.4	4.1	16.9	16.0	14.1	10.7	8.4	24.1	25.0	23.2	18.6	15.1
	28	6.3	5.1	4.2	2.9	2.2	10.4	9.3	8.0	5.9	4.6	15.4	15.1	13.5	10.5	8.3
10	16	47.5	43.1	36.9	27.4	21.4	73.7	74.2	67.7	53.4	42.8	98.0	108.8	106.3	90.4	75.1
	18	31.9	28.0	23.7	17.3	13.4	52.1	50.3	44.7	34.3	27.2	72.5	76.7	72.5	59.3	48.3
	20	23.1	19.5	16.1	11.6	8.9	37.4	34.9	30.4	22.9	18.0	54.2	55.1	50.6	40.2	32.3
	24	12.5	10.0	8.0	5.7	4.3	21.2	18.5	15.6	11.4	8.8	31.1	29.8	26.4	20.2	15.9
	28	7.2	5.6	4.4	3.1	2.3	12.9	10.7	8.8	6.2	4.8	19.6	17.7	15.2	11.2	8.7
12	16	56.2	48.0	40.0	28.9	22.3	91.3	86.1	75.5	57.3	45.1	130.1	134.2	124.5	99.9	80.7
	18	38.4	31.5	25.7	18.3	14.1	61.7	56.5	48.7	36.3	28.4	92.3	91.3	82.5	64.3	51.3
	20	26.7	21.4	17.2	12.1	9.3	45.3	39.6	33.3	24.3	18.9	66.3	63.5	56.3	43.1	34.1
	24	13.9	10.7	8.5	5.9	4.5	25.0	20.6	16.8	12.0	9.2	38.4	34.3	29.2	21.5	16.8
	28	7.9	5.9	4.6	3.2	2.4	14.6	11.6	9.3	6.5	5.0	23.5	19.9	16.5	11.9	9.1

h	m	ℓ_1 = 24					ℓ_1 = 28					ℓ_1 = 32				
		γ=0.50	γ=0.75	γ=1.00	γ=1.50	γ=2.00	γ=0.50	γ=0.75	γ=1.00	γ=1.50	γ=2.00	γ=0.50	γ=0.75	γ=1.00	γ=1.50	γ=2.00
10	16	118.4	142.2	148.2	136.9	118.7	139.1	180.2	201.7	207.5	192.2	154.0	209.5	247.2	279.0	276.2
	18	90.6	104.1	104.6	92.0	77.7	110.0	137.5	148.7	144.7	129.4	124.6	164.9	188.9	202.0	192.1
	20	70.0	77.2	75.2	63.6	52.7	88.0	106.1	111.0	103.0	89.6	101.9	131.2	145.9	148.4	136.4
	24	42.8	44.1	40.8	32.7	26.4	57.4	64.7	64.1	55.3	46.3	69.8	84.8	89.3	83.6	73.0
	28	26.7	26.2	23.6	18.3	14.5	38.2	40.7	38.6	31.8	25.9	48.8	56.1	56.4	49.6	41.9
12	16	166.0	185.9	183.0	157.0	130.8	205.9	251.6	266.4	251.6	220.9	236.3	307.5	345.8	358.3	333.5
	18	122.4	131.1	124.7	102.9	84.2	157.9	185.2	189.1	170.0	145.2	186.5	234.5	254.8	250.1	224.7
	20	91.2	94.0	87.1	69.9	56.4	122.4	138.0	136.6	118.0	98.7	148.8	180.7	190.2	178.2	155.7
	24	51.8	50.5	45.3	35.0	27.8	75.0	79.2	74.7	61.1	49.7	96.6	110.1	109.8	95.8	80.5
	28	33.7	30.7	26.4	19.6	15.3	46.7	47.2	43.2	34.2	27.4	63.9	69.0	66.2	55.0	45.2
14	16	211.6	223.7	211.0	172.4	140.3	276.5	320.1	323.5	286.8	243.2	329.6	409.8	440.7	425.9	379.0
	18	150.5	152.8	140.2	111.3	89.4	205.6	228.1	222.9	189.6	157.4	253.5	303.2	314.8	289.3	250.1
	20	108.1	106.5	95.9	74.6	59.4	154.4	165.1	157.0	129.4	105.8	196.9	227.0	228.6	201.7	170.6
	24	64.0	58.5	50.4	37.6	29.4	88.7	89.9	82.4	65.3	52.4	121.1	131.2	126.0	105.0	86.3
	28	39.4	34.1	28.5	20.7	16.1	58.6	55.4	48.6	36.9	29.1	76.9	79.4	73.7	59.2	47.8
16	16	252.0	255.0	233.7	185.1	148.6	346.1	382.5	372.7	315.9	261.8	428.4	510.4	528.2	483.4	416.9
	18	172.8	169.5	152.2	118.1	94.0	249.6	265.0	250.8	205.6	167.7	321.0	367.3	367.6	322.0	271.3
	20	129.6	121.1	105.7	79.8	62.7	181.6	186.9	173.1	138.6	111.9	242.7	268.0	260.9	221.0	183.1
	24	73.8	64.6	54.3	39.7	30.8	108.8	104.1	92.0	70.3	55.6	140.9	147.6	138.4	112.3	91.2
	28	44.1	36.9	30.4	21.8	16.8	68.3	61.7	52.8	39.1	30.5	94.4	92.2	82.6	63.9	50.8

Kec

TABLE 8-7 VALUE OF K(EC) COLUMNS ABOVE l_c (FT) = 10.0 PAGE 3

h	m	ℓ_1 = 15					ℓ_1 = 18					ℓ_1 = 21				
		γ=0.50	γ=0.75	γ=1.00	γ=1.50	γ=2.00	γ=0.50	γ=0.75	γ=1.00	γ=1.50	γ=2.00	γ=0.50	γ=0.75	γ=1.00	γ=1.50	γ=2.00
6	16	28.2	36.2	40.2	40.8	37.5	32.7	44.8	53.4	61.3	61.6	35.7	50.7	63.0	79.1	85.9
	18	22.1	27.4	29.3	28.2	25.1	26.4	35.3	40.9	44.6	43.0	29.3	41.0	50.0	60.1	62.7
	20	17.4	20.9	21.7	20.0	17.3	21.5	28.1	31.7	32.9	30.7	24.4	33.6	40.0	46.1	46.4
	24	11.1	12.5	12.3	10.6	8.8	14.7	18.2	19.4	18.6	16.5	17.4	23.0	26.3	27.9	26.4
	28	7.2	7.7	7.3	6.0	4.9	10.2	12.0	12.3	11.1	9.5	12.7	16.1	17.6	17.5	15.8
8	16	49.5	57.8	58.7	52.5	44.7	62.7	80.1	88.4	88.9	81.0	72.3	97.8	114.7	127.9	125.5
	18	36.6	41.0	40.4	34.6	28.9	48.5	59.8	63.9	61.0	53.9	57.6	75.9	86.4	91.5	86.5
	20	27.2	29.5	28.3	23.6	19.4	37.9	45.2	46.8	42.9	37.0	46.4	59.4	65.9	66.6	60.9
	24	16.2	16.5	15.1	12.0	9.7	23.5	26.4	26.1	22.5	18.8	30.7	37.3	39.3	36.9	32.2
	28	10.5	10.0	8.9	6.8	5.3	15.2	16.3	15.5	12.8	10.5	20.6	23.9	24.3	21.6	18.3
10	16	69.1	74.8	71.8	59.7	49.1	96.2	114.6	118.6	108.5	93.6	117.8	150.8	167.0	168.6	154.2
	18	48.0	50.4	47.3	38.4	31.2	70.9	81.5	81.8	71.9	60.7	90.4	112.1	120.1	115.5	102.5
	20	36.7	36.7	33.3	26.2	20.9	52.6	58.7	57.5	49.1	40.8	70.0	84.1	87.7	81.1	70.3
	24	21.5	20.1	17.5	13.2	10.3	32.2	33.6	31.3	25.3	20.5	42.5	48.6	48.6	42.5	35.7
	28	13.1	11.6	9.9	7.2	5.6	21.0	20.5	18.4	14.3	11.4	28.7	30.9	29.6	24.5	20.1
12	16	87.9	89.6	82.4	65.5	52.7	127.3	143.6	142.1	122.7	102.7	166.4	202.4	213.5	200.5	175.5
	18	63.4	61.5	54.9	42.3	33.6	89.0	97.5	94.4	79.4	65.5	122.6	144.3	147.8	133.4	114.1
	20	46.2	43.1	37.6	28.4	22.3	69.0	72.0	67.4	54.5	44.2	90.9	104.0	104.2	91.3	76.9
	24	25.6	22.6	19.1	14.0	10.9	41.2	40.1	35.9	27.7	22.0	57.0	60.9	57.9	47.7	39.0
	28	15.2	12.9	10.7	7.7	5.9	25.5	23.6	20.5	15.4	12.0	37.2	37.5	34.3	27.0	21.7

h	m	ℓ_1 = 24					ℓ_1 = 28					ℓ_1 = 32				
		γ=0.50	γ=0.75	γ=1.00	γ=1.50	γ=2.00	γ=0.50	γ=0.75	γ=1.00	γ=1.50	γ=2.00	γ=0.50	γ=0.75	γ=1.00	γ=1.50	γ=2.00
10	16	134.3	180.8	210.8	232.6	226.4	150.4	211.0	258.0	312.8	328.8	161.9	232.7	293.1	379.7	425.1
	18	106.0	138.9	157.4	165.1	155.0	121.6	167.4	200.2	231.9	234.4	133.0	188.6	233.7	291.7	314.9
	20	84.5	107.8	118.9	119.3	108.7	99.5	134.3	157.0	174.1	170.0	110.6	154.7	188.4	226.4	236.2
	24	54.7	66.4	69.8	65.3	57.1	68.1	88.2	98.7	101.4	93.8	78.7	106.6	125.1	139.9	137.5
	28	35.7	41.6	42.4	37.9	32.2	47.4	59.0	63.6	61.6	54.9	57.1	74.8	84.8	89.0	83.6
12	16	198.3	255.9	285.4	291.6	268.8	231.0	314.4	371.1	419.0	415.0	255.0	358.9	440.2	537.6	568.8
	18	151.6	189.6	205.2	200.0	178.9	182.3	242.2	278.4	299.4	285.9	205.6	284.1	341.3	398.8	406.1
	20	116.7	141.9	149.6	140.3	122.7	145.3	188.4	211.1	217.5	201.6	167.7	227.4	267.3	299.5	294.9
	24	69.7	81.1	82.3	73.3	62.3	93.8	116.3	124.8	120.0	106.5	113.9	148.6	167.7	174.6	163.0
	28	48.8	53.2	51.3	43.0	35.4	60.9	72.9	75.8	69.8	60.4	78.5	98.8	107.6	105.9	95.4
14	16	265.3	329.4	353.8	341.2	303.3	322.5	425.6	485.7	516.2	488.7	366.2	503.5	601.1	694.4	700.4
	18	195.6	235.5	246.0	228.0	198.0	247.8	318.2	353.1	358.0	328.4	289.1	388.9	453.1	499.6	485.8
	20	144.7	169.9	173.7	156.4	133.7	191.8	240.1	259.8	253.5	226.9	230.5	303.3	345.2	365.0	344.4
	24	92.8	101.7	98.4	82.8	68.4	115.5	138.9	144.9	134.0	116.3	148.7	187.9	205.3	203.1	183.3
	28	60.9	63.1	58.6	47.2	38.1	81.9	92.7	91.9	79.7	66.7	98.3	120.0	126.9	119.6	104.9
16	16	329.2	396.1	413.7	383.1	332.7	419.0	537.5	595.8	603.2	552.7	490.3	658.9	766.9	843.7	819.3
	18	232.8	273.0	278.8	250.6	214.0	312.4	389.8	420.6	408.5	364.6	378.3	496.2	562.8	591.5	555.8
	20	183.7	206.3	203.5	175.0	146.1	233.7	284.9	300.9	283.2	248.1	294.0	376.9	417.6	422.4	386.8
	24	113.3	118.8	111.5	90.6	73.6	150.6	172.6	173.1	152.0	128.2	178.3	220.4	235.7	225.8	200.0
	28	71.8	71.6	64.9	50.9	40.7	100.8	109.4	105.1	87.7	72.2	127.8	149.3	152.0	136.1	115.9

Kec

TABLE 8-7 VALUE OF K(EC) COLUMNS ABOVE l_c(FT) = 12.0 PAGE 4

h	m	ℓ_1 = 15 γ=0.50	γ=0.75	γ=1.00	γ=1.50	γ=2.00	ℓ_1 = 18 γ=0.50	γ=0.75	γ=1.00	γ=1.50	γ=2.00	ℓ_1 = 21 γ=0.50	γ=0.75	γ=1.00	γ=1.50	γ=2.00
6	16	27.4	34.3	37.2	36.3	32.5	32.2	43.4	50.7	56.3	54.9	35.3	49.7	61.0	74.4	78.7
	18	21.3	25.6	26.8	24.8	21.5	25.8	33.9	38.4	40.4	37.9	29.0	40.0	47.9	55.8	56.7
	20	16.7	19.4	19.6	17.4	14.7	21.0	26.7	29.4	29.4	26.7	24.1	32.5	38.1	42.3	41.4
	24	10.5	11.4	10.9	9.1	7.5	14.2	17.0	17.7	16.4	14.2	17.0	22.0	24.5	25.1	23.1
	28	6.7	6.9	6.4	5.1	4.1	9.7	11.1	11.0	9.6	8.1	12.3	15.2	16.2	15.5	13.6
8	16	47.1	53.0	52.4	45.1	37.7	60.8	75.6	81.3	78.4	69.8	70.9	94.1	108.0	116.0	110.6
	18	34.5	37.2	35.6	29.5	24.2	46.7	55.8	58.0	53.2	46.0	56.2	72.3	80.4	81.9	75.3
	20	25.5	26.5	24.8	20.0	16.2	36.2	41.7	42.0	37.1	31.4	45.0	56.2	60.6	58.9	52.5
	24	14.9	14.6	13.1	10.1	8.0	22.1	24.0	23.1	19.3	15.8	29.5	34.7	35.5	32.0	27.4
	28	9.5	8.7	7.5	5.6	4.4	14.2	14.6	13.6	10.9	8.8	19.6	21.9	21.6	18.5	15.5
10	16	64.5	66.9	62.4	50.3	40.7	91.7	105.4	105.9	93.2	78.7	114.1	142.1	153.0	148.2	132.0
	18	44.4	44.7	40.8	32.2	25.8	67.1	74.1	72.2	61.2	50.7	87.0	104.4	108.7	100.3	86.9
	20	33.6	32.1	28.5	21.8	17.2	49.4	52.9	50.3	41.5	34.0	66.9	77.6	78.5	69.8	59.2
	24	19.3	17.3	14.7	10.9	8.5	29.8	29.7	27.0	21.2	16.9	40.2	44.1	42.8	36.1	29.8
	28	11.6	9.9	8.3	6.0	4.6	19.1	17.9	15.7	11.9	9.3	26.7	27.7	25.7	20.6	16.7
12	16	80.4	78.4	70.1	54.2	43.0	119.5	129.3	124.0	103.2	84.8	159.0	186.7	190.9	171.9	146.9
	18	57.2	53.2	46.2	34.8	27.3	83.0	87.0	81.7	66.4	53.9	116.3	131.7	130.7	113.3	94.9
	20	41.2	36.9	31.4	23.2	18.1	63.5	63.4	57.6	45.2	36.2	85.6	94.1	91.3	77.0	63.7
	24	22.5	19.1	15.9	11.4	8.8	37.2	34.7	30.2	22.8	17.9	52.8	54.0	49.8	39.7	32.0
	28	13.2	10.8	8.8	6.2	4.8	22.7	20.2	17.1	12.6	9.8	34.0	32.7	29.1	22.3	17.7

h	m	ℓ_1 = 24 γ=0.50	γ=0.75	γ=1.00	γ=1.50	γ=2.00	ℓ_1 = 28 γ=0.50	γ=0.75	γ=1.00	γ=1.50	γ=2.00	ℓ_1 = 32 γ=0.50	γ=0.75	γ=1.00	γ=1.50	γ=2.00
10	16	131.5	173.3	197.5	209.3	197.7	148.5	205.4	247.0	289.4	295.3	160.6	228.7	284.7	359.0	391.4
	18	103.2	131.9	145.6	146.5	133.7	119.6	161.8	189.7	211.6	207.5	131.5	184.4	225.2	272.4	285.8
	20	81.8	101.4	108.8	104.7	92.9	97.5	128.8	147.2	156.8	148.7	109.2	150.4	180.0	208.9	211.5
	24	52.4	61.4	62.7	56.4	48.2	66.2	83.4	90.9	89.5	80.6	77.1	102.4	117.6	126.3	120.5
	28	33.9	38.1	37.6	32.3	27.0	45.7	55.1	57.7	53.6	46.6	55.6	71.0	78.5	79.0	72.1
12	16	192.2	240.9	261.2	255.4	228.9	226.4	301.9	348.4	377.4	362.4	251.8	349.2	421.2	496.5	509.3
	18	145.8	176.6	185.4	173.0	150.9	177.8	230.5	258.2	265.9	246.4	202.2	274.5	323.1	363.1	358.3
	20	111.5	130.9	133.7	120.3	102.7	141.0	177.7	193.6	190.9	171.9	164.2	218.0	250.5	269.2	256.9
	24	65.9	73.7	72.5	62.1	51.7	90.1	108.0	112.3	103.5	89.6	110.6	140.5	154.2	153.7	139.4
	28	45.4	47.4	44.4	35.9	29.1	58.0	66.9	67.4	59.5	50.4	75.6	92.2	97.5	91.9	80.6
14	16	254.2	304.7	317.0	292.1	252.9	313.5	402.8	447.2	453.8	416.5	359.3	484.4	565.8	626.5	611.3
	18	186.0	215.5	217.8	193.2	163.8	239.4	298.0	320.9	310.6	276.7	282.3	370.9	421.3	444.1	418.2
	20	136.8	154.2	152.5	131.6	110.1	184.1	222.7	233.6	217.6	189.6	223.9	286.8	317.4	320.4	293.1
	24	86.1	90.2	84.6	68.6	55.7	109.8	127.0	128.2	113.5	96.2	143.1	174.9	185.2	175.0	153.7
	28	55.7	55.0	49.7	38.7	30.9	76.7	83.1	79.7	66.4	54.6	93.8	110.3	112.9	101.8	87.1
16	16	312.1	360.6	363.7	321.8	272.5	403.8	501.3	538.4	519.1	461.4	478.0	626.4	709.7	744.8	699.0
	18	219.2	246.2	242.8	208.8	174.3	299.0	359.6	375.3	347.4	301.5	366.7	467.0	514.1	514.7	468.3
	20	171.0	183.3	174.7	144.3	118.1	222.3	260.5	265.9	238.8	203.9	283.3	351.5	377.2	363.4	322.8
	24	103.4	103.4	93.9	73.8	59.0	140.9	154.5	149.6	125.9	104.0	170.3	202.6	209.5	191.4	164.9
	28	64.5	61.3	54.0	41.1	32.4	92.9	96.1	89.3	71.8	58.1	120.2	134.5	132.2	113.3	94.4

h	m	ℓ1 = 15 γ=0.50	γ=0.75	γ=1.00	γ=1.50	γ=2.00	ℓ1 = 18 γ=0.50	γ=0.75	γ=1.00	γ=1.50	γ=2.00	ℓ1 = 21 γ=0.50	γ=0.75	γ=1.00	γ=1.50	γ=2.00
6	16	0.170	0.295	0.397	0.528	0.597	0.207	0.398	0.586	0.883	1.075	0.233	0.474	0.744	1.252	1.646
	18	0.129	0.214	0.278	0.353	0.389	0.165	0.304	0.431	0.615	0.723	0.190	0.376	0.572	0.910	1.148
	20	0.099	0.157	0.198	0.243	0.263	0.132	0.234	0.321	0.437	0.500	0.156	0.300	0.443	0.670	0.817
	24	0.060	0.089	0.106	0.124	0.132	0.086	0.143	0.185	0.234	0.258	0.108	0.196	0.272	0.379	0.438
	28	0.037	0.052	0.061	0.069	0.072	0.058	0.089	0.111	0.134	0.144	0.076	0.130	0.173	0.225	0.252
8	16	0.116	0.178	0.219	0.262	0.281	0.158	0.273	0.365	0.481	0.542	0.191	0.359	0.517	0.756	0.901
	18	0.083	0.122	0.146	0.169	0.180	0.119	0.196	0.253	0.320	0.352	0.149	0.269	0.374	0.519	0.599
	20	0.060	0.085	0.100	0.114	0.120	0.091	0.143	0.179	0.219	0.237	0.118	0.204	0.275	0.364	0.411
	24	0.034	0.045	0.051	0.056	0.059	0.054	0.079	0.095	0.111	0.118	0.075	0.121	0.154	0.192	0.210
	28	0.021	0.026	0.029	0.031	0.032	0.033	0.047	0.054	0.062	0.065	0.048	0.074	0.091	0.109	0.117
10	16	0.077	0.110	0.128	0.146	0.154	0.117	0.184	0.231	0.282	0.305	0.152	0.264	0.354	0.469	0.529
	18	0.052	0.072	0.083	0.093	0.097	0.084	0.127	0.154	0.183	0.195	0.114	0.189	0.245	0.311	0.344
	20	0.038	0.051	0.057	0.062	0.064	0.061	0.089	0.106	0.123	0.130	0.086	0.137	0.173	0.213	0.232
	24	0.021	0.026	0.029	0.031	0.031	0.035	0.048	0.055	0.061	0.064	0.050	0.075	0.092	0.108	0.115
	28	0.012	0.015	0.016	0.017	0.017	0.022	0.028	0.031	0.034	0.035	0.032	0.046	0.053	0.060	0.064
12	16	0.054	0.072	0.082	0.090	0.094	0.085	0.127	0.152	0.178	0.189	0.119	0.193	0.246	0.306	0.335
	18	0.037	0.048	0.053	0.058	0.059	0.058	0.084	0.099	0.113	0.120	0.086	0.133	0.165	0.199	0.215
	20	0.026	0.033	0.036	0.038	0.039	0.043	0.059	0.068	0.076	0.080	0.062	0.093	0.114	0.134	0.143
	24	0.014	0.016	0.018	0.018	0.019	0.024	0.031	0.035	0.038	0.039	0.037	0.051	0.060	0.068	0.071
	28	0.008	0.009	0.009	0.010	0.010	0.014	0.018	0.019	0.021	0.021	0.023	0.030	0.034	0.037	0.039

h	m	ℓ1 = 24 γ=0.50	γ=0.75	γ=1.00	γ=1.50	γ=2.00	ℓ1 = 28 γ=0.50	γ=0.75	γ=1.00	γ=1.50	γ=2.00	ℓ1 = 32 γ=0.50	γ=0.75	γ=1.00	γ=1.50	γ=2.00
10	16	0.181	0.337	0.482	0.697	0.825	0.210	0.418	0.639	1.028	1.306	0.230	0.478	0.767	1.348	1.834
	18	0.140	0.251	0.346	0.475	0.547	0.167	0.323	0.479	0.730	0.893	0.188	0.381	0.595	0.994	1.298
	20	0.109	0.189	0.252	0.332	0.374	0.135	0.253	0.363	0.527	0.626	0.155	0.306	0.466	0.742	0.936
	24	0.068	0.110	0.140	0.174	0.190	0.090	0.158	0.214	0.289	0.328	0.107	0.202	0.292	0.428	0.511
	28	0.043	0.066	0.082	0.098	0.105	0.060	0.101	0.131	0.168	0.186	0.076	0.136	0.188	0.258	0.297
12	16	0.149	0.261	0.354	0.474	0.538	0.182	0.344	0.501	0.742	0.892	0.206	0.413	0.635	1.030	1.318
	18	0.111	0.187	0.245	0.315	0.349	0.141	0.257	0.362	0.509	0.594	0.164	0.319	0.476	0.732	0.903
	20	0.084	0.136	0.173	0.215	0.236	0.110	0.195	0.265	0.358	0.407	0.132	0.249	0.361	0.530	0.633
	24	0.048	0.074	0.091	0.109	0.117	0.069	0.114	0.148	0.188	0.208	0.087	0.155	0.213	0.291	0.333
	28	0.032	0.046	0.054	0.062	0.065	0.043	0.069	0.087	0.106	0.115	0.058	0.099	0.130	0.169	0.189
14	16	0.122	0.202	0.263	0.334	0.369	0.156	0.281	0.392	0.545	0.631	0.183	0.352	0.521	0.791	0.966
	18	0.088	0.140	0.177	0.218	0.237	0.117	0.204	0.275	0.365	0.413	0.142	0.265	0.379	0.547	0.647
	20	0.064	0.099	0.122	0.147	0.158	0.089	0.149	0.196	0.252	0.280	0.112	0.201	0.279	0.386	0.446
	24	0.038	0.055	0.065	0.075	0.079	0.052	0.083	0.105	0.129	0.140	0.070	0.119	0.157	0.204	0.228
	28	0.024	0.033	0.037	0.041	0.043	0.035	0.052	0.063	0.073	0.078	0.045	0.073	0.093	0.116	0.128
16	16	0.098	0.157	0.199	0.245	0.266	0.132	0.230	0.309	0.410	0.464	0.161	0.299	0.427	0.615	0.727
	18	0.068	0.106	0.131	0.157	0.169	0.097	0.162	0.211	0.270	0.300	0.122	0.219	0.303	0.416	0.479
	20	0.052	0.077	0.092	0.107	0.114	0.071	0.115	0.147	0.184	0.201	0.093	0.162	0.218	0.289	0.326
	24	0.030	0.042	0.048	0.054	0.056	0.043	0.066	0.080	0.094	0.101	0.055	0.091	0.117	0.149	0.164
	28	0.018	0.024	0.027	0.029	0.030	0.028	0.039	0.046	0.053	0.056	0.037	0.058	0.071	0.086	0.092

Table 8-8.
SYMBOLS: $\gamma = l_1/l_2 = 1/\beta$; $m = l_1/c_1$; $c_2 = c_1 \cdot l_2/l_1$; h = Slab thickness; l_c = Story height. This Table furnishes the Ratio of Flexural Stiffness of the equivalent column to the Flexural Stiffness of slab at a joint in the direction of moments at end spans in a flat plate construction, (α_{ec}).

α_{ec}

α_{ec}

TABLE 8-8 VALUE OF α_{ec} NO COLUMNS ABOVE l_c(FT) = 12.0 PAGE 2

h	m	ℓ1 = 15					ℓ1 = 18					ℓ1 = 21				
		γ=0.50	γ=0.75	γ=1.00	γ=1.50	γ=2.00	γ=0.50	γ=0.75	γ=1.00	γ=1.50	γ=2.00	γ=0.50	γ=0.75	γ=1.00	γ=1.50	γ=2.00
6	16	0.161	0.271	0.355	0.456	0.507	0.201	0.376	0.540	0.784	0.931	0.229	0.458	0.704	1.142	1.459
	18	0.122	0.194	0.245	0.302	0.328	0.159	0.284	0.392	0.539	0.620	0.185	0.359	0.534	0.818	1.005
	20	0.093	0.141	0.173	0.206	0.221	0.126	0.216	0.289	0.379	0.426	0.152	0.284	0.409	0.595	0.708
	24	0.055	0.078	0.092	0.104	0.110	0.081	0.129	0.163	0.200	0.217	0.104	0.182	0.246	0.330	0.374
	28	0.034	0.046	0.052	0.058	0.060	0.053	0.080	0.097	0.113	0.121	0.072	0.119	0.153	0.194	0.213
8	16	0.107	0.158	0.189	0.220	0.234	0.150	0.249	0.324	0.412	0.456	0.184	0.336	0.472	0.662	0.771
	18	0.076	0.107	0.125	0.142	0.149	0.112	0.177	0.222	0.272	0.295	0.143	0.249	0.337	0.449	0.509
	20	0.054	0.074	0.085	0.095	0.099	0.084	0.128	0.156	0.185	0.198	0.112	0.187	0.244	0.313	0.347
	24	0.030	0.039	0.043	0.047	0.048	0.049	0.070	0.082	0.093	0.098	0.070	0.109	0.135	0.163	0.175
	28	0.018	0.022	0.024	0.026	0.026	0.030	0.041	0.046	0.051	0.054	0.044	0.066	0.078	0.091	0.097
10	16	0.070	0.095	0.108	0.121	0.126	0.108	0.164	0.199	0.236	0.252	0.144	0.240	0.313	0.400	0.442
	18	0.047	0.062	0.070	0.077	0.079	0.077	0.111	0.132	0.152	0.161	0.107	0.170	0.214	0.263	0.286
	20	0.034	0.043	0.047	0.051	0.053	0.055	0.077	0.090	0.102	0.107	0.080	0.122	0.150	0.179	0.192
	24	0.018	0.022	0.024	0.025	0.025	0.031	0.041	0.046	0.051	0.052	0.046	0.066	0.078	0.090	0.095
	28	0.010	0.012	0.013	0.013	0.014	0.019	0.024	0.026	0.028	0.028	0.029	0.039	0.045	0.050	0.052
12	16	0.047	0.061	0.068	0.074	0.076	0.077	0.110	0.128	0.146	0.154	0.111	0.171	0.212	0.255	0.275
	18	0.032	0.040	0.044	0.047	0.048	0.052	0.072	0.083	0.093	0.097	0.079	0.117	0.141	0.165	0.175
	20	0.023	0.027	0.029	0.031	0.031	0.038	0.051	0.057	0.062	0.064	0.056	0.081	0.096	0.111	0.117
	24	0.012	0.013	0.014	0.015	0.015	0.021	0.026	0.029	0.031	0.031	0.033	0.044	0.050	0.055	0.057
	28	0.006	0.007	0.008	0.008	0.008	0.012	0.015	0.016	0.017	0.017	0.020	0.025	0.028	0.030	0.031

h	m	ℓ1 = 24					ℓ1 = 28					ℓ1 = 32				
		γ=0.50	γ=0.75	γ=1.00	γ=1.50	γ=2.00	γ=0.50	γ=0.75	γ=1.00	γ=1.50	γ=2.00	γ=0.50	γ=0.75	γ=1.00	γ=1.50	γ=2.00
10	16	0.174	0.314	0.437	0.605	0.700	0.205	0.398	0.594	0.917	1.133	0.227	0.463	0.728	1.234	1.629
	18	0.134	0.231	0.309	0.408	0.459	0.162	0.305	0.440	0.642	0.765	0.184	0.366	0.559	0.896	1.137
	20	0.103	0.171	0.223	0.283	0.312	0.130	0.236	0.329	0.458	0.531	0.151	0.291	0.432	0.660	0.809
	24	0.063	0.098	0.121	0.146	0.157	0.085	0.144	0.190	0.247	0.275	0.104	0.189	0.265	0.373	0.435
	28	0.039	0.058	0.070	0.082	0.087	0.057	0.091	0.115	0.142	0.155	0.072	0.125	0.168	0.222	0.250
12	16	0.141	0.237	0.312	0.401	0.446	0.175	0.322	0.454	0.644	0.754	0.201	0.393	0.590	0.917	1.137
	18	0.104	0.168	0.213	0.264	0.288	0.135	0.237	0.323	0.436	0.497	0.159	0.301	0.436	0.642	0.769
	20	0.078	0.121	0.149	0.180	0.193	0.105	0.177	0.234	0.303	0.339	0.127	0.232	0.326	0.459	0.534
	24	0.044	0.065	0.078	0.090	0.096	0.064	0.102	0.128	0.157	0.171	0.083	0.142	0.189	0.247	0.277
	28	0.029	0.039	0.045	0.050	0.053	0.040	0.061	0.074	0.088	0.094	0.055	0.089	0.114	0.142	0.156
14	16	0.113	0.180	0.226	0.277	0.301	0.148	0.257	0.347	0.462	0.522	0.177	0.330	0.473	0.686	0.814
	18	0.081	0.123	0.151	0.180	0.192	0.110	0.184	0.240	0.306	0.339	0.136	0.245	0.339	0.468	0.539
	20	0.058	0.086	0.103	0.121	0.128	0.083	0.133	0.169	0.209	0.228	0.106	0.184	0.247	0.327	0.369
	24	0.034	0.047	0.054	0.061	0.063	0.048	0.073	0.089	0.106	0.113	0.065	0.106	0.136	0.170	0.187
	28	0.021	0.027	0.031	0.033	0.035	0.031	0.045	0.052	0.060	0.063	0.041	0.064	0.080	0.096	0.104
16	16	0.090	0.137	0.168	0.199	0.213	0.124	0.206	0.268	0.341	0.376	0.154	0.275	0.380	0.521	0.600
	18	0.062	0.091	0.109	0.128	0.135	0.090	0.143	0.181	0.222	0.242	0.115	0.199	0.265	0.348	0.391
	20	0.046	0.065	0.076	0.086	0.090	0.065	0.101	0.125	0.150	0.162	0.087	0.145	0.188	0.240	0.265
	24	0.026	0.035	0.039	0.043	0.044	0.039	0.056	0.066	0.076	0.080	0.051	0.080	0.100	0.122	0.132
	28	0.016	0.020	0.022	0.023	0.024	0.024	0.033	0.038	0.042	0.044	0.034	0.050	0.060	0.069	0.074

α_{ec}

TABLE 8-8 VALUE OF α_{ec} COLUMNS ABOVE l_c (FT) = 10.0 PAGE 3

h	m	ℓ1 = 15 γ=0.50	γ=0.75	γ=1.00	γ=1.50	γ=2.00	ℓ1 = 18 γ=0.50	γ=0.75	γ=1.00	γ=1.50	γ=2.00	ℓ1 = 21 γ=0.50	γ=0.75	γ=1.00	γ=1.50	γ=2.00
6	16	0.192	0.371	0.549	0.837	1.024	0.223	0.459	0.729	1.256	1.682	0.243	0.519	0.860	1.620	2.345
	18	0.151	0.281	0.402	0.581	0.687	0.181	0.363	0.561	0.917	1.180	0.201	0.422	0.685	1.235	1.719
	20	0.120	0.215	0.298	0.412	0.475	0.148	0.289	0.435	0.678	0.844	0.168	0.346	0.550	0.951	1.276
	24	0.076	0.129	0.170	0.219	0.244	0.101	0.188	0.268	0.386	0.456	0.120	0.238	0.362	0.578	0.729
	28	0.049	0.079	0.101	0.125	0.136	0.070	0.124	0.170	0.230	0.263	0.088	0.167	0.244	0.363	0.437
8	16	0.142	0.249	0.338	0.453	0.514	0.180	0.346	0.509	0.767	0.933	0.208	0.422	0.660	1.105	1.445
	18	0.105	0.177	0.233	0.300	0.334	0.140	0.259	0.369	0.529	0.623	0.166	0.329	0.499	0.793	0.999
	20	0.079	0.128	0.164	0.205	0.225	0.109	0.196	0.271	0.373	0.429	0.134	0.258	0.381	0.579	0.706
	24	0.047	0.072	0.088	0.105	0.112	0.068	0.115	0.152	0.197	0.219	0.089	0.162	0.228	0.322	0.375
	28	0.030	0.044	0.051	0.059	0.062	0.044	0.071	0.090	0.112	0.122	0.060	0.104	0.141	0.189	0.214
10	16	0.101	0.165	0.211	0.264	0.289	0.141	0.253	0.349	0.480	0.552	0.173	0.333	0.492	0.745	0.909
	18	0.071	0.111	0.139	0.170	0.184	0.104	0.180	0.242	0.319	0.359	0.133	0.248	0.355	0.513	0.607
	20	0.054	0.081	0.099	0.116	0.124	0.078	0.130	0.170	0.218	0.242	0.103	0.187	0.260	0.360	0.417
	24	0.032	0.044	0.052	0.058	0.061	0.048	0.075	0.093	0.113	0.122	0.063	0.108	0.144	0.189	0.213
	28	0.019	0.026	0.029	0.032	0.033	0.031	0.046	0.055	0.064	0.068	0.042	0.069	0.088	0.110	0.120
12	16	0.075	0.114	0.140	0.167	0.180	0.108	0.183	0.242	0.314	0.350	0.141	0.259	0.364	0.513	0.598
	18	0.054	0.079	0.094	0.108	0.115	0.076	0.125	0.161	0.203	0.224	0.105	0.185	0.253	0.342	0.391
	20	0.039	0.055	0.064	0.073	0.076	0.059	0.092	0.115	0.140	0.151	0.078	0.133	0.178	0.235	0.264
	24	0.022	0.029	0.033	0.036	0.037	0.035	0.051	0.061	0.071	0.075	0.049	0.078	0.099	0.123	0.134
	28	0.013	0.016	0.018	0.020	0.020	0.022	0.030	0.035	0.039	0.041	0.032	0.048	0.059	0.070	0.074

h	m	ℓ1 = 24 γ=0.50	γ=0.75	γ=1.00	γ=1.50	γ=2.00	ℓ1 = 28 γ=0.50	γ=0.75	γ=1.00	γ=1.50	γ=2.00	ℓ1 = 32 γ=0.50	γ=0.75	γ=1.00	γ=1.50	γ=2.00
10	16	0.198	0.399	0.621	1.028	1.335	0.221	0.466	0.760	1.383	1.938	0.238	0.514	0.864	1.679	2.506
	18	0.156	0.308	0.465	0.733	0.917	0.180	0.371	0.592	1.029	1.387	0.196	0.418	0.691	1.295	1.863
	20	0.125	0.240	0.352	0.531	0.645	0.147	0.298	0.465	0.774	1.009	0.164	0.344	0.559	1.007	1.401
	24	0.081	0.148	0.207	0.291	0.340	0.101	0.197	0.293	0.453	0.558	0.117	0.238	0.372	0.624	0.818
	28	0.053	0.093	0.126	0.169	0.192	0.070	0.132	0.189	0.275	0.327	0.085	0.167	0.253	0.398	0.498
12	16	0.169	0.327	0.486	0.746	0.917	0.197	0.402	0.633	1.072	1.416	0.217	0.459	0.751	1.375	1.940
	18	0.129	0.243	0.351	0.513	0.612	0.156	0.311	0.476	0.769	0.979	0.176	0.364	0.584	1.024	1.391
	20	0.100	0.182	0.256	0.361	0.421	0.124	0.242	0.362	0.560	0.692	0.143	0.292	0.458	0.771	1.012
	24	0.060	0.104	0.141	0.189	0.214	0.080	0.150	0.214	0.310	0.367	0.098	0.192	0.288	0.451	0.561
	28	0.042	0.068	0.088	0.111	0.122	0.052	0.094	0.130	0.180	0.208	0.067	0.128	0.185	0.274	0.329
14	16	0.142	0.265	0.380	0.549	0.651	0.173	0.343	0.521	0.831	1.050	0.196	0.405	0.645	1.119	1.505
	18	0.105	0.190	0.265	0.368	0.427	0.133	0.257	0.380	0.579	0.708	0.155	0.314	0.488	0.808	1.047
	20	0.078	0.137	0.187	0.253	0.289	0.103	0.194	0.280	0.411	0.490	0.124	0.245	0.373	0.591	0.744
	24	0.050	0.082	0.106	0.134	0.148	0.062	0.113	0.157	0.218	0.252	0.080	0.152	0.222	0.330	0.397
	28	0.033	0.051	0.063	0.077	0.083	0.044	0.075	0.100	0.129	0.145	0.053	0.097	0.137	0.195	0.228
16	16	0.118	0.213	0.297	0.413	0.478	0.150	0.290	0.428	0.651	0.795	0.176	0.355	0.552	0.910	1.179
	18	0.084	0.147	0.201	0.271	0.309	0.112	0.211	0.303	0.442	0.526	0.136	0.268	0.406	0.640	0.803
	20	0.066	0.112	0.147	0.190	0.211	0.084	0.154	0.217	0.307	0.359	0.106	0.204	0.302	0.458	0.560
	24	0.041	0.064	0.081	0.098	0.107	0.054	0.094	0.125	0.165	0.186	0.064	0.120	0.171	0.246	0.290
	28	0.026	0.039	0.047	0.055	0.059	0.036	0.059	0.076	0.095	0.105	0.046	0.081	0.110	0.148	0.168

α_{ec}

TABLE 8-8 VALUE OF α_{ec} COLUMNS ABOVE l_c(FT) = 12.0 PAGE 4

h	m	ℓ_1 = 15					ℓ_1 = 18					ℓ_1 = 21				
		γ=0.50	γ=0.75	γ=1.00	γ=1.50	γ=2.00	γ=0.50	γ=0.75	γ=1.00	γ=1.50	γ=2.00	γ=0.50	γ=0.75	γ=1.00	γ=1.50	=2.00
6	16	0.187	0.352	0.508	0.744	0.888	0.219	0.444	0.693	1.152	1.501	0.241	0.509	0.832	1.525	2.150
	18	0.146	0.263	0.367	0.510	0.591	0.177	0.348	0.527	0.830	1.039	0.198	0.411	0.657	1.148	1.554
	20	0.115	0.199	0.269	0.358	0.405	0.144	0.275	0.404	0.607	0.735	0.165	0.335	0.523	0.873	1.139
	24	0.072	0.118	0.151	0.188	0.206	0.098	0.176	0.244	0.339	0.391	0.117	0.227	0.338	0.519	0.639
	28	0.046	0.072	0.089	0.107	0.115	0.067	0.115	0.153	0.200	0.224	0.085	0.157	0.224	0.321	0.378
8	16	0.135	0.229	0.301	0.390	0.434	0.175	0.326	0.468	0.677	0.804	0.204	0.406	0.622	1.002	1.274
	18	0.099	0.161	0.206	0.256	0.280	0.135	0.242	0.335	0.461	0.532	0.162	0.313	0.465	0.709	0.870
	20	0.073	0.115	0.143	0.174	0.188	0.105	0.181	0.243	0.322	0.363	0.130	0.244	0.351	0.512	0.609
	24	0.043	0.063	0.076	0.088	0.093	0.064	0.105	0.134	0.168	0.184	0.085	0.151	0.206	0.279	0.319
	28	0.027	0.038	0.044	0.049	0.051	0.041	0.064	0.079	0.095	0.102	0.057	0.096	0.126	0.162	0.180
10	16	0.095	0.148	0.184	0.222	0.240	0.139	0.233	0.312	0.412	0.464	0.168	0.314	0.451	0.655	0.778
	18	0.065	0.099	0.120	0.142	0.152	0.099	0.164	0.213	0.271	0.300	0.128	0.231	0.321	0.445	0.514
	20	0.049	0.071	0.084	0.097	0.102	0.073	0.117	0.149	0.185	0.201	0.099	0.172	0.233	0.310	0.351
	24	0.028	0.038	0.043	0.048	0.050	0.044	0.066	0.080	0.094	0.101	0.059	0.098	0.127	0.161	0.177
	28	0.017	0.022	0.024	0.026	0.027	0.028	0.040	0.046	0.053	0.055	0.039	0.061	0.076	0.092	0.099
12	16	0.068	0.100	0.119	0.138	0.147	0.101	0.165	0.211	0.264	0.289	0.135	0.239	0.325	0.440	0.501
	18	0.049	0.068	0.079	0.089	0.093	0.071	0.111	0.139	0.170	0.184	0.099	0.169	0.223	0.291	0.325
	20	0.035	0.047	0.054	0.059	0.062	0.054	0.081	0.099	0.116	0.124	0.073	0.121	0.156	0.198	0.218
	24	0.019	0.024	0.027	0.029	0.030	0.032	0.044	0.052	0.058	0.061	0.045	0.069	0.085	0.102	0.110
	28	0.011	0.014	0.015	0.016	0.016	0.019	0.026	0.029	0.032	0.033	0.029	0.042	0.050	0.057	0.061

h	m	ℓ_1 = 24					ℓ_1 = 28					ℓ_1 = 32				
		γ=0.50	γ=0.75	γ=1.00	γ=1.50	γ=2.00	γ=0.50	γ=0.75	γ=1.00	γ=1.50	γ=2.00	γ=0.50	γ=0.75	γ=1.00	γ=1.50	γ=2.00
10	16	0.193	0.383	0.582	0.925	1.166	0.218	0.454	0.728	1.279	1.741	0.236	0.505	0.839	1.587	2.308
	18	0.152	0.292	0.431	0.650	0.791	0.177	0.359	0.561	0.939	1.228	0.194	0.409	0.666	1.209	1.691
	20	0.121	0.225	0.322	0.465	0.551	0.144	0.286	0.436	0.698	0.882	0.161	0.334	0.534	0.929	1.255
	24	0.077	0.137	0.186	0.251	0.286	0.098	0.186	0.270	0.399	0.480	0.114	0.228	0.350	0.564	0.717
	28	0.050	0.085	0.112	0.144	0.161	0.068	0.123	0.172	0.239	0.278	0.082	0.159	0.234	0.353	0.430
12	16	0.163	0.308	0.445	0.653	0.781	0.193	0.386	0.594	0.966	1.236	0.214	0.446	0.718	1.270	1.737
	18	0.124	0.226	0.317	0.444	0.516	0.152	0.296	0.442	0.683	0.843	0.173	0.352	0.553	0.932	1.227
	20	0.095	0.168	0.229	0.309	0.352	0.121	0.228	0.332	0.491	0.590	0.140	0.280	0.430	0.693	0.882
	24	0.056	0.095	0.124	0.160	0.178	0.077	0.139	0.193	0.267	0.308	0.095	0.181	0.265	0.397	0.480
	28	0.039	0.061	0.076	0.093	0.100	0.050	0.086	0.116	0.154	0.174	0.065	0.119	0.168	0.238	0.278
14	16	0.136	0.245	0.340	0.470	0.543	0.168	0.324	0.480	0.731	0.895	0.193	0.390	0.607	1.009	1.313
	18	0.100	0.174	0.234	0.312	0.353	0.129	0.241	0.346	0.502	0.596	0.152	0.299	0.454	0.718	0.901
	20	0.073	0.125	0.164	0.213	0.238	0.099	0.180	0.252	0.353	0.410	0.121	0.232	0.343	0.519	0.633
	24	0.046	0.073	0.091	0.111	0.120	0.059	0.103	0.139	0.184	0.208	0.077	0.142	0.201	0.284	0.333
	28	0.030	0.044	0.054	0.063	0.067	0.041	0.067	0.086	0.108	0.118	0.051	0.089	0.122	0.166	0.189
16	16	0.112	0.194	0.261	0.347	0.392	0.145	0.270	0.387	0.560	0.664	0.172	0.338	0.510	0.804	1.006
	18	0.079	0.133	0.175	0.226	0.251	0.108	0.194	0.271	0.376	0.435	0.132	0.253	0.371	0.557	0.676
	20	0.061	0.099	0.126	0.156	0.171	0.080	0.141	0.192	0.259	0.295	0.102	0.190	0.273	0.394	0.467
	24	0.037	0.056	0.068	0.080	0.085	0.051	0.084	0.108	0.137	0.151	0.061	0.110	0.152	0.208	0.239
	28	0.023	0.033	0.039	0.044	0.047	0.033	0.052	0.065	0.078	0.084	0.043	0.073	0.096	0.123	0.137

α_{ec}

h	m	l1 = 15 γ=0.50	γ=0.75	γ=1.00	γ=1.50	γ=2.00	l1 = 18 γ=0.50	γ=0.75	γ=1.00	γ=1.50	γ=2.00	l1 = 21 γ=0.50	γ=0.75	γ=1.00	γ=1.50	γ=2.00
6	16	0.094	0.148	0.184	0.224	0.243	0.111	0.185	0.240	0.304	0.336	0.122	0.209	0.277	0.361	0.404
	18	0.074	0.114	0.141	0.169	0.182	0.092	0.151	0.195	0.247	0.272	0.103	0.177	0.236	0.309	0.347
	20	0.058	0.088	0.107	0.127	0.135	0.075	0.123	0.158	0.197	0.216	0.088	0.150	0.199	0.260	0.292
	24	0.037	0.053	0.062	0.071	0.075	0.051	0.081	0.101	0.123	0.133	0.063	0.106	0.139	0.178	0.198
	28	0.023	0.032	0.037	0.042	0.044	0.035	0.053	0.065	0.076	0.082	0.046	0.075	0.095	0.119	0.130
8	16	0.067	0.098	0.116	0.135	0.142	0.088	0.139	0.173	0.211	0.228	0.104	0.171	0.221	0.279	0.308
	18	0.049	0.070	0.082	0.094	0.099	0.069	0.106	0.131	0.157	0.169	0.084	0.138	0.177	0.222	0.243
	20	0.037	0.051	0.059	0.066	0.069	0.054	0.081	0.099	0.116	0.124	0.068	0.110	0.140	0.173	0.189
	24	0.021	0.028	0.031	0.035	0.036	0.033	0.048	0.056	0.065	0.068	0.045	0.070	0.087	0.104	0.112
	28	0.013	0.016	0.018	0.019	0.020	0.021	0.029	0.033	0.038	0.039	0.030	0.045	0.054	0.063	0.068
10	16	0.046	0.064	0.074	0.083	0.086	0.068	0.101	0.122	0.143	0.152	0.086	0.135	0.170	0.207	0.225
	18	0.032	0.043	0.050	0.055	0.057	0.050	0.073	0.087	0.100	0.106	0.066	0.103	0.128	0.154	0.166
	20	0.024	0.031	0.035	0.038	0.039	0.037	0.053	0.062	0.071	0.074	0.051	0.078	0.096	0.114	0.122
	24	0.013	0.016	0.018	0.019	0.020	0.022	0.030	0.034	0.037	0.039	0.031	0.045	0.054	0.063	0.067
	28	0.008	0.009	0.010	0.010	0.011	0.013	0.017	0.019	0.021	0.022	0.020	0.028	0.033	0.037	0.039
12	16	0.033	0.044	0.049	0.054	0.056	0.051	0.073	0.086	0.098	0.103	0.069	0.105	0.128	0.152	0.163
	18	0.023	0.029	0.033	0.035	0.036	0.036	0.050	0.058	0.066	0.069	0.051	0.076	0.092	0.108	0.115
	20	0.016	0.020	0.022	0.024	0.024	0.027	0.036	0.041	0.046	0.048	0.038	0.055	0.066	0.077	0.081
	24	0.009	0.010	0.011	0.012	0.012	0.015	0.020	0.022	0.023	0.024	0.023	0.032	0.036	0.041	0.043
	28	0.005	0.006	0.006	0.006	0.006	0.009	0.011	0.012	0.013	0.013	0.014	0.019	0.021	0.023	0.024

h	m	l1 = 24 γ=0.50	γ=0.75	γ=1.00	γ=1.50	γ=2.00	l1 = 28 γ=0.50	γ=0.75	γ=1.00	γ=1.50	γ=2.00	l1 = 32 γ=0.50	γ=0.75	γ=1.00	γ=1.50	γ=2.00
10	16	0.099	0.164	0.211	0.267	0.293	0.112	0.191	0.253	0.329	0.368	0.121	0.210	0.282	0.373	0.420
	18	0.080	0.130	0.167	0.209	0.229	0.093	0.158	0.210	0.274	0.306	0.102	0.179	0.242	0.324	0.367
	20	0.064	0.103	0.131	0.162	0.177	0.077	0.131	0.173	0.224	0.250	0.087	0.152	0.206	0.276	0.314
	24	0.041	0.064	0.079	0.096	0.103	0.053	0.088	0.114	0.145	0.160	0.063	0.109	0.147	0.194	0.219
	28	0.027	0.040	0.049	0.058	0.062	0.037	0.059	0.075	0.093	0.102	0.046	0.078	0.103	0.133	0.149
12	16	0.084	0.134	0.170	0.209	0.227	0.100	0.166	0.217	0.276	0.306	0.111	0.190	0.252	0.329	0.369
	18	0.065	0.102	0.128	0.155	0.168	0.080	0.133	0.172	0.219	0.242	0.091	0.157	0.209	0.274	0.308
	20	0.050	0.077	0.096	0.115	0.124	0.064	0.106	0.136	0.171	0.188	0.076	0.129	0.172	0.225	0.252
	24	0.030	0.045	0.054	0.064	0.068	0.042	0.066	0.084	0.103	0.112	0.052	0.087	0.114	0.146	0.162
	28	0.020	0.028	0.033	0.038	0.039	0.027	0.042	0.052	0.062	0.067	0.036	0.058	0.075	0.094	0.103
14	16	0.070	0.109	0.135	0.163	0.175	0.087	0.142	0.183	0.229	0.251	0.100	0.169	0.222	0.287	0.319
	18	0.052	0.080	0.097	0.116	0.124	0.068	0.110	0.140	0.174	0.190	0.081	0.136	0.178	0.229	0.255
	20	0.039	0.058	0.070	0.083	0.089	0.053	0.084	0.106	0.130	0.142	0.065	0.109	0.142	0.181	0.200
	24	0.024	0.034	0.040	0.045	0.047	0.032	0.049	0.061	0.074	0.079	0.042	0.069	0.088	0.110	0.121
	28	0.015	0.020	0.023	0.026	0.027	0.022	0.032	0.038	0.044	0.047	0.028	0.044	0.055	0.068	0.073
16	16	0.058	0.088	0.107	0.127	0.136	0.076	0.121	0.153	0.189	0.206	0.090	0.149	0.194	0.247	0.273
	18	0.041	0.062	0.075	0.088	0.094	0.057	0.090	0.113	0.138	0.150	0.070	0.116	0.151	0.191	0.210
	20	0.032	0.046	0.054	0.063	0.066	0.043	0.067	0.083	0.101	0.109	0.055	0.090	0.116	0.145	0.160
	24	0.019	0.026	0.029	0.033	0.034	0.027	0.040	0.048	0.056	0.059	0.034	0.054	0.068	0.084	0.091
	28	0.011	0.015	0.017	0.018	0.019	0.017	0.024	0.029	0.032	0.034	0.023	0.035	0.043	0.051	0.055

% $M_{neg\ ext.}$

% $M_{neg\ ext.}$

Table 8-9.
SYMBOLS: $\gamma = l_1/l_2 = 1/\beta$; $m = l_1/c_1$; $c_2 = c_1 \cdot l_2/l_1$; h = Slab thickness; l_c = Story height. This Table furnishes the distribution of the total static design moment in end spans of flat plate construction into exterior negative design moments, as per Sect. 13.3.3.3 of the ACI-318-71 Code.

TABLE 8-9 ACI-13.3.3.3, FLAT PLATES. END SPAN, EXTERIOR NEG. MOMENT % M_O NO COLUMNS ABOVE l_{C}(FT) = 12.0 PAGE 2

h	m	ℓ1 = 15 γ=0.50	γ=0.75	γ=1.00	γ=1.50	γ=2.00	ℓ1 = 18 γ=0.50	γ=0.75	γ=1.00	γ=1.50	γ=2.00	ℓ1 = 21 γ=0.50	γ=0.75	γ=1.00	γ=1.50	γ=2.00
6	16	0.090	0.138	0.170	0.203	0.218	0.109	0.177	0.228	0.285	0.313	0.121	0.204	0.268	0.346	0.385
	18	0.070	0.105	0.128	0.150	0.160	0.089	0.143	0.183	0.227	0.248	0.101	0.171	0.226	0.292	0.325
	20	0.055	0.080	0.096	0.111	0.117	0.072	0.115	0.145	0.178	0.194	0.085	0.144	0.188	0.242	0.269
	24	0.034	0.047	0.054	0.061	0.064	0.049	0.074	0.091	0.108	0.116	0.061	0.100	0.128	0.161	0.177
	28	0.021	0.028	0.032	0.035	0.037	0.033	0.048	0.057	0.066	0.070	0.044	0.069	0.086	0.105	0.114
8	16	0.062	0.088	0.103	0.117	0.123	0.084	0.129	0.159	0.189	0.203	0.101	0.163	0.208	0.259	0.283
	18	0.045	0.062	0.072	0.080	0.084	0.065	0.097	0.118	0.139	0.148	0.081	0.129	0.163	0.201	0.219
	20	0.033	0.045	0.051	0.056	0.058	0.050	0.073	0.087	0.101	0.107	0.065	0.102	0.127	0.155	0.167
	24	0.019	0.024	0.027	0.029	0.030	0.030	0.042	0.049	0.055	0.058	0.042	0.063	0.077	0.091	0.097
	28	0.011	0.014	0.015	0.016	0.016	0.019	0.025	0.029	0.032	0.033	0.027	0.040	0.047	0.054	0.057
10	16	0.042	0.056	0.063	0.070	0.072	0.063	0.091	0.108	0.124	0.131	0.082	0.126	0.155	0.185	0.199
	18	0.029	0.038	0.042	0.046	0.048	0.046	0.065	0.075	0.086	0.090	0.062	0.094	0.114	0.135	0.144
	20	0.021	0.027	0.029	0.031	0.032	0.034	0.046	0.053	0.060	0.062	0.048	0.071	0.084	0.098	0.104
	24	0.011	0.014	0.015	0.016	0.016	0.019	0.025	0.028	0.031	0.032	0.028	0.040	0.047	0.053	0.056
	28	0.006	0.008	0.008	0.008	0.009	0.012	0.015	0.016	0.017	0.018	0.018	0.024	0.028	0.031	0.032
12	16	0.029	0.037	0.041	0.044	0.046	0.046	0.064	0.074	0.083	0.086	0.064	0.095	0.113	0.132	0.140
	18	0.020	0.025	0.027	0.029	0.029	0.032	0.043	0.050	0.055	0.057	0.047	0.068	0.080	0.092	0.097
	20	0.014	0.017	0.018	0.019	0.020	0.024	0.031	0.035	0.038	0.039	0.035	0.049	0.057	0.064	0.068
	24	0.007	0.008	0.009	0.009	0.009	0.013	0.016	0.018	0.019	0.020	0.020	0.027	0.031	0.034	0.035
	28	0.004	0.004	0.005	0.005	0.005	0.008	0.009	0.010	0.010	0.011	0.012	0.016	0.018	0.019	0.019

h	m	ℓ1 = 24 γ=0.50	γ=0.75	γ=1.00	γ=1.50	γ=2.00	ℓ1 = 28 γ=0.50	γ=0.75	γ=1.00	γ=1.50	γ=2.00	ℓ1 = 32 γ=0.50	γ=0.75	γ=1.00	γ=1.50	γ=2.00
10	16	0.096	0.155	0.197	0.245	0.267	0.110	0.185	0.242	0.311	0.345	0.120	0.205	0.274	0.359	0.402
	18	0.076	0.122	0.153	0.188	0.204	0.091	0.152	0.198	0.254	0.281	0.101	0.174	0.233	0.307	0.345
	20	0.061	0.095	0.118	0.143	0.154	0.075	0.124	0.161	0.204	0.225	0.085	0.146	0.196	0.258	0.290
	24	0.038	0.058	0.070	0.083	0.088	0.051	0.082	0.104	0.128	0.140	0.061	0.103	0.136	0.176	0.197
	28	0.024	0.036	0.042	0.049	0.052	0.035	0.054	0.067	0.081	0.087	0.044	0.072	0.093	0.118	0.130
12	16	0.080	0.124	0.154	0.186	0.200	0.097	0.158	0.203	0.254	0.279	0.109	0.183	0.241	0.310	0.345
	18	0.061	0.093	0.114	0.135	0.145	0.077	0.124	0.159	0.197	0.215	0.089	0.150	0.197	0.254	0.282
	20	0.047	0.070	0.084	0.099	0.105	0.061	0.098	0.123	0.151	0.164	0.073	0.122	0.160	0.204	0.226
	24	0.027	0.039	0.047	0.054	0.056	0.039	0.060	0.074	0.088	0.095	0.049	0.080	0.103	0.129	0.141
	28	0.018	0.024	0.028	0.031	0.032	0.025	0.037	0.045	0.052	0.056	0.033	0.053	0.066	0.081	0.087
14	16	0.066	0.099	0.120	0.141	0.150	0.084	0.133	0.167	0.205	0.223	0.097	0.161	0.208	0.264	0.291
	18	0.048	0.071	0.085	0.099	0.105	0.064	0.101	0.125	0.152	0.164	0.078	0.128	0.164	0.207	0.227
	20	0.035	0.051	0.061	0.070	0.074	0.050	0.076	0.094	0.112	0.121	0.062	0.101	0.128	0.160	0.175
	24	0.021	0.029	0.033	0.037	0.039	0.029	0.044	0.053	0.062	0.066	0.040	0.062	0.078	0.094	0.102
	28	0.013	0.017	0.019	0.021	0.021	0.020	0.028	0.032	0.036	0.038	0.026	0.039	0.048	0.057	0.061
16	16	0.054	0.078	0.093	0.108	0.114	0.072	0.111	0.137	0.165	0.177	0.086	0.140	0.179	0.222	0.243
	18	0.038	0.054	0.064	0.073	0.077	0.053	0.081	0.099	0.118	0.126	0.067	0.107	0.136	0.168	0.183
	20	0.029	0.040	0.046	0.051	0.054	0.040	0.059	0.072	0.085	0.090	0.052	0.082	0.103	0.125	0.136
	24	0.016	0.022	0.024	0.026	0.027	0.024	0.034	0.040	0.046	0.048	0.031	0.048	0.059	0.070	0.076
	28	0.010	0.012	0.014	0.015	0.015	0.015	0.021	0.024	0.026	0.027	0.021	0.031	0.036	0.042	0.044

% $M_{neg\ ext.}$

% $M_{neg\ ext.}$

TABLE 8-9 ACI-13.3.3.3, FLAT PLATES. END SPAN, EXTERIOR NEG. MOMENT % M_o COLUMNS ABOVE l_c(FT) = 10.0 PAGE 3

h	m	ℓ1 = 15					ℓ1 = 18					ℓ1 = 21				
		γ=0.50	γ=0.75	γ=1.00	γ=1.50	γ=2.00	γ=0.50	γ=0.75	γ=1.00	γ=1.50	γ=2.00	γ=0.50	γ=0.75	γ=1.00	γ=1.50	γ=2.00
6	16	0.105	0.176	0.230	0.296	0.328	0.118	0.204	0.274	0.361	0.407	0.127	0.222	0.300	0.401	0.455
	18	0.085	0.142	0.186	0.238	0.264	0.099	0.173	0.233	0.311	0.351	0.108	0.192	0.264	0.359	0.410
	20	0.069	0.115	0.149	0.189	0.209	0.083	0.146	0.197	0.262	0.297	0.093	0.167	0.230	0.316	0.364
	24	0.046	0.074	0.094	0.117	0.127	0.059	0.103	0.137	0.181	0.203	0.069	0.125	0.173	0.238	0.274
	28	0.030	0.048	0.059	0.072	0.078	0.043	0.072	0.094	0.121	0.135	0.052	0.093	0.127	0.173	0.197
8	16	0.081	0.129	0.164	0.202	0.220	0.099	0.167	0.219	0.282	0.313	0.112	0.193	0.258	0.341	0.384
	18	0.062	0.098	0.123	0.150	0.162	0.079	0.133	0.175	0.224	0.249	0.092	0.160	0.216	0.287	0.324
	20	0.047	0.073	0.091	0.110	0.119	0.064	0.106	0.138	0.176	0.195	0.077	0.133	0.179	0.238	0.269
	24	0.029	0.043	0.052	0.061	0.065	0.041	0.067	0.085	0.107	0.117	0.053	0.091	0.121	0.158	0.177
	28	0.019	0.027	0.032	0.036	0.038	0.027	0.043	0.054	0.065	0.071	0.036	0.061	0.080	0.103	0.114
10	16	0.060	0.092	0.113	0.135	0.145	0.080	0.131	0.168	0.210	0.231	0.096	0.162	0.214	0.277	0.309
	18	0.043	0.065	0.079	0.094	0.101	0.061	0.099	0.126	0.157	0.171	0.076	0.129	0.170	0.220	0.245
	20	0.033	0.049	0.058	0.067	0.071	0.047	0.075	0.094	0.116	0.126	0.061	0.102	0.134	0.172	0.191
	24	0.020	0.027	0.032	0.036	0.037	0.029	0.045	0.055	0.066	0.070	0.038	0.063	0.082	0.103	0.114
	28	0.012	0.016	0.018	0.020	0.021	0.019	0.028	0.033	0.039	0.041	0.026	0.042	0.052	0.064	0.069
12	16	0.045	0.066	0.080	0.093	0.099	0.063	0.100	0.126	0.155	0.168	0.080	0.133	0.173	0.220	0.243
	18	0.033	0.047	0.055	0.063	0.067	0.046	0.072	0.090	0.110	0.119	0.061	0.101	0.131	0.165	0.182
	20	0.024	0.034	0.039	0.044	0.046	0.036	0.055	0.067	0.080	0.085	0.047	0.076	0.098	0.123	0.135
	24	0.014	0.018	0.020	0.022	0.023	0.022	0.032	0.037	0.043	0.045	0.030	0.047	0.058	0.071	0.077
	28	0.008	0.010	0.011	0.012	0.013	0.014	0.019	0.022	0.024	0.026	0.020	0.030	0.036	0.042	0.045

h	m	ℓ1 = 24					ℓ1 = 28					ℓ1 = 32				
		γ=0.50	γ=0.75	γ=1.00	γ=1.50	γ=2.00	γ=0.50	γ=0.75	γ=1.00	γ=1.50	γ=2.00	γ=0.50	γ=0.75	γ=1.00	γ=1.50	γ=2.00
10	16	0.107	0.185	0.249	0.329	0.371	0.118	0.206	0.280	0.377	0.428	0.125	0.220	0.301	0.407	0.464
	18	0.088	0.153	0.206	0.274	0.310	0.099	0.176	0.241	0.329	0.377	0.106	0.191	0.265	0.366	0.423
	20	0.072	0.125	0.169	0.225	0.254	0.083	0.149	0.206	0.283	0.326	0.091	0.166	0.233	0.326	0.379
	24	0.048	0.083	0.111	0.146	0.164	0.059	0.107	0.147	0.202	0.233	0.068	0.124	0.176	0.249	0.292
	28	0.032	0.055	0.072	0.094	0.104	0.042	0.075	0.103	0.140	0.160	0.051	0.093	0.131	0.185	0.216
12	16	0.094	0.160	0.212	0.277	0.311	0.107	0.186	0.251	0.336	0.381	0.116	0.204	0.278	0.376	0.428
	18	0.074	0.127	0.169	0.220	0.247	0.087	0.154	0.209	0.282	0.321	0.097	0.173	0.239	0.328	0.378
	20	0.059	0.100	0.132	0.172	0.192	0.072	0.126	0.172	0.233	0.265	0.081	0.147	0.204	0.283	0.327
	24	0.036	0.061	0.080	0.103	0.114	0.048	0.084	0.115	0.153	0.174	0.058	0.104	0.145	0.202	0.233
	28	0.026	0.041	0.052	0.065	0.070	0.032	0.056	0.075	0.099	0.112	0.041	0.073	0.101	0.139	0.161
14	16	0.081	0.136	0.179	0.230	0.256	0.095	0.166	0.222	0.295	0.332	0.106	0.187	0.255	0.343	0.390
	18	0.062	0.104	0.136	0.175	0.194	0.076	0.133	0.179	0.238	0.269	0.087	0.155	0.213	0.290	0.332
	20	0.047	0.078	0.102	0.131	0.145	0.061	0.105	0.142	0.189	0.213	0.072	0.128	0.176	0.241	0.277
	24	0.031	0.049	0.062	0.077	0.084	0.038	0.066	0.088	0.116	0.130	0.048	0.086	0.118	0.161	0.185
	28	0.020	0.031	0.038	0.046	0.049	0.027	0.045	0.059	0.074	0.082	0.032	0.057	0.078	0.106	0.120
16	16	0.068	0.114	0.149	0.190	0.210	0.085	0.146	0.195	0.256	0.288	0.097	0.170	0.231	0.309	0.351
	18	0.050	0.083	0.108	0.138	0.153	0.065	0.113	0.151	0.199	0.224	0.078	0.137	0.187	0.253	0.289
	20	0.040	0.065	0.083	0.103	0.113	0.050	0.087	0.116	0.152	0.171	0.062	0.110	0.150	0.204	0.233
	24	0.025	0.039	0.048	0.058	0.062	0.033	0.055	0.072	0.092	0.102	0.039	0.069	0.095	0.128	0.146
	28	0.016	0.024	0.029	0.034	0.036	0.023	0.036	0.046	0.056	0.061	0.028	0.049	0.064	0.084	0.093

% $M_{ext.}^{neg.}$

% $M_{ext.}^{neg}$

TABLE 8-9 ACI-13.3.3.3, FLAT PLATES. END SPAN, EXTERIOR NEG. MOMENT % M_o COLUMNS ABOVE l_c(FT) = 12.0 PAGE 4

h	m	ℓ1 = 15 γ=0.50	γ=0.75	γ=1.00	γ=1.50	γ=2.00	ℓ1 = 18 γ=0.50	γ=0.75	γ=1.00	γ=1.50	γ=2.00	ℓ1 = 21 γ=0.50	γ=0.75	γ=1.00	γ=1.50	γ=2.00
6	16	0.102	0.169	0.219	0.277	0.305	0.117	0.200	0.266	0.348	0.390	0.126	0.219	0.295	0.392	0.443
	18	0.082	0.135	0.174	0.219	0.241	0.097	0.168	0.224	0.294	0.331	0.107	0.189	0.257	0.347	0.395
	20	0.067	0.108	0.137	0.171	0.187	0.082	0.140	0.187	0.245	0.275	0.092	0.163	0.223	0.303	0.346
	24	0.043	0.068	0.085	0.103	0.111	0.058	0.097	0.127	0.164	0.182	0.068	0.120	0.164	0.222	0.253
	28	0.028	0.043	0.053	0.062	0.067	0.041	0.067	0.086	0.108	0.118	0.051	0.088	0.119	0.158	0.178
8	16	0.077	0.121	0.150	0.182	0.196	0.096	0.160	0.207	0.262	0.289	0.110	0.187	0.249	0.325	0.364
	18	0.058	0.090	0.111	0.132	0.142	0.077	0.126	0.163	0.205	0.225	0.090	0.155	0.206	0.269	0.302
	20	0.044	0.067	0.081	0.096	0.103	0.061	0.099	0.127	0.158	0.173	0.075	0.127	0.169	0.220	0.246
	24	0.027	0.038	0.046	0.052	0.055	0.039	0.061	0.077	0.093	0.101	0.051	0.085	0.111	0.142	0.157
	28	0.017	0.024	0.027	0.030	0.032	0.025	0.039	0.047	0.056	0.060	0.035	0.057	0.072	0.090	0.099
10	16	0.056	0.083	0.101	0.118	0.125	0.077	0.122	0.154	0.189	0.206	0.093	0.155	0.202	0.257	0.284
	18	0.040	0.058	0.070	0.081	0.086	0.058	0.091	0.114	0.138	0.150	0.074	0.122	0.158	0.200	0.220
	20	0.030	0.043	0.050	0.057	0.060	0.044	0.068	0.084	0.101	0.109	0.058	0.095	0.122	0.154	0.169
	24	0.018	0.024	0.027	0.030	0.031	0.027	0.040	0.048	0.056	0.059	0.036	0.058	0.073	0.090	0.098
	28	0.011	0.014	0.015	0.017	0.017	0.018	0.025	0.029	0.032	0.034	0.024	0.037	0.046	0.055	0.058
12	16	0.041	0.059	0.069	0.079	0.083	0.060	0.092	0.113	0.135	0.145	0.077	0.125	0.159	0.198	0.217
	18	0.030	0.041	0.047	0.053	0.055	0.043	0.065	0.079	0.094	0.101	0.058	0.094	0.118	0.146	0.159
	20	0.022	0.029	0.033	0.036	0.038	0.033	0.049	0.058	0.067	0.071	0.044	0.070	0.088	0.107	0.116
	24	0.012	0.015	0.017	0.018	0.019	0.020	0.027	0.032	0.036	0.037	0.028	0.042	0.051	0.060	0.064
	28	0.007	0.008	0.009	0.010	0.010	0.012	0.016	0.018	0.020	0.021	0.018	0.026	0.031	0.035	0.037

h	m	ℓ1 = 24 γ=0.50	γ=0.75	γ=1.00	γ=1.50	γ=2.00	ℓ1 = 28 γ=0.50	γ=0.75	γ=1.00	γ=1.50	γ=2.00	ℓ1 = 32 γ=0.50	γ=0.75	γ=1.00	γ=1.50	γ=2.00
10	16	0.105	0.180	0.239	0.312	0.349	0.116	0.203	0.273	0.364	0.412	0.124	0.218	0.296	0.398	0.453
	18	0.086	0.147	0.195	0.256	0.287	0.097	0.171	0.233	0.314	0.358	0.105	0.188	0.259	0.355	0.408
	20	0.070	0.119	0.158	0.206	0.231	0.082	0.144	0.197	0.267	0.304	0.090	0.163	0.226	0.313	0.361
	24	0.047	0.078	0.102	0.130	0.144	0.058	0.102	0.138	0.185	0.210	0.066	0.120	0.168	0.234	0.271
	28	0.031	0.051	0.065	0.082	0.090	0.041	0.071	0.095	0.125	0.141	0.049	0.089	0.123	0.169	0.195
12	16	0.091	0.153	0.200	0.256	0.285	0.105	0.181	0.242	0.319	0.359	0.114	0.200	0.271	0.363	0.412
	18	0.072	0.120	0.156	0.200	0.221	0.085	0.148	0.199	0.263	0.297	0.095	0.169	0.231	0.313	0.358
	20	0.056	0.093	0.121	0.153	0.169	0.070	0.121	0.162	0.214	0.241	0.080	0.142	0.195	0.266	0.304
	24	0.034	0.056	0.072	0.089	0.098	0.046	0.079	0.105	0.137	0.153	0.056	0.099	0.136	0.184	0.210
	28	0.024	0.037	0.046	0.055	0.059	0.031	0.051	0.067	0.086	0.096	0.039	0.069	0.093	0.124	0.141
14	16	0.078	0.128	0.165	0.208	0.228	0.093	0.159	0.210	0.274	0.307	0.105	0.182	0.245	0.326	0.369
	18	0.059	0.096	0.123	0.154	0.169	0.074	0.126	0.167	0.217	0.242	0.085	0.149	0.203	0.271	0.308
	20	0.044	0.072	0.092	0.114	0.125	0.058	0.099	0.131	0.169	0.189	0.070	0.122	0.166	0.222	0.252
	24	0.029	0.044	0.054	0.065	0.070	0.036	0.060	0.079	0.101	0.112	0.046	0.080	0.108	0.144	0.162
	28	0.019	0.027	0.033	0.038	0.040	0.026	0.041	0.051	0.063	0.069	0.031	0.053	0.071	0.092	0.103
16	16	0.065	0.105	0.134	0.167	0.183	0.082	0.138	0.181	0.233	0.259	0.095	0.164	0.219	0.289	0.326
	18	0.047	0.076	0.097	0.119	0.130	0.063	0.106	0.138	0.177	0.197	0.076	0.131	0.176	0.232	0.262
	20	0.037	0.058	0.073	0.088	0.094	0.048	0.080	0.104	0.133	0.148	0.060	0.104	0.139	0.183	0.207
	24	0.023	0.034	0.041	0.048	0.051	0.031	0.050	0.063	0.078	0.085	0.037	0.064	0.085	0.112	0.125
	28	0.014	0.021	0.024	0.027	0.029	0.021	0.032	0.039	0.047	0.050	0.027	0.044	0.057	0.071	0.078

% $M^{neg.}_{ext.}$

% M^{neg}_{ext}

TABLE 8-10 ACI-13.3.3.3, FLAT PLATES. END SPAN, POSITIVE MOMENT % M_O NO COLUMNS ABOVE ℓ_C (FT) = 10.0 PAGE 1

h	m	ℓ1 = 15 γ=0.50	γ=0.75	γ=1.00	γ=1.50	γ=2.00	ℓ1 = 18 γ=0.50	γ=0.75	γ=1.00	γ=1.50	γ=2.00	ℓ1 = 21 γ=0.50	γ=0.75	γ=1.00	γ=1.50	γ=2.00
6	16	0.589	0.566	0.550	0.533	0.525	0.581	0.550	0.526	0.498	0.484	0.577	0.539	0.510	0.474	0.455
	18	0.597	0.580	0.569	0.556	0.551	0.590	0.564	0.545	0.523	0.512	0.585	0.553	0.528	0.496	0.480
	20	0.604	0.591	0.583	0.575	0.571	0.597	0.576	0.561	0.544	0.536	0.592	0.565	0.543	0.517	0.504
	24	0.614	0.607	0.602	0.599	0.597	0.607	0.594	0.586	0.576	0.572	0.602	0.584	0.569	0.553	0.544
	28	0.619	0.615	0.613	0.611	0.611	0.614	0.606	0.601	0.596	0.594	0.610	0.597	0.588	0.578	0.573
8	16	0.600	0.587	0.579	0.571	0.568	0.591	0.569	0.555	0.538	0.531	0.584	0.555	0.534	0.509	0.497
	18	0.608	0.599	0.594	0.589	0.587	0.600	0.584	0.573	0.562	0.557	0.593	0.570	0.553	0.534	0.525
	20	0.614	0.607	0.604	0.601	0.599	0.606	0.594	0.587	0.579	0.576	0.600	0.582	0.569	0.555	0.548
	24	0.620	0.617	0.616	0.614	0.614	0.615	0.609	0.605	0.601	0.600	0.610	0.599	0.592	0.584	0.581
	28	0.624	0.622	0.622	0.621	0.621	0.620	0.617	0.615	0.613	0.612	0.616	0.610	0.606	0.602	0.600
10	16	0.609	0.602	0.598	0.594	0.592	0.600	0.586	0.577	0.568	0.564	0.592	0.571	0.556	0.540	0.533
	18	0.615	0.611	0.608	0.606	0.605	0.608	0.598	0.592	0.586	0.584	0.601	0.585	0.574	0.563	0.558
	20	0.619	0.616	0.614	0.613	0.612	0.613	0.607	0.603	0.599	0.597	0.607	0.596	0.588	0.580	0.577
	24	0.624	0.622	0.622	0.621	0.621	0.620	0.617	0.615	0.613	0.613	0.616	0.610	0.606	0.602	0.600
	28	0.626	0.625	0.625	0.625	0.625	0.623	0.622	0.621	0.620	0.620	0.621	0.617	0.615	0.613	0.613
12	16	0.615	0.611	0.608	0.606	0.605	0.607	0.598	0.592	0.587	0.585	0.600	0.584	0.574	0.564	0.559
	18	0.619	0.617	0.615	0.614	0.614	0.614	0.608	0.604	0.601	0.599	0.607	0.597	0.590	0.583	0.580
	20	0.622	0.621	0.620	0.619	0.619	0.618	0.614	0.611	0.610	0.609	0.613	0.605	0.601	0.596	0.594
	24	0.626	0.625	0.625	0.624	0.624	0.623	0.621	0.620	0.619	0.619	0.619	0.616	0.614	0.612	0.611
	28	0.627	0.627	0.627	0.627	0.627	0.625	0.625	0.624	0.624	0.624	0.623	0.621	0.620	0.619	0.619

h	m	ℓ1 = 24 γ=0.50	γ=0.75	γ=1.00	γ=1.50	γ=2.00	ℓ1 = 28 γ=0.50	γ=0.75	γ=1.00	γ=1.50	γ=2.00	ℓ1 = 32 γ=0.50	γ=0.75	γ=1.00	γ=1.50	γ=2.00
10	16	0.586	0.559	0.538	0.514	0.503	0.581	0.547	0.520	0.488	0.471	0.577	0.539	0.508	0.469	0.448
	18	0.595	0.573	0.557	0.539	0.530	0.589	0.561	0.539	0.511	0.497	0.585	0.552	0.525	0.490	0.471
	20	0.602	0.585	0.573	0.560	0.553	0.596	0.573	0.555	0.533	0.522	0.592	0.564	0.540	0.510	0.494
	24	0.612	0.602	0.595	0.588	0.585	0.606	0.591	0.580	0.567	0.560	0.602	0.582	0.566	0.546	0.535
	28	0.618	0.612	0.608	0.604	0.603	0.613	0.604	0.597	0.589	0.586	0.610	0.596	0.585	0.572	0.565
12	16	0.593	0.571	0.556	0.539	0.532	0.586	0.558	0.536	0.510	0.497	0.582	0.548	0.521	0.487	0.470
	18	0.601	0.585	0.574	0.562	0.557	0.595	0.572	0.555	0.535	0.525	0.590	0.562	0.539	0.511	0.497
	20	0.608	0.596	0.588	0.580	0.576	0.602	0.584	0.571	0.556	0.548	0.597	0.574	0.555	0.532	0.521
	24	0.616	0.610	0.606	0.602	0.600	0.611	0.601	0.593	0.585	0.581	0.607	0.592	0.580	0.566	0.559
	28	0.621	0.617	0.615	0.613	0.612	0.618	0.611	0.607	0.603	0.600	0.614	0.604	0.597	0.589	0.585
14	16	0.599	0.582	0.571	0.559	0.554	0.592	0.568	0.551	0.531	0.521	0.586	0.556	0.534	0.506	0.492
	18	0.607	0.595	0.587	0.579	0.576	0.600	0.582	0.569	0.555	0.548	0.595	0.571	0.553	0.530	0.519
	20	0.613	0.604	0.599	0.593	0.591	0.607	0.593	0.584	0.573	0.568	0.601	0.583	0.568	0.551	0.543
	24	0.619	0.615	0.612	0.610	0.609	0.616	0.608	0.603	0.597	0.595	0.611	0.600	0.591	0.582	0.577
	28	0.623	0.621	0.619	0.618	0.618	0.620	0.616	0.613	0.610	0.609	0.617	0.610	0.606	0.600	0.598
16	16	0.604	0.591	0.583	0.574	0.571	0.597	0.577	0.563	0.548	0.541	0.591	0.565	0.546	0.523	0.512
	18	0.611	0.603	0.597	0.591	0.589	0.605	0.590	0.581	0.570	0.565	0.599	0.579	0.564	0.547	0.539
	20	0.616	0.609	0.606	0.602	0.601	0.611	0.600	0.593	0.586	0.583	0.606	0.590	0.579	0.567	0.561
	24	0.621	0.618	0.617	0.615	0.615	0.618	0.612	0.609	0.605	0.604	0.615	0.606	0.600	0.593	0.590
	28	0.624	0.623	0.622	0.621	0.621	0.622	0.619	0.617	0.615	0.615	0.619	0.614	0.611	0.607	0.606

% Mpos

% Mpos

Table 8-10.
SYMBOLS: $\gamma = l_1/l_2 = 1/\beta$; $m = l_1/c_1$; $c_2 = c_1 \cdot l_2/l_1$; h = Slab thickness; l_c = Story height. This Table furnishes the distribution of the total static design moment in end spans of flat plate construction to positive design moments, as per Section 13.3.3.3 of the ACI 318-71 Code.

%Mpos

TABLE 8-10 ACI-13.3.3.3, FLAT PLATES. END SPAN, POSITIVE MOMENT %M_O NO COLUMNS ABOVE ℓ_C(FT) = 12.0 PAGE 2

h	m	ℓ1 = 15					ℓ1 = 18					ℓ1 = 21				
		γ=0.50	γ=0.75	γ=1.00	γ=1.50	γ=2.00	γ=0.50	γ=0.75	γ=1.00	γ=1.50	γ=2.00	γ=0.50	γ=0.75	γ=1.00	γ=1.50	γ=2.00
6	16	0.590	0.570	0.556	0.542	0.535	0.582	0.553	0.531	0.506	0.494	0.577	0.542	0.514	0.480	0.463
	18	0.599	0.584	0.574	0.564	0.560	0.591	0.568	0.551	0.531	0.522	0.586	0.555	0.532	0.503	0.489
	20	0.606	0.595	0.588	0.582	0.579	0.598	0.580	0.567	0.552	0.546	0.593	0.567	0.548	0.525	0.513
	24	0.615	0.609	0.606	0.603	0.602	0.608	0.597	0.590	0.583	0.579	0.603	0.586	0.574	0.560	0.553
	28	0.620	0.617	0.616	0.614	0.614	0.615	0.609	0.605	0.601	0.599	0.611	0.600	0.592	0.584	0.580
8	16	0.602	0.591	0.585	0.579	0.576	0.593	0.574	0.561	0.548	0.542	0.586	0.559	0.540	0.518	0.508
	18	0.610	0.602	0.598	0.595	0.593	0.601	0.587	0.579	0.570	0.566	0.594	0.574	0.559	0.543	0.535
	20	0.615	0.610	0.607	0.605	0.604	0.608	0.598	0.592	0.586	0.583	0.601	0.585	0.574	0.563	0.557
	24	0.621	0.619	0.618	0.617	0.616	0.616	0.611	0.608	0.606	0.604	0.611	0.602	0.596	0.590	0.588
	28	0.624	0.623	0.623	0.622	0.622	0.621	0.618	0.617	0.616	0.615	0.617	0.612	0.609	0.606	0.605
10	16	0.611	0.605	0.602	0.599	0.598	0.602	0.590	0.583	0.576	0.573	0.594	0.575	0.563	0.549	0.544
	18	0.617	0.613	0.611	0.609	0.609	0.609	0.601	0.597	0.592	0.591	0.602	0.589	0.580	0.571	0.567
	20	0.620	0.618	0.617	0.616	0.615	0.615	0.609	0.606	0.604	0.602	0.609	0.599	0.593	0.587	0.584
	24	0.624	0.623	0.623	0.623	0.622	0.621	0.618	0.617	0.616	0.615	0.617	0.612	0.609	0.606	0.605
	28	0.626	0.626	0.626	0.626	0.626	0.624	0.623	0.622	0.622	0.622	0.622	0.619	0.617	0.616	0.616
12	16	0.617	0.613	0.612	0.610	0.610	0.609	0.602	0.598	0.594	0.592	0.602	0.588	0.580	0.572	0.569
	18	0.621	0.619	0.618	0.617	0.617	0.615	0.611	0.608	0.606	0.605	0.609	0.600	0.595	0.590	0.588
	20	0.623	0.622	0.621	0.621	0.621	0.619	0.616	0.614	0.613	0.612	0.614	0.608	0.605	0.602	0.600
	24	0.626	0.626	0.625	0.625	0.625	0.624	0.622	0.622	0.621	0.621	0.621	0.618	0.616	0.615	0.614
	28	0.628	0.627	0.627	0.627	0.627	0.626	0.625	0.625	0.625	0.625	0.624	0.622	0.622	0.621	0.621

h	m	ℓ1 = 24					ℓ1 = 28					ℓ1 = 32				
		γ=0.50	γ=0.75	γ=1.00	γ=1.50	γ=2.00	γ=0.50	γ=0.75	γ=1.00	γ=1.50	γ=2.00	γ=0.50	γ=0.75	γ=1.00	γ=1.50	γ=2.00
10	16	0.588	0.563	0.544	0.524	0.514	0.582	0.550	0.525	0.495	0.481	0.578	0.541	0.511	0.475	0.456
	18	0.596	0.577	0.563	0.548	0.541	0.590	0.564	0.544	0.520	0.508	0.586	0.554	0.529	0.497	0.481
	20	0.603	0.588	0.578	0.568	0.563	0.597	0.576	0.560	0.541	0.532	0.593	0.566	0.545	0.518	0.504
	24	0.613	0.604	0.599	0.594	0.591	0.607	0.594	0.585	0.574	0.569	0.603	0.585	0.571	0.553	0.545
	28	0.619	0.614	0.611	0.608	0.607	0.614	0.606	0.601	0.595	0.592	0.610	0.598	0.589	0.579	0.573
12	16	0.595	0.576	0.563	0.549	0.543	0.588	0.561	0.542	0.520	0.509	0.583	0.550	0.526	0.496	0.480
	18	0.603	0.589	0.580	0.571	0.567	0.596	0.576	0.561	0.544	0.537	0.591	0.565	0.544	0.520	0.508
	20	0.609	0.599	0.593	0.587	0.584	0.603	0.587	0.576	0.564	0.559	0.598	0.577	0.561	0.541	0.532
	24	0.618	0.612	0.609	0.606	0.605	0.613	0.603	0.598	0.591	0.589	0.608	0.595	0.585	0.574	0.569
	28	0.622	0.619	0.617	0.616	0.615	0.619	0.613	0.610	0.607	0.605	0.615	0.607	0.601	0.595	0.592
14	16	0.601	0.587	0.578	0.569	0.565	0.593	0.572	0.557	0.541	0.533	0.587	0.560	0.540	0.516	0.504
	18	0.608	0.599	0.593	0.587	0.584	0.602	0.586	0.575	0.564	0.559	0.596	0.574	0.559	0.540	0.531
	20	0.614	0.607	0.603	0.599	0.598	0.608	0.596	0.589	0.581	0.577	0.603	0.586	0.574	0.560	0.554
	24	0.620	0.617	0.615	0.613	0.613	0.617	0.610	0.607	0.603	0.601	0.612	0.602	0.596	0.589	0.585
	28	0.624	0.622	0.621	0.620	0.620	0.621	0.617	0.615	0.614	0.613	0.618	0.612	0.609	0.605	0.603
16	16	0.606	0.596	0.589	0.583	0.580	0.598	0.582	0.570	0.558	0.553	0.592	0.569	0.552	0.533	0.524
	18	0.613	0.606	0.602	0.598	0.596	0.606	0.594	0.587	0.578	0.575	0.600	0.583	0.571	0.557	0.551
	20	0.617	0.612	0.610	0.607	0.606	0.612	0.604	0.598	0.593	0.590	0.607	0.594	0.585	0.575	0.571
	24	0.622	0.620	0.619	0.618	0.617	0.619	0.614	0.612	0.610	0.609	0.616	0.609	0.604	0.599	0.597
	28	0.625	0.624	0.623	0.623	0.623	0.623	0.620	0.619	0.618	0.618	0.620	0.616	0.614	0.611	0.610

%Mpos

TABLE 8-10 ACI-13.3.3.3, FLAT PLATES. END SPAN, POSITIVE MOMENT % M_O COLUMNS ABOVE ℓ_c(FT) = 10.0 PAGE 3

h	m	$\ell 1$ = 15					$\ell 1$ = 18					$\ell 1$ = 21				
		γ=0.50	γ=0.75	γ=1.00	γ=1.50	γ=2.00	γ=0.50	γ=0.75	γ=1.00	γ=1.50	γ=2.00	γ=0.50	γ=0.75	γ=1.00	γ=1.50	γ=2.00
6	16	0.584	0.554	0.530	0.502	0.488	0.578	0.541	0.511	0.474	0.454	0.575	0.534	0.500	0.456	0.433
	18	0.593	0.568	0.549	0.527	0.515	0.587	0.555	0.529	0.496	0.478	0.583	0.546	0.516	0.475	0.452
	20	0.599	0.580	0.565	0.548	0.539	0.593	0.567	0.545	0.516	0.501	0.589	0.557	0.530	0.493	0.473
	24	0.610	0.597	0.589	0.579	0.574	0.604	0.585	0.570	0.551	0.542	0.599	0.576	0.555	0.527	0.511
	28	0.616	0.609	0.604	0.598	0.596	0.611	0.598	0.589	0.577	0.571	0.607	0.589	0.575	0.555	0.544
8	16	0.595	0.574	0.559	0.542	0.534	0.587	0.557	0.535	0.508	0.494	0.581	0.546	0.518	0.482	0.464
	18	0.603	0.587	0.576	0.565	0.559	0.595	0.572	0.554	0.533	0.522	0.590	0.560	0.536	0.506	0.490
	20	0.609	0.598	0.590	0.582	0.578	0.602	0.584	0.570	0.553	0.545	0.596	0.572	0.552	0.527	0.514
	24	0.617	0.611	0.607	0.603	0.601	0.612	0.601	0.592	0.583	0.579	0.607	0.590	0.577	0.561	0.553
	28	0.621	0.618	0.616	0.614	0.613	0.618	0.611	0.606	0.601	0.599	0.614	0.603	0.595	0.585	0.580
10	16	0.604	0.590	0.581	0.571	0.567	0.595	0.573	0.557	0.539	0.530	0.588	0.559	0.537	0.510	0.496
	18	0.611	0.601	0.595	0.589	0.586	0.603	0.587	0.575	0.562	0.555	0.596	0.574	0.556	0.535	0.524
	20	0.615	0.608	0.604	0.600	0.599	0.609	0.597	0.589	0.579	0.575	0.603	0.585	0.572	0.555	0.547
	24	0.621	0.617	0.616	0.614	0.613	0.617	0.610	0.606	0.601	0.599	0.613	0.602	0.594	0.585	0.580
	28	0.624	0.622	0.621	0.621	0.620	0.621	0.617	0.615	0.613	0.612	0.618	0.611	0.607	0.602	0.599
12	16	0.610	0.601	0.595	0.589	0.587	0.602	0.586	0.575	0.563	0.557	0.595	0.572	0.555	0.535	0.525
	18	0.615	0.609	0.605	0.602	0.601	0.610	0.598	0.591	0.582	0.578	0.603	0.586	0.573	0.558	0.551
	20	0.619	0.615	0.612	0.610	0.610	0.614	0.606	0.600	0.595	0.593	0.609	0.596	0.587	0.576	0.571
	24	0.623	0.622	0.621	0.620	0.619	0.620	0.616	0.613	0.611	0.610	0.616	0.609	0.604	0.599	0.596
	28	0.626	0.625	0.624	0.624	0.624	0.623	0.621	0.620	0.619	0.618	0.621	0.617	0.614	0.611	0.610

h	m	$\ell 1$ = 24					$\ell 1$ = 28					$\ell 1$ = 32				
		γ=0.50	γ=0.75	γ=1.00	γ=1.50	γ=2.00	γ=0.50	γ=0.75	γ=1.00	γ=1.50	γ=2.00	γ=0.50	γ=0.75	γ=1.00	γ=1.50	γ=2.00
10	16	0.583	0.550	0.522	0.488	0.469	0.579	0.540	0.509	0.467	0.445	0.576	0.534	0.500	0.454	0.429
	18	0.592	0.564	0.541	0.511	0.496	0.587	0.554	0.525	0.487	0.467	0.583	0.547	0.515	0.472	0.447
	20	0.598	0.575	0.556	0.532	0.520	0.593	0.565	0.541	0.507	0.489	0.590	0.558	0.529	0.489	0.466
	24	0.608	0.593	0.581	0.566	0.558	0.604	0.583	0.566	0.542	0.529	0.600	0.576	0.554	0.522	0.503
	28	0.615	0.606	0.598	0.589	0.584	0.611	0.597	0.585	0.569	0.560	0.608	0.589	0.573	0.550	0.536
12	16	0.589	0.560	0.538	0.510	0.496	0.583	0.549	0.521	0.485	0.465	0.579	0.541	0.509	0.467	0.445
	18	0.597	0.575	0.557	0.534	0.523	0.592	0.563	0.539	0.508	0.491	0.588	0.555	0.526	0.488	0.467
	20	0.604	0.586	0.572	0.555	0.546	0.598	0.575	0.555	0.529	0.515	0.594	0.566	0.541	0.508	0.489
	24	0.614	0.603	0.595	0.585	0.580	0.609	0.593	0.580	0.563	0.554	0.604	0.584	0.567	0.542	0.529
	28	0.618	0.611	0.607	0.601	0.599	0.616	0.605	0.597	0.587	0.581	0.612	0.598	0.586	0.569	0.560
14	16	0.595	0.571	0.552	0.530	0.519	0.588	0.558	0.533	0.502	0.486	0.583	0.549	0.520	0.482	0.461
	18	0.603	0.585	0.571	0.554	0.546	0.596	0.572	0.552	0.527	0.513	0.592	0.562	0.538	0.504	0.486
	20	0.609	0.596	0.585	0.573	0.567	0.603	0.584	0.568	0.548	0.537	0.598	0.574	0.553	0.525	0.510
	24	0.616	0.608	0.602	0.596	0.593	0.613	0.601	0.591	0.579	0.573	0.609	0.592	0.578	0.560	0.550
	28	0.621	0.616	0.613	0.609	0.608	0.618	0.610	0.604	0.597	0.594	0.615	0.605	0.596	0.584	0.577
16	16	0.600	0.580	0.565	0.548	0.539	0.593	0.567	0.545	0.519	0.505	0.587	0.556	0.530	0.496	0.478
	18	0.608	0.593	0.583	0.570	0.563	0.601	0.581	0.564	0.544	0.533	0.596	0.570	0.549	0.520	0.505
	20	0.612	0.601	0.594	0.585	0.581	0.608	0.592	0.579	0.564	0.555	0.603	0.582	0.564	0.541	0.529
	24	0.618	0.612	0.608	0.604	0.602	0.615	0.605	0.598	0.590	0.586	0.612	0.599	0.589	0.574	0.566
	28	0.622	0.619	0.617	0.615	0.614	0.620	0.614	0.610	0.605	0.603	0.617	0.608	0.602	0.593	0.589

% Mpos

%Mpos

%Mpos

TABLE 8-10 ACI-13.3.3.3, FLAT PLATES. END SPAN, POSITIVE MOMENT % M_o COLUMNS ABOVE ℓ_c (FT) = 12.0 PAGE 4

		ℓ1 = 15					ℓ1 = 18					ℓ1 = 21				
h	m	γ=0.50	γ=0.75	γ=1.00	γ=1.50	γ=2.00	γ=0.50	γ=0.75	γ=1.00	γ=1.50	γ=2.00	γ=0.50	γ=0.75	γ=1.00	γ=1.50	γ=2.00
6	16	0.585	0.557	0.535	0.510	0.498	0.579	0.543	0.515	0.480	0.461	0.575	0.535	0.502	0.460	0.438
	18	0.594	0.571	0.554	0.535	0.525	0.587	0.557	0.533	0.502	0.487	0.583	0.548	0.518	0.480	0.459
	20	0.601	0.583	0.570	0.556	0.549	0.594	0.569	0.549	0.524	0.511	0.590	0.559	0.533	0.499	0.480
	24	0.611	0.600	0.593	0.585	0.581	0.604	0.588	0.574	0.559	0.551	0.600	0.578	0.559	0.534	0.520
	28	0.617	0.611	0.607	0.602	0.601	0.612	0.601	0.592	0.583	0.578	0.607	0.591	0.578	0.561	0.553
8	16	0.596	0.577	0.565	0.551	0.545	0.588	0.561	0.540	0.516	0.505	0.582	0.549	0.522	0.489	0.473
	18	0.604	0.591	0.582	0.572	0.568	0.596	0.575	0.559	0.541	0.532	0.590	0.563	0.541	0.513	0.499
	20	0.610	0.600	0.594	0.588	0.585	0.603	0.586	0.575	0.561	0.555	0.597	0.575	0.557	0.535	0.523
	24	0.618	0.613	0.610	0.607	0.605	0.613	0.603	0.596	0.589	0.586	0.607	0.593	0.582	0.568	0.562
	28	0.622	0.619	0.618	0.616	0.616	0.618	0.613	0.609	0.605	0.603	0.614	0.605	0.598	0.590	0.587
10	16	0.605	0.593	0.586	0.579	0.575	0.596	0.577	0.563	0.548	0.541	0.589	0.563	0.542	0.519	0.507
	18	0.612	0.604	0.599	0.594	0.592	0.604	0.590	0.580	0.570	0.565	0.598	0.577	0.561	0.543	0.534
	20	0.616	0.611	0.608	0.605	0.603	0.610	0.600	0.593	0.586	0.582	0.604	0.588	0.577	0.563	0.557
	24	0.622	0.619	0.618	0.616	0.616	0.618	0.612	0.609	0.605	0.604	0.614	0.604	0.598	0.591	0.587
	28	0.625	0.623	0.623	0.622	0.622	0.622	0.619	0.617	0.615	0.615	0.619	0.613	0.610	0.606	0.604
12	16	0.612	0.604	0.600	0.595	0.594	0.604	0.590	0.581	0.571	0.567	0.596	0.575	0.561	0.544	0.536
	18	0.616	0.612	0.609	0.607	0.606	0.611	0.601	0.595	0.589	0.586	0.604	0.589	0.578	0.566	0.561
	20	0.620	0.617	0.615	0.614	0.613	0.615	0.608	0.604	0.600	0.599	0.610	0.599	0.592	0.583	0.579
	24	0.624	0.623	0.622	0.621	0.621	0.621	0.617	0.616	0.614	0.613	0.617	0.611	0.607	0.603	0.602
	28	0.626	0.626	0.625	0.625	0.625	0.624	0.622	0.621	0.621	0.620	0.622	0.618	0.616	0.614	0.613

		ℓ1 = 24					ℓ1 = 28					ℓ1 = 32				
h	m	γ=0.50	γ=0.75	γ=1.00	γ=1.50	γ=2.00	γ=0.50	γ=0.75	γ=1.00	γ=1.50	γ=2.00	γ=0.50	γ=0.75	γ=1.00	γ=1.50	γ=2.00
10	16	0.584	0.552	0.526	0.495	0.479	0.579	0.542	0.511	0.472	0.452	0.576	0.535	0.502	0.458	0.434
	18	0.592	0.566	0.545	0.519	0.506	0.587	0.555	0.529	0.494	0.475	0.584	0.548	0.518	0.476	0.454
	20	0.599	0.578	0.561	0.540	0.530	0.594	0.567	0.544	0.514	0.498	0.590	0.559	0.532	0.495	0.474
	24	0.609	0.596	0.585	0.573	0.567	0.604	0.586	0.570	0.550	0.539	0.601	0.577	0.557	0.529	0.513
	28	0.616	0.607	0.601	0.594	0.591	0.612	0.599	0.588	0.575	0.569	0.608	0.591	0.576	0.556	0.545
12	16	0.590	0.564	0.543	0.519	0.507	0.584	0.551	0.525	0.492	0.475	0.580	0.543	0.512	0.473	0.452
	18	0.598	0.578	0.562	0.543	0.534	0.593	0.566	0.544	0.516	0.501	0.588	0.557	0.530	0.494	0.475
	20	0.605	0.589	0.577	0.563	0.556	0.599	0.577	0.560	0.537	0.526	0.595	0.568	0.545	0.515	0.498
	24	0.614	0.605	0.598	0.591	0.587	0.609	0.595	0.584	0.570	0.563	0.605	0.586	0.571	0.550	0.539
	28	0.619	0.613	0.610	0.606	0.604	0.616	0.607	0.600	0.592	0.588	0.612	0.600	0.589	0.576	0.569
14	16	0.596	0.574	0.558	0.540	0.531	0.589	0.561	0.539	0.511	0.497	0.584	0.551	0.524	0.489	0.471
	18	0.604	0.588	0.576	0.563	0.556	0.597	0.575	0.558	0.536	0.525	0.593	0.565	0.542	0.512	0.497
	20	0.610	0.598	0.590	0.580	0.576	0.604	0.587	0.573	0.556	0.548	0.599	0.577	0.558	0.534	0.521
	24	0.617	0.610	0.606	0.601	0.599	0.614	0.603	0.595	0.586	0.581	0.609	0.595	0.583	0.567	0.559
	28	0.621	0.617	0.615	0.613	0.612	0.618	0.612	0.607	0.602	0.600	0.616	0.606	0.599	0.590	0.585
16	16	0.601	0.584	0.571	0.557	0.551	0.594	0.570	0.551	0.529	0.518	0.588	0.559	0.535	0.505	0.489
	18	0.609	0.597	0.588	0.578	0.573	0.602	0.584	0.570	0.553	0.545	0.597	0.573	0.554	0.529	0.517
	20	0.613	0.604	0.598	0.592	0.589	0.609	0.595	0.584	0.572	0.566	0.603	0.585	0.569	0.550	0.540
	24	0.619	0.615	0.612	0.609	0.607	0.616	0.608	0.602	0.596	0.593	0.613	0.602	0.593	0.581	0.575
	28	0.623	0.620	0.619	0.617	0.617	0.620	0.616	0.612	0.609	0.608	0.618	0.610	0.605	0.599	0.596

%Mpos

%Mneg

h	m	$\ell_1 = 15$					$\ell_1 = 18$					$\ell_1 = 21$				
		γ=0.50	γ=0.75	γ=1.00	γ=1.50	γ=2.00	γ=0.50	γ=0.75	γ=1.00	γ=1.50	γ=2.00	γ=0.50	γ=0.75	γ=1.00	γ=1.50	γ=2.00
6	16	0.735	0.727	0.721	0.715	0.712	0.732	0.721	0.713	0.703	0.698	0.731	0.717	0.707	0.694	0.687
	18	0.738	0.732	0.728	0.723	0.721	0.735	0.726	0.719	0.711	0.708	0.734	0.722	0.713	0.702	0.696
	20	0.740	0.736	0.733	0.730	0.729	0.738	0.731	0.725	0.719	0.716	0.736	0.726	0.719	0.709	0.705
	24	0.744	0.741	0.740	0.738	0.738	0.742	0.737	0.734	0.731	0.729	0.740	0.733	0.728	0.722	0.719
	28	0.746	0.744	0.744	0.743	0.743	0.744	0.741	0.739	0.738	0.737	0.742	0.738	0.735	0.731	0.729
8	16	0.739	0.734	0.732	0.729	0.728	0.736	0.728	0.723	0.717	0.714	0.733	0.723	0.715	0.706	0.702
	18	0.742	0.739	0.737	0.735	0.734	0.739	0.733	0.729	0.725	0.723	0.736	0.728	0.722	0.715	0.712
	20	0.744	0.742	0.740	0.739	0.739	0.741	0.737	0.734	0.732	0.730	0.739	0.732	0.728	0.723	0.720
	24	0.746	0.745	0.745	0.744	0.744	0.744	0.742	0.741	0.739	0.739	0.742	0.739	0.736	0.733	0.732
	28	0.747	0.747	0.747	0.746	0.746	0.746	0.745	0.744	0.744	0.743	0.745	0.743	0.741	0.740	0.739
10	16	0.742	0.740	0.738	0.737	0.736	0.739	0.734	0.731	0.727	0.726	0.736	0.729	0.723	0.718	0.715
	18	0.744	0.743	0.742	0.741	0.741	0.742	0.738	0.736	0.734	0.733	0.739	0.734	0.730	0.726	0.724
	20	0.746	0.745	0.744	0.744	0.743	0.744	0.741	0.740	0.739	0.738	0.742	0.737	0.735	0.732	0.731
	24	0.747	0.747	0.747	0.746	0.746	0.746	0.745	0.744	0.744	0.743	0.745	0.742	0.741	0.740	0.739
	28	0.748	0.748	0.748	0.748	0.748	0.747	0.747	0.746	0.746	0.746	0.746	0.745	0.744	0.744	0.743
12	16	0.744	0.743	0.742	0.741	0.741	0.742	0.738	0.736	0.734	0.734	0.739	0.733	0.730	0.726	0.724
	18	0.746	0.745	0.744	0.744	0.744	0.744	0.742	0.740	0.739	0.739	0.742	0.738	0.735	0.733	0.732
	20	0.747	0.746	0.746	0.746	0.746	0.745	0.744	0.743	0.742	0.742	0.744	0.741	0.739	0.738	0.737
	24	0.748	0.748	0.748	0.748	0.748	0.747	0.746	0.746	0.746	0.746	0.746	0.745	0.744	0.743	0.743
	28	0.749	0.749	0.749	0.748	0.748	0.748	0.748	0.748	0.747	0.747	0.747	0.747	0.746	0.746	0.746

h	m	$\ell_1 = 24$					$\ell_1 = 28$					$\ell_1 = 32$				
		γ=0.50	γ=0.75	γ=1.00	γ=1.50	γ=2.00	γ=0.50	γ=0.75	γ=1.00	γ=1.50	γ=2.00	γ=0.50	γ=0.75	γ=1.00	γ=1.50	γ=2.00
10	16	0.734	0.724	0.717	0.708	0.704	0.732	0.720	0.710	0.699	0.693	0.731	0.717	0.706	0.692	0.685
	18	0.737	0.729	0.724	0.717	0.714	0.735	0.725	0.717	0.707	0.702	0.734	0.722	0.712	0.700	0.693
	20	0.740	0.734	0.729	0.725	0.722	0.738	0.729	0.723	0.715	0.711	0.736	0.726	0.718	0.707	0.701
	24	0.743	0.740	0.737	0.735	0.734	0.741	0.736	0.732	0.727	0.725	0.740	0.733	0.727	0.720	0.716
	28	0.745	0.743	0.742	0.741	0.740	0.744	0.740	0.738	0.735	0.734	0.742	0.737	0.734	0.729	0.727
12	16	0.736	0.729	0.723	0.717	0.715	0.734	0.724	0.716	0.707	0.702	0.732	0.720	0.711	0.699	0.693
	18	0.739	0.734	0.730	0.726	0.724	0.737	0.729	0.723	0.716	0.712	0.735	0.725	0.717	0.707	0.702
	20	0.742	0.738	0.735	0.732	0.730	0.740	0.733	0.729	0.723	0.721	0.738	0.730	0.723	0.715	0.711
	24	0.745	0.743	0.741	0.740	0.739	0.743	0.739	0.737	0.734	0.732	0.741	0.736	0.732	0.727	0.724
	28	0.746	0.745	0.744	0.744	0.743	0.745	0.743	0.741	0.740	0.739	0.744	0.740	0.738	0.735	0.734
14	16	0.739	0.733	0.729	0.724	0.722	0.736	0.728	0.721	0.714	0.711	0.734	0.723	0.715	0.705	0.700
	18	0.741	0.737	0.734	0.732	0.730	0.739	0.733	0.728	0.723	0.720	0.737	0.729	0.722	0.714	0.710
	20	0.743	0.740	0.739	0.737	0.736	0.741	0.736	0.733	0.729	0.728	0.739	0.733	0.728	0.722	0.719
	24	0.746	0.744	0.743	0.742	0.742	0.745	0.742	0.740	0.738	0.737	0.743	0.739	0.736	0.732	0.731
	28	0.747	0.746	0.746	0.745	0.745	0.746	0.745	0.744	0.743	0.742	0.745	0.743	0.741	0.739	0.738
16	16	0.740	0.736	0.733	0.730	0.728	0.738	0.731	0.726	0.720	0.718	0.736	0.726	0.720	0.711	0.707
	18	0.743	0.740	0.738	0.736	0.735	0.741	0.736	0.732	0.728	0.726	0.739	0.732	0.726	0.720	0.717
	20	0.745	0.742	0.741	0.740	0.739	0.743	0.739	0.737	0.734	0.733	0.741	0.736	0.732	0.727	0.725
	24	0.747	0.745	0.745	0.744	0.744	0.745	0.743	0.742	0.741	0.740	0.744	0.741	0.739	0.737	0.735
	28	0.748	0.747	0.747	0.747	0.746	0.747	0.746	0.745	0.744	0.744	0.746	0.744	0.743	0.742	0.741

Table 8-11.
SYMBOLS: $\gamma = l_1/l_2 = 1/\beta$; $m = l_1/c_1$; $c_2 = c_1 \cdot l_2/l_1$; h = Slab thickness; l_c = Story height. This Table furnishes the distribution of the total static design moment in end spans of flat plate construction to interior negative design moments, as per Section 13.3.3.3 of ACI 318-71 Code.

%Mneg

TABLE 8-11 ACI-13.3.3.3, FLAT PLATES. END SPAN, INTERIOR NEG. MOMENT % M_O NO COLUMNS ABOVE ℓ_c(FT) = 12.0 PAGE 2

h	m	ℓ_1 = 15 γ=0.50	γ=0.75	γ=1.00	γ=1.50	γ=2.00	ℓ_1 = 18 γ=0.50	γ=0.75	γ=1.00	γ=1.50	γ=2.00	ℓ_1 = 21 γ=0.50	γ=0.75	γ=1.00	γ=1.50	γ=2.00
6	16	0.736	0.728	0.723	0.718	0.716	0.733	0.722	0.714	0.706	0.701	0.731	0.718	0.708	0.696	0.690
	18	0.739	0.733	0.730	0.726	0.725	0.736	0.727	0.721	0.714	0.711	0.734	0.723	0.715	0.704	0.699
	20	0.741	0.737	0.735	0.732	0.731	0.738	0.732	0.727	0.722	0.720	0.736	0.727	0.720	0.712	0.708
	24	0.744	0.742	0.741	0.740	0.740	0.742	0.738	0.735	0.733	0.732	0.740	0.734	0.730	0.725	0.722
	28	0.746	0.745	0.745	0.744	0.744	0.744	0.742	0.741	0.739	0.739	0.743	0.739	0.736	0.733	0.732
8	16	0.740	0.736	0.734	0.731	0.730	0.736	0.730	0.725	0.720	0.718	0.734	0.724	0.717	0.710	0.706
	18	0.742	0.740	0.738	0.737	0.737	0.739	0.734	0.731	0.728	0.727	0.737	0.730	0.724	0.718	0.716
	20	0.744	0.743	0.742	0.741	0.740	0.742	0.738	0.736	0.734	0.733	0.739	0.734	0.730	0.726	0.724
	24	0.747	0.746	0.745	0.745	0.745	0.745	0.743	0.742	0.741	0.741	0.743	0.740	0.738	0.735	0.735
	28	0.748	0.747	0.747	0.747	0.747	0.747	0.746	0.745	0.745	0.744	0.745	0.743	0.742	0.741	0.741
10	16	0.743	0.741	0.740	0.739	0.738	0.740	0.735	0.733	0.730	0.729	0.737	0.730	0.726	0.721	0.719
	18	0.745	0.744	0.743	0.742	0.742	0.742	0.739	0.738	0.736	0.736	0.740	0.735	0.732	0.729	0.727
	20	0.746	0.745	0.745	0.745	0.744	0.744	0.742	0.741	0.740	0.740	0.742	0.739	0.736	0.734	0.733
	24	0.748	0.747	0.747	0.747	0.747	0.746	0.746	0.745	0.745	0.744	0.745	0.743	0.742	0.741	0.741
	28	0.748	0.748	0.748	0.748	0.748	0.748	0.747	0.747	0.747	0.747	0.747	0.746	0.745	0.745	0.745
12	16	0.745	0.744	0.743	0.743	0.742	0.742	0.740	0.738	0.737	0.736	0.740	0.735	0.732	0.729	0.728
	18	0.746	0.746	0.745	0.745	0.745	0.744	0.743	0.742	0.741	0.741	0.742	0.739	0.737	0.735	0.735
	20	0.747	0.747	0.747	0.746	0.746	0.746	0.745	0.744	0.744	0.743	0.744	0.742	0.741	0.740	0.739
	24	0.748	0.748	0.748	0.748	0.748	0.747	0.747	0.747	0.746	0.746	0.746	0.745	0.745	0.744	0.744
	28	0.749	0.749	0.749	0.749	0.749	0.748	0.748	0.748	0.748	0.748	0.748	0.747	0.747	0.747	0.746

h	m	ℓ_1 = 24 γ=0.50	γ=0.75	γ=1.00	γ=1.50	γ=2.00	ℓ_1 = 28 γ=0.50	γ=0.75	γ=1.00	γ=1.50	γ=2.00	ℓ_1 = 32 γ=0.50	γ=0.75	γ=1.00	γ=1.50	γ=2.00
10	16	0.735	0.726	0.719	0.712	0.708	0.732	0.721	0.712	0.702	0.696	0.731	0.718	0.707	0.694	0.688
	18	0.738	0.731	0.726	0.720	0.718	0.735	0.726	0.719	0.710	0.706	0.734	0.723	0.714	0.702	0.696
	20	0.740	0.735	0.731	0.727	0.726	0.738	0.730	0.725	0.718	0.715	0.736	0.727	0.719	0.710	0.705
	24	0.744	0.741	0.739	0.737	0.736	0.742	0.737	0.733	0.730	0.728	0.740	0.734	0.728	0.722	0.719
	28	0.746	0.744	0.743	0.742	0.741	0.744	0.741	0.739	0.737	0.736	0.743	0.738	0.735	0.731	0.729
12	16	0.737	0.730	0.726	0.721	0.719	0.735	0.725	0.718	0.710	0.707	0.733	0.721	0.712	0.702	0.696
	18	0.740	0.735	0.732	0.729	0.727	0.738	0.730	0.725	0.719	0.716	0.736	0.726	0.719	0.710	0.706
	20	0.742	0.739	0.736	0.734	0.733	0.740	0.734	0.730	0.726	0.724	0.738	0.731	0.725	0.718	0.715
	24	0.745	0.743	0.742	0.741	0.741	0.743	0.740	0.738	0.736	0.735	0.742	0.737	0.734	0.730	0.728
	28	0.747	0.746	0.745	0.745	0.744	0.746	0.744	0.743	0.741	0.741	0.744	0.741	0.739	0.737	0.736
14	16	0.739	0.734	0.731	0.728	0.726	0.737	0.729	0.724	0.718	0.715	0.734	0.725	0.717	0.709	0.705
	18	0.742	0.738	0.736	0.734	0.733	0.740	0.734	0.730	0.726	0.724	0.737	0.730	0.724	0.718	0.714
	20	0.744	0.742	0.740	0.739	0.738	0.742	0.738	0.735	0.732	0.731	0.740	0.734	0.730	0.725	0.723
	24	0.746	0.745	0.744	0.744	0.743	0.745	0.743	0.741	0.740	0.739	0.743	0.740	0.737	0.735	0.734
	28	0.747	0.747	0.746	0.746	0.746	0.746	0.745	0.744	0.744	0.744	0.745	0.743	0.742	0.741	0.740
16	16	0.741	0.737	0.735	0.733	0.732	0.738	0.732	0.728	0.724	0.722	0.736	0.728	0.722	0.715	0.712
	18	0.744	0.741	0.740	0.738	0.738	0.741	0.737	0.734	0.731	0.730	0.739	0.733	0.729	0.724	0.721
	20	0.745	0.743	0.742	0.742	0.741	0.743	0.740	0.738	0.736	0.736	0.741	0.737	0.734	0.730	0.729
	24	0.747	0.746	0.746	0.745	0.745	0.746	0.744	0.743	0.742	0.742	0.745	0.742	0.740	0.739	0.738
	28	0.748	0.748	0.747	0.747	0.747	0.747	0.746	0.746	0.745	0.745	0.746	0.745	0.744	0.743	0.743

%Mneg

%Mneg

TABLE 8-11 ACI-13.3.3.3, FLAT PLATES. END SPAN, INTERIOR NEG. MOMENT % M_O COLUMNS ABOVE ℓ_c (FT) = 10.0 PAGE 3

h	m	ℓ_1 = 15 γ=0.50	γ=0.75	γ=1.00	γ=1.50	γ=2.00	ℓ_1 = 18 γ=0.50	γ=0.75	γ=1.00	γ=1.50	γ=2.00	ℓ_1 = 21 γ=0.50	γ=0.75	γ=1.00	γ=1.50	γ=2.00
6	16	0.733	0.722	0.714	0.704	0.599	0.731	0.718	0.707	0.694	0.687	0.730	0.715	0.703	0.688	0.679
	18	0.736	0.728	0.721	0.713	0.709	0.734	0.723	0.714	0.702	0.695	0.733	0.720	0.709	0.694	0.686
	20	0.739	0.732	0.727	0.720	0.717	0.737	0.727	0.719	0.709	0.704	0.735	0.724	0.714	0.701	0.693
	24	0.742	0.738	0.735	0.731	0.730	0.740	0.734	0.728	0.722	0.718	0.739	0.730	0.723	0.713	0.707
	28	0.745	0.742	0.740	0.738	0.737	0.743	0.738	0.735	0.731	0.729	0.741	0.735	0.730	0.723	0.719
8	16	0.737	0.730	0.724	0.718	0.716	0.734	0.724	0.716	0.706	0.701	0.732	0.720	0.710	0.697	0.690
	18	0.740	0.734	0.731	0.726	0.724	0.737	0.729	0.723	0.715	0.711	0.735	0.725	0.716	0.705	0.700
	20	0.742	0.738	0.735	0.732	0.731	0.740	0.733	0.728	0.722	0.719	0.738	0.729	0.722	0.713	0.708
	24	0.745	0.743	0.741	0.740	0.739	0.743	0.739	0.736	0.733	0.731	0.741	0.735	0.731	0.725	0.722
	28	0.747	0.745	0.745	0.744	0.744	0.745	0.743	0.741	0.739	0.739	0.744	0.740	0.737	0.734	0.732
10	16	0.740	0.735	0.732	0.729	0.727	0.737	0.729	0.724	0.717	0.714	0.735	0.724	0.717	0.707	0.702
	18	0.743	0.739	0.737	0.735	0.734	0.740	0.734	0.730	0.725	0.723	0.738	0.730	0.723	0.716	0.712
	20	0.744	0.742	0.740	0.739	0.738	0.742	0.738	0.735	0.732	0.730	0.740	0.734	0.729	0.723	0.720
	24	0.746	0.745	0.745	0.744	0.744	0.745	0.743	0.741	0.739	0.739	0.744	0.740	0.737	0.734	0.732
	28	0.748	0.747	0.747	0.746	0.746	0.746	0.745	0.744	0.743	0.743	0.745	0.743	0.741	0.740	0.739
12	16	0.743	0.739	0.737	0.735	0.734	0.740	0.734	0.730	0.726	0.724	0.737	0.729	0.723	0.716	0.712
	18	0.744	0.742	0.741	0.740	0.739	0.742	0.738	0.736	0.733	0.731	0.740	0.734	0.729	0.724	0.721
	20	0.746	0.744	0.743	0.743	0.742	0.744	0.741	0.739	0.737	0.736	0.742	0.738	0.734	0.730	0.729
	24	0.747	0.747	0.746	0.746	0.746	0.746	0.745	0.744	0.743	0.742	0.745	0.742	0.740	0.739	0.738
	28	0.748	0.748	0.748	0.748	0.747	0.747	0.747	0.746	0.746	0.745	0.746	0.745	0.744	0.743	0.743

h	m	ℓ_1 = 24 γ=0.50	γ=0.75	γ=1.00	γ=1.50	γ=2.00	ℓ_1 = 28 γ=0.50	γ=0.75	γ=1.00	γ=1.50	γ=2.00	ℓ_1 = 32 γ=0.50	γ=0.75	γ=1.00	γ=1.50	γ=2.00
10	16	0.733	0.721	0.711	0.699	0.692	0.731	0.718	0.706	0.691	0.684	0.730	0.716	0.703	0.687	0.678
	18	0.736	0.726	0.718	0.707	0.702	0.734	0.722	0.712	0.699	0.691	0.733	0.720	0.709	0.693	0.684
	20	0.738	0.730	0.723	0.715	0.710	0.737	0.726	0.718	0.706	0.699	0.735	0.724	0.714	0.699	0.691
	24	0.742	0.737	0.732	0.727	0.724	0.740	0.733	0.727	0.718	0.714	0.739	0.730	0.722	0.711	0.704
	28	0.744	0.741	0.738	0.735	0.733	0.743	0.738	0.734	0.728	0.725	0.742	0.735	0.729	0.721	0.716
12	16	0.735	0.725	0.717	0.707	0.702	0.733	0.721	0.711	0.698	0.691	0.732	0.718	0.707	0.692	0.684
	18	0.738	0.730	0.723	0.716	0.711	0.736	0.726	0.717	0.706	0.700	0.735	0.723	0.713	0.699	0.691
	20	0.740	0.734	0.729	0.723	0.720	0.738	0.730	0.723	0.714	0.709	0.737	0.727	0.718	0.706	0.699
	24	0.744	0.740	0.737	0.734	0.732	0.742	0.736	0.732	0.726	0.723	0.741	0.733	0.727	0.718	0.714
	28	0.745	0.743	0.741	0.739	0.739	0.745	0.741	0.738	0.734	0.732	0.743	0.738	0.734	0.728	0.725
14	16	0.737	0.729	0.722	0.714	0.710	0.735	0.724	0.715	0.704	0.698	0.733	0.721	0.710	0.697	0.689
	18	0.740	0.733	0.729	0.723	0.720	0.738	0.729	0.722	0.713	0.708	0.736	0.726	0.717	0.705	0.698
	20	0.742	0.737	0.734	0.729	0.727	0.740	0.733	0.728	0.720	0.717	0.738	0.730	0.722	0.712	0.707
	24	0.745	0.742	0.740	0.738	0.737	0.744	0.739	0.736	0.732	0.729	0.742	0.736	0.731	0.725	0.721
	28	0.746	0.745	0.744	0.742	0.742	0.745	0.742	0.740	0.738	0.737	0.744	0.741	0.737	0.733	0.731
16	16	0.739	0.732	0.727	0.720	0.717	0.736	0.727	0.719	0.710	0.705	0.734	0.723	0.714	0.702	0.695
	18	0.742	0.737	0.733	0.728	0.726	0.739	0.732	0.726	0.719	0.715	0.737	0.728	0.721	0.710	0.705
	20	0.743	0.739	0.737	0.734	0.732	0.742	0.736	0.732	0.726	0.723	0.740	0.733	0.726	0.718	0.714
	24	0.746	0.743	0.742	0.741	0.740	0.744	0.741	0.738	0.735	0.734	0.743	0.739	0.735	0.730	0.727
	28	0.747	0.746	0.745	0.744	0.744	0.746	0.744	0.742	0.741	0.740	0.745	0.742	0.740	0.737	0.735

% Mneg

%Mneg

TABLE 8-11 ACI-13.3.3.3, FLAT PLATES. END SPAN, INTERIOR NEG. MOMENT % M_o COLUMNS ABOVE ℓ_c (FT) = 12.0 PAGE 4

h	m	ℓ_1 = 15					ℓ_1 = 18					ℓ_1 = 21				
		γ=0.50	γ=0.75	γ=1.00	γ=1.50	γ=2.00	γ=0.50	γ=0.75	γ=1.00	γ=1.50	γ=2.00	γ=0.50	γ=0.75	γ=1.00	γ=1.50	γ=2.00
6	16	0.734	0.723	0.716	0.707	0.702	0.731	0.719	0.709	0.696	0.689	0.730	0.716	0.704	0.689	0.681
	18	0.737	0.729	0.723	0.716	0.712	0.734	0.724	0.715	0.704	0.699	0.733	0.720	0.710	0.696	0.689
	20	0.739	0.733	0.728	0.723	0.721	0.737	0.728	0.721	0.712	0.707	0.735	0.724	0.715	0.703	0.696
	24	0.743	0.739	0.736	0.734	0.732	0.741	0.735	0.730	0.724	0.721	0.739	0.731	0.724	0.715	0.711
	28	0.745	0.743	0.741	0.740	0.739	0.743	0.739	0.736	0.733	0.731	0.742	0.736	0.731	0.725	0.722
8	16	0.738	0.731	0.726	0.721	0.719	0.735	0.725	0.718	0.709	0.705	0.733	0.721	0.711	0.699	0.693
	18	0.740	0.736	0.732	0.729	0.728	0.738	0.730	0.724	0.718	0.715	0.736	0.726	0.718	0.708	0.703
	20	0.743	0.739	0.737	0.735	0.734	0.740	0.734	0.730	0.725	0.723	0.738	0.730	0.723	0.715	0.712
	24	0.745	0.744	0.742	0.741	0.741	0.743	0.740	0.738	0.735	0.734	0.742	0.736	0.732	0.728	0.725
	28	0.747	0.746	0.745	0.745	0.745	0.746	0.743	0.742	0.741	0.740	0.744	0.741	0.738	0.735	0.734
10	16	0.741	0.737	0.734	0.731	0.730	0.738	0.731	0.726	0.720	0.718	0.735	0.726	0.718	0.710	0.706
	18	0.743	0.740	0.739	0.737	0.736	0.740	0.735	0.732	0.728	0.726	0.738	0.731	0.725	0.719	0.716
	20	0.745	0.743	0.742	0.741	0.740	0.743	0.739	0.736	0.734	0.733	0.740	0.735	0.731	0.726	0.723
	24	0.747	0.746	0.745	0.745	0.745	0.745	0.743	0.742	0.741	0.740	0.744	0.741	0.738	0.735	0.734
	28	0.748	0.747	0.747	0.747	0.747	0.747	0.746	0.745	0.744	0.744	0.746	0.744	0.742	0.741	0.740
12	16	0.743	0.740	0.739	0.737	0.737	0.740	0.735	0.732	0.729	0.727	0.738	0.730	0.725	0.719	0.716
	18	0.745	0.743	0.742	0.741	0.741	0.743	0.739	0.737	0.735	0.734	0.740	0.735	0.731	0.727	0.725
	20	0.746	0.745	0.744	0.744	0.744	0.744	0.742	0.740	0.739	0.738	0.743	0.739	0.736	0.733	0.732
	24	0.748	0.747	0.747	0.747	0.747	0.746	0.745	0.745	0.744	0.744	0.745	0.743	0.742	0.740	0.740
	28	0.748	0.748	0.748	0.748	0.748	0.748	0.747	0.747	0.746	0.746	0.747	0.745	0.745	0.744	0.744

h	m	ℓ_1 = 24					ℓ_1 = 28					ℓ_1 = 32				
		γ=0.50	γ=0.75	γ=1.00	γ=1.50	γ=2.00	γ=0.50	γ=0.75	γ=1.00	γ=1.50	γ=2.00	γ=0.50	γ=0.75	γ=1.00	γ=1.50	γ=2.00
10	16	0.733	0.722	0.713	0.701	0.696	0.732	0.718	0.707	0.693	0.686	0.730	0.716	0.704	0.688	0.680
	18	0.736	0.727	0.719	0.710	0.705	0.734	0.723	0.714	0.701	0.694	0.733	0.720	0.710	0.695	0.687
	20	0.739	0.731	0.725	0.718	0.714	0.737	0.727	0.719	0.708	0.703	0.736	0.724	0.715	0.701	0.694
	24	0.742	0.737	0.734	0.729	0.727	0.741	0.734	0.728	0.721	0.717	0.739	0.731	0.724	0.713	0.708
	28	0.745	0.742	0.739	0.737	0.736	0.743	0.739	0.735	0.730	0.728	0.742	0.736	0.731	0.723	0.719
12	16	0.735	0.726	0.719	0.710	0.706	0.733	0.722	0.712	0.700	0.694	0.732	0.719	0.708	0.694	0.686
	18	0.738	0.731	0.725	0.719	0.715	0.736	0.727	0.719	0.709	0.704	0.735	0.723	0.714	0.701	0.694
	20	0.741	0.735	0.731	0.726	0.723	0.739	0.731	0.725	0.717	0.712	0.737	0.728	0.719	0.709	0.703
	24	0.744	0.741	0.738	0.736	0.734	0.742	0.737	0.733	0.728	0.726	0.741	0.734	0.729	0.721	0.717
	28	0.746	0.744	0.742	0.741	0.740	0.745	0.742	0.739	0.736	0.735	0.743	0.739	0.735	0.730	0.728
14	16	0.737	0.730	0.724	0.717	0.714	0.735	0.725	0.717	0.707	0.702	0.733	0.721	0.712	0.699	0.693
	18	0.740	0.735	0.730	0.726	0.723	0.738	0.730	0.724	0.716	0.712	0.736	0.726	0.718	0.708	0.702
	20	0.743	0.738	0.735	0.732	0.730	0.740	0.734	0.729	0.723	0.720	0.739	0.731	0.724	0.715	0.711
	24	0.745	0.743	0.741	0.739	0.739	0.744	0.740	0.737	0.734	0.732	0.742	0.737	0.733	0.727	0.724
	28	0.747	0.745	0.744	0.744	0.743	0.745	0.743	0.742	0.740	0.739	0.745	0.741	0.739	0.735	0.734
16	16	0.739	0.733	0.729	0.724	0.721	0.737	0.728	0.722	0.714	0.710	0.735	0.724	0.716	0.705	0.699
	18	0.742	0.738	0.735	0.731	0.729	0.740	0.733	0.728	0.722	0.719	0.738	0.729	0.722	0.714	0.709
	20	0.744	0.740	0.738	0.736	0.735	0.742	0.737	0.733	0.729	0.727	0.740	0.733	0.728	0.721	0.718
	24	0.746	0.744	0.743	0.742	0.742	0.745	0.742	0.740	0.737	0.736	0.744	0.740	0.736	0.732	0.730
	28	0.747	0.746	0.746	0.745	0.745	0.746	0.745	0.743	0.742	0.742	0.745	0.743	0.741	0.738	0.737

%Mneg

%Mneg

TABLE 8-12 - α, PER ACI CODE, SECTION 13.0 PAGE 1

SLAB THICKNESS = 6.0   ℓ = SPAN (FT.)   I = MOMENT OF INERTIA OF BEAM PER SECT. 13.1.5

BEAM		L SHAPE										T SHAPE									
h	b	I	ℓ=12	ℓ=16	ℓ=20	ℓ=24	ℓ=28	ℓ=32	ℓ=36	ℓ=40	ℓ=44	I	ℓ=12	ℓ=16	ℓ=20	ℓ=24	ℓ=28	ℓ=32	ℓ=36	ℓ=40	ℓ=44
12	12	2095.2	1.6	1.2	1.0	0.8	0.7	0.6	0.5	0.5	0.4	2376.0	0.9	0.7	0.6	0.5	0.4	0.3	0.3	0.3	0.3
	18	2977.7	2.3	1.7	1.4	1.1	1.0	0.9	0.8	0.7	0.6	3294.0	1.3	1.0	0.8	0.6	0.5	0.5	0.4	0.4	0.3
	24	3852.0	3.0	2.2	1.8	1.5	1.3	1.1	1.0	0.9	0.8	4190.4	1.6	1.2	1.0	0.8	0.7	0.6	0.5	0.5	0.4
	30	4722.5	3.6	2.7	2.2	1.8	1.6	1.4	1.2	1.1	1.0	5076.0	2.0	1.5	1.2	1.0	0.8	0.7	0.7	0.6	0.5
	36	5591.0	4.3	3.2	2.6	2.2	1.8	1.6	1.4	1.3	1.2	5955.4	2.3	1.7	1.4	1.1	1.0	0.9	0.8	0.7	0.6
16	12	5418.8	4.2	3.1	2.5	2.1	1.8	1.6	1.4	1.3	1.1	6302.1	2.4	1.8	1.5	1.2	1.0	0.9	0.8	0.7	0.7
	18	7565.3	5.8	4.4	3.5	2.9	2.5	2.2	1.9	1.8	1.6	8621.6	3.3	2.5	2.0	1.7	1.4	1.2	1.1	1.0	0.9
	24	9669.2	7.5	5.6	4.5	3.7	3.2	2.8	2.5	2.2	2.0	10837.7	4.2	3.1	2.5	2.1	1.8	1.6	1.4	1.3	1.1
	30	11753.3	9.1	6.8	5.4	4.5	3.9	3.4	3.0	2.7	2.5	13000.0	5.0	3.8	3.0	2.5	2.1	1.9	1.7	1.5	1.4
	36	13826.4	10.7	8.0	6.4	5.3	4.6	4.0	3.6	3.2	2.9	15130.7	5.8	4.4	3.5	2.9	2.5	2.2	1.9	1.8	1.6
20	12	11300.8	8.7	6.5	5.2	4.4	3.7	3.3	2.9	2.6	2.4	13346.3	5.1	3.9	3.1	2.6	2.2	1.9	1.7	1.5	1.4
	18	15589.2	12.0	9.0	7.2	6.0	5.2	4.5	4.0	3.6	3.3	18116.7	7.0	5.2	4.2	3.5	3.0	2.6	2.3	2.1	1.9
	24	19754.9	15.2	11.4	9.1	7.6	6.5	5.7	5.1	4.6	4.2	22601.7	8.7	6.5	5.2	4.4	3.7	3.3	2.9	2.6	2.4
	30	23862.5	18.4	13.8	11.0	9.2	7.9	6.9	6.1	5.5	5.0	26935.2	10.4	7.8	6.2	5.2	4.5	3.9	3.5	3.1	2.8
	36	27937.9	21.6	16.2	12.9	10.8	9.2	8.1	7.2	6.5	5.9	31178.5	12.0	9.0	7.2	6.0	5.2	4.5	4.0	3.6	3.3
24	12	20510.1	15.8	11.9	9.5	7.9	6.3	5.9	5.3	4.7	4.3	24469.7	9.4	7.1	5.7	4.7	4.0	3.5	3.1	2.8	2.6
	18	28058.4	21.6	16.2	13.0	10.8	9.3	8.1	7.2	6.5	5.9	33048.0	12.8	9.6	7.7	6.4	5.5	4.8	4.3	3.8	3.5
	24	35338.7	27.3	20.5	16.4	13.6	11.7	10.2	9.1	8.2	7.4	41020.3	15.8	11.9	9.5	7.9	6.8	5.9	5.3	4.7	4.3
	30	42490.9	32.8	24.6	19.7	16.4	14.1	12.3	10.9	9.8	8.9	48666.4	18.8	14.1	11.3	9.4	8.0	7.0	6.3	5.6	5.1
	36	49572.0	38.3	28.7	23.0	19.1	16.4	14.3	12.8	11.5	10.4	56116.8	21.6	16.2	13.0	10.8	9.3	8.1	7.2	6.5	5.9
28	12	33815.0	26.1	19.6	15.7	13.0	11.2	9.8	8.7	7.8	7.1	40632.6	15.7	11.8	9.4	7.8	6.7	5.9	5.2	4.7	4.3
	18	45981.0	35.5	26.6	21.3	17.7	15.2	13.3	11.8	10.6	9.7	54683.2	21.1	15.8	12.7	10.5	9.0	7.9	7.0	6.3	5.8
	24	57649.7	44.5	33.4	26.7	22.2	19.[illegible]	16.7	14.8	13.3	12.1	67630.1	26.1	19.6	15.7	13.0	11.2	9.8	8.7	7.8	7.1
	30	69078.9	53.3	40.0	32.0	26.7	22.8	20.0	17.8	16.0	14.5	79977.2	30.9	23.1	18.5	15.4	13.2	11.6	10.3	9.3	8.4
	36	80374.6	62.0	46.5	37.2	31.0	26.6	23.3	20.7	18.6	16.9	91962.1	35.5	26.6	21.3	17.7	15.2	13.3	11.8	10.6	9.7
32	12	50898.9	39.3	29.5	23.6	19.6	16.8	14.7	13.1	11.8	10.7	61444.5	23.7	17.8	14.2	11.9	10.2	8.9	7.9	7.1	6.5
	18	69052.8	53.3	40.0	32.0	26.6	22.8	20.0	17.8	16.0	14.5	82464.0	31.8	23.9	19.1	15.9	13.6	11.9	10.6	9.5	8.7
	24	86461.4	66.7	50.0	40.0	33.4	28.6	25.0	22.2	20.0	18.2	101797.8	39.3	29.5	23.6	19.6	16.8	14.7	13.1	11.8	10.7
	30	103513.7	79.9	59.9	47.9	39.9	34.2	30.0	26.6	24.0	21.8	120224.0	46.4	34.8	27.8	23.2	19.9	17.4	15.5	13.9	12.6
	36	120368.0	92.9	69.7	55.7	46.4	39.8	34.8	31.0	27.9	25.3	138105.6	53.3	40.0	32.0	26.6	22.8	20.0	17.8	16.0	14.5
36	12	71388.0	55.1	41.3	33.0	27.5	23.6	20.7	18.4	16.5	15.0	86400.0	33.3	25.0	20.0	16.7	14.3	12.5	11.1	10.0	9.1
	18	96925.0	74.8	56.1	44.9	37.4	32.1	28.0	24.9	22.4	20.4	115709.5	44.6	33.5	26.8	22.3	19.1	16.7	14.9	13.4	12.2
	24	121515.4	93.8	70.3	56.3	46.9	40.2	35.2	31.3	28.1	25.6	142776.0	55.1	41.3	33.0	27.5	23.6	20.7	18.4	16.5	15.0
	30	145660.2	112.4	84.3	67.4	56.2	48.2	42.1	37.5	33.7	30.7	168661.8	65.1	48.8	39.0	32.5	27.9	24.4	21.7	19.5	17.7
	36	169560.0	130.8	98.1	78.5	65.4	56.1	49.1	43.6	39.3	35.7	193850.1	74.8	56.1	44.9	37.4	32.1	28.0	24.9	22.4	20.4

TABLE 8-12 - α, PER ACI CODE, SECTION 13.0

SLAB THICKNESS = 7.0 ℓ = SPAN (FT.) I = MOMENT OF INERTIA OF BEAM PER SECT. 13.1.5

L SHAPE

BEAM h	b	I	ℓ=12	ℓ=16	ℓ=20	ℓ=24	ℓ=28	ℓ=32	ℓ=36	ℓ=40	ℓ=44
12	12	2046.8	1.0	0.7	0.6	0.5	0.4	0.4	0.3	0.3	0.3
	18	2923.1	1.4	1.1	0.9	0.7	0.6	0.5	0.5	0.4	0.4
	24	3793.9	1.8	1.4	1.1	0.9	0.8	0.7	0.6	0.6	0.5
	30	4662.2	2.3	1.7	1.4	1.1	1.0	0.8	0.8	0.7	0.6
	36	5529.2	2.7	2.0	1.6	1.3	1.2	1.0	0.9	0.8	0.7
16	12	5313.8	2.6	1.9	1.5	1.3	1.1	1.0	0.9	0.8	0.7
	18	7448.0	3.6	2.7	2.2	1.8	1.6	1.4	1.2	1.1	1.0
	24	9545.1	4.6	3.5	2.8	2.3	2.0	1.7	1.5	1.4	1.3
	30	11624.9	5.6	4.2	3.4	2.8	2.4	2.1	1.9	1.7	1.5
	36	13695.2	6.7	5.0	4.0	3.3	2.9	2.5	2.2	2.0	1.8
20	12	11159.3	5.4	4.1	3.3	2.7	2.3	2.0	1.8	1.6	1.5
	18	15440.5	7.5	5.6	4.5	3.8	3.2	2.8	2.5	2.3	2.0
	24	19603.5	9.5	7.1	5.7	4.8	4.1	3.6	3.2	2.9	2.6
	30	23710.0	11.5	8.6	6.9	5.8	4.9	4.3	3.8	3.5	3.1
	36	27784.9	13.5	10.1	8.1	6.8	5.8	5.1	4.5	4.1	3.7
24	12	20393.8	9.9	7.4	5.9	5.0	4.2	3.7	3.3	3.0	2.7
	18	27962.8	13.6	10.2	8.2	6.8	5.8	5.1	4.5	4.1	3.7
	24	35259.5	17.1	12.8	10.3	8.6	7.3	6.4	5.7	5.1	4.7
	30	42424.2	20.6	15.5	12.4	10.3	8.8	7.7	6.9	6.2	5.6
	36	49514.8	24.1	18.0	14.4	12.0	10.3	9.0	8.0	7.2	6.6
28	12	33826.5	16.4	12.3	9.9	8.2	7.0	6.2	5.5	4.9	4.5
	18	46075.4	22.4	16.8	13.4	11.2	9.6	8.4	7.5	6.7	6.1
	24	57802.0	28.1	21.1	16.9	14.0	12.0	10.5	9.4	8.4	7.7
	30	69273.2	33.7	25.2	20.2	16.8	14.4	12.6	11.2	10.1	9.2
	36	80600.3	39.2	29.4	23.5	19.6	16.8	14.7	13.1	11.7	10.7
32	12	52266.1	25.4	19.0	15.2	12.7	10.9	9.5	8.5	7.6	6.9
	18	70838.6	34.4	25.8	20.7	17.2	14.8	12.9	11.5	10.3	9.4
	24	88519.9	43.0	32.3	25.8	21.5	18.4	16.1	14.3	12.9	11.7
	30	105762.3	51.4	38.5	30.8	25.7	22.0	19.3	17.1	15.4	14.0
	36	122756.3	59.6	44.7	35.8	29.8	25.6	22.4	19.9	17.9	16.3
36	12	75803.9	36.8	27.6	22.1	18.4	15.8	13.8	12.3	11.1	10.0
	18	102423.4	49.8	37.3	29.9	24.9	21.3	18.7	16.6	14.9	13.6
	24	127701.5	62.1	46.5	37.2	31.0	26.6	23.3	20.7	18.6	16.9
	30	152319.4	74.0	55.5	44.4	37.0	31.7	27.8	24.7	22.2	20.2
	36	176563.8	85.8	64.3	51.5	42.9	36.8	32.2	28.6	25.7	23.4

T SHAPE

BEAM h	b	I	ℓ=12	ℓ=16	ℓ=20	ℓ=24	ℓ=28	ℓ=32	ℓ=36	ℓ=40	ℓ=44
12	12	2308.2	0.6	0.4	0.3	0.3	0.2	0.2	0.2	0.2	0.2
	18	3208.2	0.8	0.6	0.5	0.4	0.3	0.3	0.3	0.2	0.2
	24	4093.7	1.0	0.7	0.6	0.5	0.4	0.4	0.3	0.3	0.3
	30	4972.1	1.2	0.9	0.7	0.6	0.5	0.5	0.4	0.4	0.3
	36	5846.3	1.4	1.1	0.9	0.7	0.6	0.5	0.5	0.4	0.4
16	12	6151.0	1.5	1.1	0.9	0.7	0.6	0.6	0.5	0.4	0.4
	18	8433.4	2.0	1.5	1.2	1.0	0.9	0.8	0.7	0.6	0.6
	24	10627.6	2.6	1.9	1.5	1.3	1.1	1.0	0.9	0.8	0.7
	30	12775.4	3.1	2.3	1.9	1.6	1.3	1.2	1.0	0.9	0.8
	36	14896.0	3.6	2.7	2.2	1.8	1.6	1.4	1.2	1.1	1.0
20	12	13116.3	3.2	2.4	1.9	1.6	1.4	1.2	1.1	1.0	0.9
	18	17850.5	4.3	3.3	2.6	2.2	1.9	1.6	1.4	1.3	1.2
	24	22318.6	5.4	4.1	3.3	2.7	2.3	2.0	1.8	1.6	1.5
	30	26643.0	6.5	4.9	3.9	3.2	2.8	2.4	2.2	1.9	1.8
	36	30881.1	7.5	5.6	4.5	3.8	3.2	2.8	2.5	2.3	2.0
24	12	24210.8	5.9	4.4	3.5	2.9	2.5	2.2	2.0	1.8	1.6
	18	32795.0	8.0	6.0	4.8	4.0	3.4	3.0	2.7	2.4	2.2
	24	40787.6	9.9	7.4	5.9	5.0	4.2	3.7	3.3	3.0	2.7
	30	48455.3	11.8	8.8	7.1	5.9	5.0	4.4	3.9	3.5	3.2
	36	55925.6	13.6	10.2	8.2	6.8	5.8	5.1	4.5	4.1	3.7
28	12	40439.7	9.8	7.4	5.9	4.9	4.2	3.7	3.3	2.9	2.7
	18	54600.1	13.3	9.9	8.0	6.6	5.7	5.0	4.4	4.0	3.6
	24	67653.0	16.4	12.3	9.9	8.2	7.0	6.2	5.5	4.9	4.5
	30	80090.5	19.5	14.6	11.7	9.7	8.3	7.3	6.5	5.8	5.3
	36	92150.8	22.4	16.8	13.4	11.2	9.6	8.4	7.5	6.7	6.1
32	12	62807.5	15.3	11.4	9.2	7.6	6.5	5.7	5.1	4.6	4.2
	18	84598.4	20.6	15.4	12.3	10.3	8.8	7.7	6.9	6.2	5.6
	24	104532.2	25.4	19.0	15.2	12.7	10.9	9.5	8.5	7.6	6.9
	30	123425.5	30.0	22.5	18.0	15.0	12.9	11.2	10.0	9.0	8.2
	36	141677.2	34.4	25.8	20.7	17.2	14.8	12.9	11.5	10.3	9.4
36	12	91466.1	22.2	16.7	13.3	11.1	9.5	8.3	7.4	6.7	6.1
	18	122937.4	29.9	22.4	17.9	14.9	12.8	11.2	10.0	9.0	8.1
	24	151607.8	36.8	27.6	22.1	18.4	15.8	13.8	12.3	11.1	10.0
	30	178710.3	43.4	32.6	26.1	21.7	18.6	16.3	14.5	13.0	11.8
	36	204846.9	49.8	37.3	29.9	24.9	21.3	18.7	16.6	14.9	13.6

TABLE 8-12 - α, PER ACI CODE, SECTION 13.0

SLAB THICKNESS = 8.0 ℓ = SPAN (FT.) I = MOMENT OF INERTIA OF BEAM PER SECT. 13.1.5

BEAM		L SHAPE										T SHAPE									
h	b	I	ℓ=12	ℓ=16	ℓ=20	ℓ=24	ℓ=28	ℓ=32	ℓ=36	ℓ=40	ℓ=44	I	ℓ=12	ℓ=16	ℓ=20	ℓ=24	ℓ=28	ℓ=32	ℓ=36	ℓ=40	ℓ=44
12	12	2003.3	0.7	0.5	0.4	0.3	0.3	0.2	0.2	0.2	0.2	2246.5	0.4	0.3	0.2	0.2	0.2	0.1	0.1	0.1	0.1
	18	2874.1	0.9	0.7	0.6	0.5	0.4	0.4	0.3	0.3	0.3	3130.8	0.5	0.4	0.3	0.3	0.2	0.2	0.2	0.2	0.1
	24	3741.8	1.2	0.9	0.7	0.6	0.5	0.5	0.4	0.4	0.3	4006.7	0.7	0.5	0.4	0.3	0.3	0.2	0.2	0.2	0.2
	30	4608.2	1.5	1.1	0.9	0.8	0.6	0.6	0.5	0.5	0.4	4878.6	0.8	0.6	0.5	0.4	0.3	0.3	0.3	0.2	0.2
	36	5473.8	1.8	1.3	1.1	0.9	0.8	0.7	0.6	0.5	0.5	5748.3	0.9	0.7	0.6	0.5	0.4	0.4	0.3	0.3	0.3
16	12	5205.3	1.7	1.3	1.0	0.8	0.7	0.6	0.6	0.5	0.5	6007.4	1.0	0.7	0.6	0.5	0.4	0.4	0.3	0.3	0.3
	18	7323.1	2.4	1.8	1.4	1.2	1.0	0.9	0.8	0.7	0.7	8244.5	1.3	1.0	0.8	0.7	0.6	0.5	0.4	0.4	0.4
	24	9411.0	3.1	2.3	1.8	1.5	1.3	1.1	1.0	0.9	0.8	10410.6	1.7	1.3	1.0	0.8	0.7	0.6	0.6	0.5	0.5
	30	11484.8	3.7	2.8	2.2	1.9	1.6	1.4	1.2	1.1	1.0	12539.5	2.0	1.5	1.2	1.0	0.9	0.8	0.7	0.6	0.6
	36	13550.9	4.4	3.3	2.6	2.2	1.9	1.7	1.5	1.3	1.2	14646.3	2.4	1.8	1.4	1.2	1.0	0.9	0.8	0.7	0.7
20	12	10980.5	3.6	2.7	2.1	1.8	1.5	1.3	1.2	1.1	1.0	12864.0	2.1	1.6	1.3	1.0	0.9	0.8	0.7	0.6	0.6
	18	15240.4	5.0	3.7	3.0	2.5	2.1	1.9	1.7	1.5	1.4	17531.8	2.9	2.1	1.7	1.4	1.2	1.1	1.0	0.9	0.8
	24	19392.0	6.3	4.7	3.8	3.2	2.7	2.4	2.1	1.9	1.7	21961.1	3.6	2.7	2.1	1.8	1.5	1.3	1.2	1.1	1.0
	30	23491.3	7.6	5.7	4.6	3.8	3.3	2.9	2.5	2.3	2.1	26260.3	4.3	3.2	2.6	2.1	1.8	1.6	1.4	1.3	1.2
	36	27561.4	9.0	6.7	5.4	4.5	3.3	3.4	3.0	2.7	2.4	30480.8	5.0	3.7	3.0	2.5	2.1	1.9	1.7	1.5	1.4
24	12	20178.0	6.6	4.9	3.9	3.3	2.8	2.5	2.2	2.0	1.8	23863.2	3.9	2.9	2.3	1.9	1.7	1.5	1.3	1.2	1.1
	18	27738.2	9.0	6.8	5.4	4.5	3.9	3.4	3.0	2.7	2.5	32388.9	5.3	4.0	3.2	2.6	2.3	2.0	1.8	1.6	1.4
	24	35033.2	11.4	8.6	6.8	5.7	4.9	4.3	3.8	3.4	3.1	40356.1	6.6	4.9	3.9	3.3	2.8	2.5	2.2	2.0	1.8
	30	42198.1	13.7	10.3	8.2	6.9	5.9	5.2	4.6	4.1	3.7	48011.8	7.8	5.9	4.7	3.9	3.3	2.9	2.6	2.3	2.1
	36	49289.6	16.0	12.0	9.6	8.0	6.9	6.0	5.3	4.8	4.4	55476.4	9.0	6.8	5.4	4.5	3.9	3.4	3.0	2.7	2.5
28	12	33644.0	11.0	8.2	6.6	5.5	4.7	4.1	3.7	3.3	3.0	40048.9	6.5	4.9	3.9	3.3	2.8	2.4	2.2	2.0	1.8
	18	45925.9	14.9	11.2	9.0	7.5	6.4	5.6	5.0	4.5	4.1	54207.4	8.8	6.6	5.3	4.4	3.8	3.3	2.9	2.6	2.4
	24	57680.4	18.8	14.1	11.3	9.4	8.0	7.0	6.3	5.6	5.1	67288.0	11.0	8.2	6.6	5.5	4.7	4.1	3.7	3.3	3.0
	30	69173.3	22.5	16.9	13.5	11.3	9.7	8.4	7.5	6.8	6.1	79759.0	13.0	9.7	7.8	6.5	5.6	4.9	4.3	3.9	3.5
	36	80517.5	26.2	19.7	15.7	13.1	11.2	9.8	8.7	7.9	7.1	91851.8	14.9	11.2	9.0	7.5	6.4	5.6	5.0	4.5	4.1
32	12	52224.0	17.0	12.8	10.2	8.5	7.3	6.4	5.7	5.1	4.6	62464.0	10.2	7.6	6.1	5.1	4.4	3.8	3.4	3.0	2.8
	18	70912.0	23.1	17.3	13.9	11.5	9.9	8.7	7.7	6.9	6.3	84377.6	13.7	10.3	8.2	6.9	5.9	5.1	4.6	4.1	3.7
	24	88678.4	28.9	21.7	17.3	14.4	12.4	10.8	9.6	8.7	7.9	104448.0	17.0	12.8	10.2	8.5	7.3	6.4	5.7	5.1	4.6
	30	105984.0	34.5	25.9	20.7	17.3	14.8	12.9	11.5	10.3	9.4	123465.1	20.1	15.1	12.1	10.0	8.6	7.5	6.7	6.0	5.5
	36	123026.2	40.0	30.0	24.0	20.0	17.2	15.0	13.3	12.0	10.9	141824.0	23.1	17.3	13.9	11.5	9.9	8.7	7.7	6.9	6.3
36	12	76763.0	25.0	18.7	15.0	12.5	10.7	9.4	8.3	7.5	6.8	92151.0	15.0	11.2	9.0	7.5	6.4	5.6	5.0	4.5	4.1
	18	103804.5	33.8	25.3	20.3	16.9	14.5	12.7	11.3	10.1	9.2	124289.0	20.2	15.2	12.1	10.1	8.7	7.6	6.7	6.1	5.5
	24	129371.6	42.1	31.6	25.3	21.1	18.0	15.8	14.0	12.6	11.5	153526.1	25.0	18.7	15.0	12.5	10.7	9.4	8.3	7.5	6.8
	30	154196.8	50.2	37.6	30.1	25.1	21.5	18.8	16.7	15.1	13.7	181092.5	29.5	22.1	17.7	14.7	12.6	11.1	9.8	8.8	8.0
	36	178596.6	58.1	43.6	34.9	29.1	24.9	21.8	19.4	17.4	15.9	207609.1	33.8	25.3	20.3	16.9	14.5	12.7	11.3	10.1	9.2

α

α

TABLE 8-12 - α, PER ACI CODE, SECTION 13.0

SLAB THICKNESS = 10.0 ℓ = SPAN (FT.) I = MOMENT OF INERTIA OF BEAM PER SECT. 13.1.5

BEAM		L SHAPE										T SHAPE									
h	b	I	ℓ=12	ℓ=16	ℓ=20	ℓ=24	ℓ=28	ℓ=32	ℓ=36	ℓ=40	ℓ=44	I	ℓ=12	ℓ=16	ℓ=20	ℓ=24	ℓ=28	ℓ=32	ℓ=36	ℓ=40	ℓ=44
12	12	1912.2	0.3	0.2	0.2	0.2	0.1	0.1	0.1	0.1	0.1	2092.6	0.2	0.1	0.1	0.1	0.1	0.1	0.1	0.1	0.0
	18	2776.9	0.5	0.3	0.3	0.2	0.2	0.2	0.2	0.1	0.1	2959.0	0.2	0.2	0.1	0.1	0.1	0.1	0.1	0.1	0.1
	24	3641.3	0.6	0.5	0.4	0.3	0.3	0.2	0.2	0.2	0.2	3824.4	0.3	0.2	0.2	0.2	0.1	0.1	0.1	0.1	0.1
	30	4505.6	0.8	0.6	0.5	0.4	0.3	0.3	0.3	0.2	0.2	4689.3	0.4	0.3	0.2	0.2	0.2	0.1	0.1	0.1	0.1
	36	5369.7	0.9	0.7	0.5	0.4	0.4	0.3	0.3	0.3	0.2	5553.9	0.5	0.3	0.3	0.2	0.2	0.2	0.2	0.1	0.1
16	12	5007.4	0.8	0.6	0.5	0.4	0.4	0.3	0.3	0.3	0.2	5760.6	0.5	0.4	0.3	0.2	0.2	0.2	0.2	0.1	0.1
	18	7090.8	1.2	0.9	0.7	0.6	0.5	0.4	0.4	0.4	0.3	7906.3	0.7	0.5	0.4	0.3	0.3	0.2	0.2	0.2	0.2
	24	9159.0	1.5	1.1	0.9	0.8	0.7	0.6	0.5	0.5	0.4	10014.8	0.8	0.6	0.5	0.4	0.4	0.3	0.3	0.3	0.2
	30	11220.0	1.9	1.4	1.1	0.9	0.8	0.7	0.6	0.6	0.5	12104.0	1.0	0.8	0.6	0.5	0.4	0.4	0.3	0.3	0.3
	36	13277.0	2.2	1.7	1.3	1.1	0.9	0.8	0.7	0.7	0.6	14181.7	1.2	0.9	0.7	0.6	0.5	0.4	0.4	0.4	0.3
20	12	10598.0	1.8	1.3	1.1	0.9	0.8	0.7	0.6	0.5	0.5	12393.9	1.0	0.8	0.6	0.5	0.4	0.4	0.3	0.3	0.3
	18	14789.8	2.5	1.8	1.5	1.2	1.1	0.9	0.8	0.7	0.7	16880.9	1.4	1.1	0.8	0.7	0.6	0.5	0.5	0.4	0.4
	24	18902.2	3.2	2.4	1.9	1.6	1.4	1.2	1.1	0.9	0.9	21196.0	1.8	1.3	1.1	0.9	0.8	0.7	0.6	0.5	0.5
	30	22976.1	3.8	2.9	2.3	1.9	1.6	1.4	1.3	1.1	1.0	25416.6	2.1	1.6	1.3	1.1	0.9	0.8	0.7	0.6	0.6
	36	27028.4	4.5	3.4	2.7	2.3	1.9	1.7	1.5	1.4	1.2	29579.7	2.5	1.8	1.5	1.2	1.1	0.9	0.8	0.7	0.7
24	12	19606.7	3.3	2.5	2.0	1.6	1.4	1.2	1.1	1.0	0.9	23113.9	1.9	1.4	1.2	1.0	0.8	0.7	0.6	0.6	0.5
	18	27083.6	4.5	3.4	2.7	2.3	1.9	1.7	1.5	1.4	1.2	31393.8	2.6	2.0	1.6	1.3	1.1	1.0	0.9	0.8	0.7
	24	34333.3	5.7	4.3	3.4	2.9	2.5	2.1	1.9	1.7	1.6	39213.4	3.3	2.5	2.0	1.6	1.4	1.2	1.1	1.0	0.9
	30	41469.9	6.9	5.2	4.1	3.5	3.0	2.6	2.3	2.1	1.9	46771.7	3.9	2.9	2.3	1.9	1.7	1.5	1.3	1.2	1.1
	36	48542.0	8.1	6.1	4.9	4.0	3.5	3.0	2.7	2.4	2.2	54167.2	4.5	3.4	2.7	2.3	1.9	1.7	1.5	1.4	1.2
28	12	32945.9	5.5	4.1	3.3	2.7	2.4	2.1	1.8	1.6	1.5	39029.2	3.3	2.4	2.0	1.6	1.4	1.2	1.1	1.0	0.9
	18	45171.1	7.5	5.6	4.5	3.8	3.2	2.8	2.5	2.3	2.1	52938.0	4.4	3.3	2.6	2.2	1.9	1.7	1.5	1.3	1.2
	24	56903.7	9.5	7.1	5.7	4.7	4.1	3.6	3.2	2.8	2.6	65891.9	5.5	4.1	3.3	2.7	2.4	2.1	1.8	1.6	1.5
	30	68387.0	11.4	8.5	6.8	5.7	4.9	4.3	3.8	3.4	3.1	78292.0	6.5	4.9	3.9	3.3	2.8	2.4	2.2	2.0	1.8
	36	79726.9	13.3	10.0	8.0	6.6	5.7	5.0	4.4	4.0	3.6	90342.3	7.5	5.6	4.5	3.8	3.2	2.8	2.5	2.3	2.1
32	12	51525.3	8.6	6.4	5.2	4.3	3.7	3.2	2.9	2.6	2.3	61245.5	5.1	3.8	3.1	2.6	2.2	1.9	1.7	1.5	1.4
	18	70248.0	11.7	8.8	7.0	5.9	5.0	4.4	3.9	3.5	3.2	83001.9	6.9	5.2	4.2	3.5	3.0	2.6	2.3	2.1	1.9
	24	88061.8	14.7	11.0	8.8	7.3	6.3	5.5	4.9	4.4	4.0	103050.6	8.6	6.4	5.2	4.3	3.7	3.2	2.9	2.6	2.3
	30	105410.2	17.6	13.2	10.5	8.8	7.5	6.6	5.9	5.3	4.8	122094.0	10.2	7.6	6.1	5.1	4.4	3.8	3.4	3.1	2.8
	36	122488.8	20.4	15.3	12.2	10.2	8.7	7.7	6.8	6.1	5.6	140496.0	11.7	8.8	7.0	5.9	5.0	4.4	3.9	3.5	3.2
36	12	76253.4	12.7	9.5	7.6	6.4	5.4	4.8	4.2	3.8	3.5	90867.6	7.6	5.7	4.5	3.8	3.2	2.8	2.5	2.3	2.1
	18	103508.7	17.3	12.9	10.4	8.6	7.4	6.5	5.8	5.2	4.7	123072.6	10.3	7.7	6.2	5.1	4.4	3.8	3.4	3.1	2.8
	24	129254.6	21.5	16.2	12.9	10.8	9.2	8.1	7.2	6.5	5.9	152506.8	12.7	9.5	7.6	6.4	5.4	4.8	4.2	3.8	3.5
	30	154220.9	25.7	19.3	15.4	12.9	11.0	9.6	8.6	7.7	7.0	180292.3	15.0	11.3	9.0	7.5	6.4	5.6	5.0	4.5	4.1
	36	178732.5	29.8	22.3	17.9	14.9	12.8	11.2	9.9	8.9	8.1	207017.4	17.3	12.9	10.4	8.6	7.4	6.5	5.8	5.2	4.7

TABLE 8-12 – α, PER ACI CODE, SECTION 13.0 PAGE 5

SLAB THICKNESS = 12.0 ℓ = SPAN (FT.) I = MOMENT OF INERTIA OF BEAM PER SECT. 13.1.5

BEAM		L SHAPE									
h	b	I	ℓ=12	ℓ=16	ℓ=20	ℓ=24	ℓ=28	ℓ=32	ℓ=36	ℓ=40	ℓ=44
16	12	4825.6	0.5	0.3	0.3	0.2	0.2	0.2	0.2	0.1	0.1
	18	6884.5	0.7	0.5	0.4	0.3	0.3	0.2	0.2	0.2	0.2
	24	8938.6	0.9	0.6	0.5	0.4	0.4	0.3	0.3	0.3	0.2
	30	10990.5	1.1	0.8	0.6	0.5	0.5	0.4	0.4	0.3	0.3
	36	13041.2	1.3	0.9	0.8	0.6	0.5	0.5	0.4	0.4	0.3
20	12	10249.1	1.0	0.7	0.6	0.5	0.4	0.4	0.3	0.3	0.3
	18	14364.6	1.4	1.0	0.8	0.7	0.6	0.5	0.5	0.4	0.4
	24	18432.0	1.8	1.3	1.1	0.9	0.8	0.7	0.6	0.5	0.5
	30	22476.1	2.2	1.6	1.3	1.1	0.9	0.8	0.7	0.7	0.6
	36	26507.2	2.6	1.9	1.5	1.3	1.1	1.0	0.9	0.8	0.7
24	12	19008.0	1.8	1.4	1.1	0.9	0.8	0.7	0.6	0.6	0.5
	18	26352.0	2.5	1.9	1.5	1.3	1.1	1.0	0.8	0.8	0.7
	24	33523.2	3.2	2.4	1.9	1.6	1.4	1.2	1.1	1.0	0.9
	30	40608.0	3.9	2.9	2.3	2.0	1.7	1.5	1.3	1.2	1.1
	36	47643.4	4.6	3.4	2.8	2.3	2.0	1.7	1.5	1.4	1.3
28	12	32075.6	3.1	2.3	1.9	1.5	1.3	1.2	1.0	0.9	0.8
	18	44130.2	4.3	3.2	2.6	2.1	1.8	1.6	1.4	1.3	1.2
	24	55765.3	5.4	4.0	3.2	2.7	2.3	2.0	1.8	1.6	1.5
	30	67185.8	6.5	4.9	3.9	3.2	2.8	2.4	2.2	1.9	1.8
	36	78481.9	7.6	5.7	4.5	3.8	3.2	2.8	2.5	2.3	2.1
32	12	50417.2	4.9	3.6	2.9	2.4	2.1	1.8	1.6	1.5	1.3
	18	68973.1	6.7	5.0	4.0	3.3	2.9	2.5	2.2	2.0	1.8
	24	86701.7	8.4	6.3	5.0	4.2	3.6	3.1	2.8	2.5	2.3
	30	104000.0	10.0	7.5	6.0	5.0	4.3	3.8	3.3	3.0	2.7
	36	121046.0	11.7	8.8	7.0	5.8	5.0	4.4	3.9	3.5	3.2
36	12	74995.2	7.2	5.4	4.3	3.6	3.1	2.7	2.4	2.2	2.0
	18	102151.3	9.9	7.4	5.9	4.9	4.2	3.7	3.3	3.0	2.7
	24	127872.0	12.3	9.3	7.4	6.2	5.3	4.6	4.1	3.7	3.4
	30	152837.0	14.7	11.1	8.8	7.4	6.3	5.5	4.9	4.4	4.0
	36	177355.6	17.1	12.8	10.3	8.6	7.3	6.4	5.7	5.1	4.7

BEAM		T SHAPE									
h	b	I	ℓ=12	ℓ=16	ℓ=20	ℓ=24	ℓ=28	ℓ=32	ℓ=36	ℓ=40	ℓ=44
16	12	5504.0	0.3	0.2	0.2	0.1	0.1	0.1	0.1	0.1	0.1
	18	7584.0	0.4	0.3	0.2	0.2	0.2	0.1	0.1	0.1	0.1
	24	9651.2	0.5	0.3	0.3	0.2	0.2	0.2	0.2	0.1	0.1
	30	11712.0	0.6	0.4	0.3	0.3	0.2	0.2	0.2	0.2	0.2
	36	13769.1	0.7	0.5	0.4	0.3	0.3	0.2	0.2	0.2	0.2
20	12	12010.6	0.6	0.4	0.3	0.3	0.2	0.2	0.2	0.2	0.2
	18	16307.4	0.8	0.6	0.5	0.4	0.3	0.3	0.3	0.2	0.2
	24	20498.2	1.0	0.7	0.6	0.5	0.4	0.4	0.3	0.3	0.3
	30	24631.2	1.2	0.9	0.7	0.6	0.5	0.4	0.4	0.4	0.3
	36	28729.2	1.4	1.0	0.8	0.7	0.6	0.5	0.5	0.4	0.4
24	12	22464.0	1.1	0.8	0.6	0.5	0.5	0.4	0.4	0.3	0.3
	18	30412.8	1.5	1.1	0.9	0.7	0.6	0.5	0.5	0.4	0.4
	24	38016.0	1.8	1.4	1.1	0.9	0.8	0.7	0.6	0.6	0.5
	30	45421.7	2.2	1.6	1.3	1.1	0.9	0.8	0.7	0.7	0.6
	36	52704.0	2.5	1.9	1.5	1.3	1.1	1.0	0.8	0.8	0.7
28	12	38028.8	1.8	1.4	1.1	0.9	0.8	0.7	0.6	0.6	0.5
	18	51484.5	2.5	1.9	1.5	1.2	1.1	0.9	0.8	0.7	0.7
	24	64151.2	3.1	2.3	1.9	1.5	1.3	1.2	1.0	0.9	0.8
	30	76353.8	3.7	2.8	2.2	1.8	1.6	1.4	1.2	1.1	1.0
	36	88260.4	4.3	3.2	2.6	2.1	1.8	1.6	1.4	1.3	1.2
32	12	59861.3	2.9	2.2	1.7	1.4	1.2	1.1	1.0	0.9	0.8
	18	81093.8	3.9	2.9	2.3	2.0	1.7	1.5	1.3	1.2	1.1
	24	100834.4	4.9	3.6	2.9	2.4	2.1	1.8	1.6	1.5	1.3
	30	119680.0	5.8	4.3	3.5	2.9	2.5	2.2	1.9	1.7	1.6
	36	137946.3	6.7	5.0	4.0	3.3	2.9	2.5	2.2	2.0	1.8
36	12	89115.4	4.3	3.2	2.6	2.1	1.8	1.6	1.4	1.3	1.2
	18	120807.5	5.8	4.4	3.5	2.9	2.5	2.2	1.9	1.7	1.6
	24	149990.4	7.2	5.4	4.3	3.6	3.1	2.7	2.4	2.2	2.0
	30	177645.9	8.6	6.4	5.1	4.3	3.7	3.2	2.9	2.6	2.3
	36	204302.7	9.9	7.4	5.9	4.9	4.2	3.7	3.3	3.0	2.7

Step 10. Distribute the design moments to column and middle strips.

(1) Compute the ratio of torsional stiffness of edge beam to the flexural stiffness of the slab having a width of center to center of supports,

$$\beta_t = C/2I_s$$

where C = torsional coefficient of edge beam per step 5-3, or Table 8-5. For flat slabs without edge beams: $\beta_t = 0$.

(2) Distribute the design moments (computed by steps 6, 7, and 8) to column and middle strips (ACI-13.3.4). For flat slabs $\alpha_1 = 0$, therefore $\alpha_1 l_2/l_1 = 0$, unless the edge beam is in the direction of the moments which are being considered. For the tabulation below, $\alpha_1 = 0$ and $\beta_t = 0$ are assumed.

	% of design moment	
Description	*Column strip*	*Middle strip*
Exterior negative moment	100	–
Positive moments	60	40
Interior negative moment	75	25

(3) For flat slabs with edge beams, use Tables 8-15 to 8-19 for distribution of design moments to column and middle strips. The procedure is described in the next section, Steps for Two-way Slabs on Beams.

Step 11. Establish shear constants at columns (Fig. 8-2):

Shear perimeter, for edge column: $b_v = 2c_1 + c_2 + 2d$; for interior column: $b_v = 2(c_1 + c_2\ 2d)$

Shear surface area, $A_c = b_v d$

Compute c_{AB} and c_{CD} by establishing center of twist.

Using ACI Commentary, Chapter 11, compute torsional moment of inertia:

$$J_c = \frac{(c_1 + \frac{1}{2}d)d^3}{6} + \frac{d(c_1 + \frac{1}{2}d)^3}{6} + \frac{2d}{3}(c_{AB}^3 + c_{CD}^3) + (c_2 + d)d(c_{AB})^2 \quad \text{(for exterior columns)}$$

where

$$c_{AB} = (c_1 + d/2)^2/(2c_1 + c_2 + 2d)$$
$$c_{CD} = c_1 + d/2 - c_{AB}$$

$$J_c = \frac{d(c_1 + d)^3}{6} + \frac{(c_1 + d)d^3}{6} + \frac{d(c_2 + d)(c_1 + d)^2}{6} \quad \text{(for interior columns)}$$

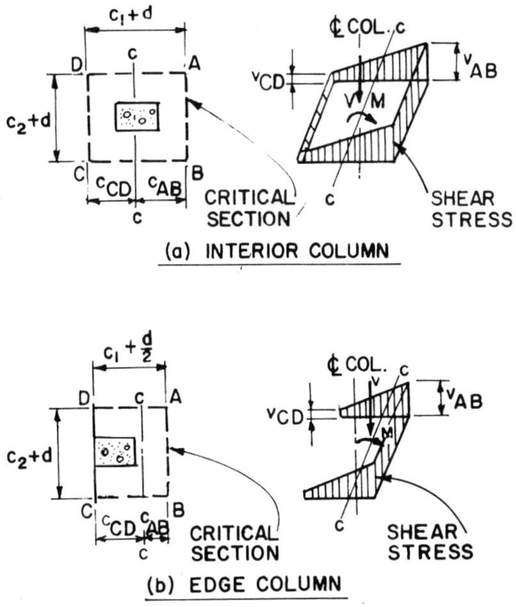

Fig. 8-2.

Compute the fraction (F_e) of the exterior negative design moment (M_{ext}, computed by step 6) to be transferred between slab and exterior column by shear:

$$F_e = 1 - \frac{1}{1 + \frac{2}{3}\sqrt{\frac{c_1 + \frac{1}{2}d}{c_2 + d}}}$$

Compute the fraction (F_i) of interior column moment (M computed by step 9) to be transferred between slab and interior column by shear (ACI-11.13.2):

$$F_i = 1 - \frac{1}{1 + \frac{2}{3}\sqrt{\frac{c_1 + d}{c_2 + d}}}$$

Step 12. Check shear in slab due to the moment transfer between exterior column and slab. Establish shear correction to center of twist:

$$\Delta V = (M_{\text{1st int. col.}} - M_{\text{ext. col.}})/l_n$$

Shear at exterior column: $V = 0.5\, w(l_1 l_2 - c_1 c_2)$
Shear at center of twist: $V_{cs} = V - \Delta V$
Moment at center of twist: $M_{cs} = M_{\text{face}} + V_{cs} c_{AB}$
Moment to be transferred by shear: $M_{sh} = F_e M_{cs}$

Design moments in column and middle strips

Location	% *Column strip*	% *Middle strip*
Interior negative design moment:		
(a) No beams:	75	25
(b) With beams:		
$l_2/l_1 = 0.5$	90	0
$l_2/l_1 = 1.0$	75	25
$l_2/l_1 = 2.0$	45	55
Exterior negative design moment:		
(a) No beams, no spandrel:	100	0
(b) No beams, with spandrel:	75	25
(c) With beams, no spandrel:	100	0
(d) With beams, with spandrel:		
$l_2/l_1 = 0.5$	90	10
$l_2/l_1 = 1.0$	75	25
$l_2/l_1 = 2.0$	45	55
Positive design moment:		
(a) No beams:	60	40
(b) With beams:		
$l_2/l_1 = 0.5$	90	10
$l_2/l_1 = 1.0$	75	25
$l_2/l_1 = 2.0$	45	55

where

c_{AB} = distance from face to center of twist, and
F_e = fraction, both from step 11

Check for maximum shear, at exterior column, at inside face of critical section:

$$v_{AB} = \frac{V_{cs}}{\phi A_c} + \frac{M_{sh} c_{AB}}{\phi J_c} \leqslant 4\sqrt{f'_c}$$

where

ϕ = 0.85, capacity reduction factor
A_c = shear area, from step 11
J_c = torsional moment of inertia, from step 11

Step 13. Shear and transfer of moments to slab at interior columns, use data from steps 11 and 12.

(1) Check shear in slab at interior columns for total panel shear:

$$V_t = w(l_1 l_2 - c_1 c_2)$$

Increase shear for first interior column by ΔV (from step 12), and compute shear stress:

$$v_u = \frac{V_t + \Delta V}{\phi A_c} \leqslant 4\sqrt{f_c'}$$

(2) Check for the combination of direct shear and shear due to transfer of unbalanced column moment (M from step 9). Full panel load on exterior span and dead load only on interior span are assumed. Add ΔV to first interior column only.

$$V_m = (w_d + 0.5 w_1) \cdot (l_1 l_2 - c_1 c_2) + \Delta V$$

$$v_{AB} = \frac{V_m}{\phi A_c} + \frac{F_i M c_{AB}}{\phi J_c} \leqslant 4\sqrt{f_c'}$$

Step 14. The portion of column moments, not transferred to slabs by shear, shall be resisted by flexure. The effective width of slab is the column width (transverse to the moment) plus the effective slab depth, and this slab section shall resist the transferred moment.

Moment transferred from column to slab at exterior column:

$$M_f = (1 - F_e) M_{cs}$$

where M_{cs} and F_e are from step 11.

The flexural transfer at interior column does not govern, since the unbalanced column moment is small, compared to the exterior one; furthermore the slab moments are distributed to the slabs at the two faces of the column in proportion to the stiffness of the adjacent spans.

$$M_f = (l - F_i) M$$

where M is from step 9 and F_i is from step 11.

Step 15. Minimum steel area (ACI-7.13, 13.5.3) see Table 3-9.

$$\rho_{\min} = 0.002\, bh \text{ for } f_y \leqslant 50 \text{ ksi}$$
$$\rho_{\min} = 0.0018\, bh \text{ for } f_y = 60 \text{ ksi}$$

Minimum top reinforcement recommended, $A_{s\,\text{neg. min}}$: #4 @ 12″

Step 16. Maximum reinforcing steel area (ACI-8.1, 10.3.2, 10.3.3, 13.3.4.6):

$$\rho_{\max} \leqslant 0.75\, \rho_b$$

(for added ductility $\rho_{\max} = 0.50\, \rho_b$ may be recommended):

$$A_{s_{\max}} \leqslant 0.75\, \frac{0.85\, \beta_1 f_c'}{f_y}\, \frac{87}{87 + f_y}\, bd$$

Step 17. Establish effective depths $d_1 = (h - \text{cover} - \frac{1}{2}\text{ bar})$, and $d_2 = (h - \text{cover} - 1\frac{1}{2}\text{ bar})$; check concrete moment capacity; compute a_u (see Fig.

8-3) and calculate required steel reinforcement:

$$A_s = \frac{M}{a_u d} \geqslant A_{s_{\min}}$$

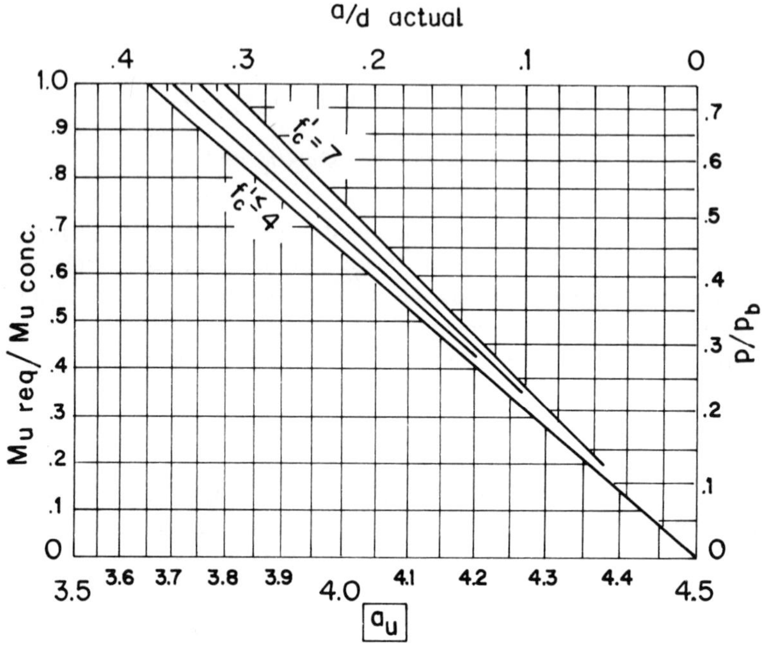

Fig. 8-3.

Step 18. Compute anchorage lengths of bars; for hardrock concrete:

$l_d = 0.04\, A_b f_y / \sqrt{f_c'} \geqslant 0.0004\, d_b f_y \geqslant 12''$, bottom bars

l_d top bars = 1.4 times requirement for bottom bars

Standard hooks (90°) at edges are to be computed according to Table 12.8.1 of ACI Code. Reinforcing bar lengths and layout is to be according to Fig. 13.5.6 of the ACI Code.

Step 19. *Crack control*. Compute

$$z = f_s \sqrt[3]{d_c A} = \begin{matrix} 175 \text{ for interior exposure} \\ 145 \text{ for exterior exposure} \end{matrix} \qquad \text{(ACI-10.6.3)}$$

where

$f_s = 0.60 f_y$

d_c = concrete cover from center of bar

A = concrete area surrounding bars

Table 3-9 gives the minimum reinforcement for crack control.

Step 20. Deflection and camber. Deflection need not be computed if the prescribed thicknesses are employed. See Table 8-1 for minimum slab thickness.

If deflections must be estimated, then the procedure of ACI-9.5.2.2 is to be used. Long-term additional deflections may be estimated by multiplying the immediate deflection by the factor, $[2 - 1.2(A_s'/A_s)] \geqslant 0.6$. It is desirable to camber all flat surfaces to 1/360 of the span, or to the computed long-term load deflection.

Step 21. As tests indicate, failure of flat slabs and flat plates occur due to shear failure around the perimeter of supports. If openings around supports are unavoidable, then the procedure described in ACI-11.12 is to be followed. See also Fig. 11-8 of Commentary of the Code.

If structural steel shear heads are employed, then the provisions of ACI-11.11 are to be followed. See also Fig. 11-7 of the Commentary.

Steps for Design of Two-Way Slabs on Beams Using Direct Design Method

The steps described in the previous section shall be utilized with the additional steps described below.

Step 2/a. Compare the beam-to-slab stiffness ratios in the two perpendicular directions.

1) Compute the $\alpha = I_b/I_s$ ratios in each direction. I_b equals the moment of inertia of the beam, where the beam size includes the slab width of $(h - t)$ or $4t$, whichever is least. I_s equals the moment of inertia of the slab bounded laterally by centerlines of the adjacent panels on each side of the beam. Values for α and I_b are given in Tables 8-12 and 8-13, respectively.

2) Compute: $R = \alpha_1 l_2^2 / \alpha_2 l_1^2$

3) Limitations: $0.2 \leqslant R \leqslant 5.0$

Step 5/e. Use the modified equations,

$$\alpha_{ec} = \frac{K_{ec}}{\sum(K_s + K_b)}$$

Step 10/b. The distribution of design moments to column and middle strips requires additional parameters: $\alpha_1 l_2/l_1$ and $\beta_t = C/2I_s$. The following tables give the design coefficients, per Section 13.3.4 of ACI Code, for varying beam depths, widths, slab thicknesses, ratios of l_1/l_2, and spans.

Table 8-15: Percentage of interior negative moment, ACI-13.3.4.1

Table 8-16: Percentage of positive moments, ACI-13.3.4.3

Tables 8-17 and 8-18: Percentage of exterior negative moment, ACI-13.3.4.2

In Table 8-17 the size of the edge beam is the same as the beam in direction of l_1, and in Table 8-18 the width of edge beam equals the beam width (in direction of l_1) multiplied by l_2/l_1.

Table 8-13

MOMENT OF INERTIA OF T/BEAMS PER ACI CODE, SECTION 13.1.5

PAGE 1

		L SHAPE							T SHAPE						
h_b	h_s	b =12	b =16	b =20	b =24	b =28	b =32	b =36	b =12	b =16	b =20	b =24	b =28	b =32	b =36
8	0	512.	682.	853.	1024.	1194.	1365.	1536.	512.	682.	853.	1024.	1194.	1365.	1536.
	4	588.	760.	932.	1104.	1275.	1446.	1617.	650.	827.	1002.	1176.	1349.	1521.	1693.
	6	558.	729.	900.	1071.	1242.	1412.	1583.	603.	774.	946.	1117.	1288.	1459.	1630.
10	0	1000.	1333.	1666.	2000.	2333.	2666.	3000.	1000.	1333.	1666.	2000.	2333.	2666.	3000.
	4	1212.	1553.	1891.	2228.	2564.	2899.	3234.	1372.	1729.	2079.	2424.	2766.	3106.	3445.
	6	1152.	1488.	1824.	2159.	2493.	2827.	3162.	1281.	1625.	1965.	2304.	2641.	2977.	3313.
	8	1099.	1433.	1766.	2100.	2433.	2767.	3100.	1195.	1530.	1864.	2198.	2532.	2866.	3200.
12	0	1728.	2304.	2880.	3456.	4032.	4608.	5184.	1728.	2304.	2880.	3456.	4032.	4608.	5184.
	4	2189.	2785.	3374.	3959.	4542.	5123.	5703.	2522.	3157.	3773.	4379.	4977.	5571.	6161.
	6	2095.	2684.	3269.	3852.	4432.	5012.	5591.	2376.	2991.	3594.	4190.	4781.	5369.	5955.
	8	2003.	2584.	3163.	3741.	4319.	4896.	5473.	2246.	2837.	3423.	4006.	4588.	5168.	5748.
	10	1912.	2488.	3065.	3641.	4217.	4793.	5369.	2092.	2670.	3247.	3824.	4401.	4977.	5553.
14	0	2744.	3658.	4573.	5488.	6402.	7317.	8232.	2744.	3658.	4573.	5488.	6402.	7317.	8232.
	4	3605.	4560.	5501.	6434.	7363.	8288.	9211.	4205.	5239.	6235.	7210.	8170.	9120.	10064.
	6	3485.	4435.	5372.	6304.	7230.	8155.	9077.	4009.	5021.	6005.	6970.	7924.	8870.	9810.
	8	3336.	4270.	5198.	6122.	7043.	7963.	8882.	3805.	4775.	5728.	6672.	7608.	8540.	9469.
	10	3206.	4127.	5046.	5964.	6881.	7797.	8713.	3627.	4561.	5488.	6413.	7335.	8255.	9174.
	12	3053.	3968.	4883.	5798.	6713.	7628.	8542.	3357.	4274.	5190.	6106.	7021.	7936.	8851.
16	0	4096.	5461.	6826.	8192.	9557.	10922.	12288.	4096.	5461.	6826.	8192.	9557.	10922.	12288.
	4	5542.	6980.	8393.	9792.	11182.	12566.	13947.	6528.	8102.	9613.	11084.	12531.	13960.	15378.
	6	5418.	6856.	8269.	9669.	11060.	12445.	13826.	6302.	7863.	9368.	10837.	12283.	13713.	15130.
	8	5205.	6621.	8021.	9411.	10794.	12174.	13550.	6007.	7509.	8972.	10410.	11832.	13243.	14646.
	10	5007.	6398.	7781.	9159.	10533.	11906.	13277.	5760.	7196.	8612.	10014.	11409.	12797.	14181.
	12	4825.	6199.	7569.	8938.	10306.	11674.	13041.	5504.	6892.	8274.	9651.	11025.	12398.	13769.
18	0	5832.	7776.	9720.	11664.	13608.	15552.	17496.	5832.	7776.	9720.	11664.	13608.	15552.	17496.
	4	8085.	10147.	12169.	14167.	16152.	18127.	20096.	9595.	11876.	14055.	16171.	18247.	20295.	22324.
	6	7992.	10065.	12096.	14101.	16092.	18072.	20044.	9374.	11664.	13854.	15984.	18072.	20131.	22169.
	8	7718.	9767.	11783.	13778.	15760.	17734.	19702.	8983.	11200.	13342.	15436.	17497.	19535.	21557.
	10	7432.	9444.	11433.	13410.	15379.	17342.	19302.	8635.	10755.	12825.	14865.	16884.	18888.	20882.
	12	7182.	9158.	11124.	13083.	15039.	16992.	18943.	8337.	10368.	12373.	14364.	16344.	18316.	20284.

Table 8-13

MOMENT OF INERTIA OF T/BEAMS PER ACI CODE, SECTION 13.1.5

		L SHAPE							T SHAPE						
h_b	h_s	b = 12	b = 16	b = 20	b = 24	b = 28	b = 32	b = 36	b = 12	b = 16	b = 20	b = 24	b = 28	b = 32	b = 36
20	0	8000.	10666.	13333.	16000.	18666.	21333.	24000.	8000.	10666.	13333.	16000.	18666.	21333.	24000.
	4	11319.	14165.	16949.	19699.	22427.	25142.	27846.	13513.	16688.	19710.	22638.	25505.	28330.	31126.
	6	11300.	14178.	16986.	19754.	22497.	25223.	27937.	13346.	16568.	19634.	22601.	25502.	28357.	31178.
	8	10980.	13837.	16632.	19392.	22128.	24850.	27561.	12664.	16010.	19027.	21961.	24837.	27674.	30480.
	10	10598.	13404.	16166.	18902.	21621.	24328.	27028.	12393.	15410.	18333.	21196.	24017.	26809.	29579.
	12	10249.	13000.	15724.	18432.	21129.	23820.	26507.	12010.	14890.	17713.	20498.	23256.	26000.	28729.
22	0	10648.	14197.	17746.	21296.	24845.	28394.	31944.	10648.	14197.	17746.	21296.	24845.	28394.	31944.
	4	14905.	18669.	22357.	26004.	29626.	33232.	36825.	17801.	21971.	25948.	29811.	33600.	37338.	41040.
	6	15441.	19312.	23078.	26782.	30448.	34089.	37711.	18338.	22724.	26877.	30883.	34789.	38625.	42410.
	8	15098.	18957.	22718.	26420.	30086.	33726.	37349.	17780.	22099.	26214.	30197.	34089.	37915.	41694.
	10	14618.	18419.	22140.	25816.	29460.	33085.	36695.	17173.	21334.	25337.	29236.	33062.	36838.	40574.
	12	14150.	17874.	21543.	25180.	28796.	32397.	35989.	16670.	20644.	24509.	28301.	32043.	35749.	39428.
24	0	13824.	18432.	23040.	27648.	32256.	36864.	41472.	13824.	18432.	23040.	27648.	32256.	36864.	41472.
	4	19145.	24003.	28772.	33493.	38184.	42857.	47515.	22856.	28202.	33315.	38291.	43178.	48006.	52791.
	6	20510.	25583.	30505.	35338.	40116.	44857.	49572.	24469.	30277.	35754.	41020.	46144.	51167.	56116.
	8	20178.	25258.	30190.	35033.	39819.	44568.	49289.	23863.	29627.	35090.	40356.	45485.	50517.	55476.
	10	19606.	24625.	29517.	34333.	39099.	43832.	48542.	23113.	28699.	34038.	39213.	44274.	49251.	54167.
	12	19008.	23930.	28755.	33523.	38253.	42957.	47643.	22464.	27812.	32976.	38016.	42969.	47860.	52704.
26	0	17576.	23434.	29293.	35152.	41010.	46869.	52728.	17576.	23434.	29293.	35152.	41010.	46869.	52728.
	4	24087.	30231.	36274.	42260.	48214.	54145.	60061.	28729.	35449.	41892.	48174.	54353.	60462.	66523.
	6	26602.	33108.	39403.	45576.	51672.	57716.	63724.	31861.	39374.	46434.	53205.	59780.	66216.	72550.
	8	26324.	32867.	39195.	45397.	51516.	57580.	63604.	31242.	38755.	45842.	52649.	59262.	65734.	72103.
	10	25678.	32163.	38457.	44635.	50738.	56791.	62806.	30353.	37676.	44638.	51356.	57903.	64327.	70656.
	12	24942.	31314.	37531.	43653.	49715.	55734.	61723.	29535.	36574.	43326.	49885.	56307.	62629.	68875.
28	0	21952.	29269.	36586.	43904.	51221.	58538.	65856.	21952.	29269.	36586.	43904.	51221.	58538.	65856.
	4	29778.	37418.	44942.	52403.	59827.	67225.	74607.	35469.	43776.	51760.	59557.	67237.	74837.	82381.
	6	33815.	42002.	49907.	57649.	65287.	72855.	80374.	40632.	50160.	59088.	67630.	75910.	84004.	91962.
	8	33644.	41912.	49884.	57680.	65362.	72967.	80517.	40048.	49642.	58656.	67288.	75652.	83824.	91851.
	10	32945.	41170.	49120.	56903.	64578.	72179.	79726.	39029.	48437.	57336.	65891.	74205.	82340.	90342.
	12	32075.	40174.	48041.	55765.	63396.	70962.	78481.	38028.	47110.	55773.	64151.	72325.	80349.	88260.
30	0	27000.	36000.	45000.	54000.	63000.	72000.	81000.	27000.	36000.	45000.	54000.	63000.	72000.	81000.
	4	36268.	45628.	54858.	64018.	73135.	82225.	91296.	43128.	53248.	62999.	72537.	81942.	91257.	100510.
	6	42243.	52382.	62154.	71712.	81133.	90463.	99728.	50904.	62784.	73885.	84486.	94747.	104765.	114605.
	8	42241.	52521.	62404.	72051.	81545.	90935.	100250.	50412.	62449.	73721.	84483.	94891.	105042.	114999.
	10	41523.	51784.	61666.	71318.	80820.	90218.	99541.	49280.	61151.	72333.	83047.	93430.	103568.	113522.
	12	40527.	50658.	60456.	70050.	79509.	88874.	98172.	48089.	59600.	70528.	81054.	91291.	101316.	111178.

Table 8-13

MOMENT OF INERTIA OF T/BEAMS PER ACI CODE, SECTION 13.1.5

		L SHAPE							T SHAPE						
h_b	h_s	b =12	b =16	b =20	b =24	b =28	b =32	b =36	b =12	b =16	b =20	b =24	b =28	b =32	b =36
32	0	32768.	43690.	54613.	65536.	76458.	87381.	98304.	32768.	43690.	54613.	65536.	76458.	87381.	98304.
	4	43605.	54926.	66102.	77200.	88251.	99272.	110273.	51754.	63931.	75690.	87210.	98581.	109852.	121053.
	6	50898.	63116.	74911.	86461.	97857.	109149.	120368.	61444.	75704.	89044.	101797.	114155.	126233.	138105.
	8	52224.	64822.	76905.	88678.	100251.	111687.	123026.	62464.	77336.	91221.	104448.	117213.	129644.	141824.
	10	51525.	64143.	76256.	88061.	99664.	111126.	122488.	61245.	75990.	89829.	103050.	115831.	128286.	140496.
	12	50417.	62911.	74947.	86701.	98268.	109704.	121046.	59861.	74224.	87801.	100834.	113474.	125822.	137946.
34	0	39304.	52405.	65506.	78608.	91709.	104810.	117912.	39304.	52405.	65506.	78608.	91709.	104810.	117912.
	4	51836.	65374.	78753.	92046.	105287.	118496.	131681.	61397.	75890.	89914.	103673.	117266.	130749.	144156.
	6	60597.	75153.	89230.	103030.	116657.	130167.	143597.	73258.	90177.	106024.	121194.	135910.	150307.	164472.
	8	63695.	78943.	93534.	107729.	121667.	135430.	149067.	76333.	94462.	111344.	127391.	142853.	157887.	172601.
	10	63064.	78385.	93051.	107313.	121310.	135124.	148806.	75061.	93125.	110025.	126128.	141660.	156771.	171561.
	12	61866.	77080.	91685.	105911.	119886.	133685.	147356.	73488.	91162.	107806.	123733.	139140.	154160.	168882.
36	0	46656.	62208.	77760.	93312.	108864.	124416.	139968.	46656.	62208.	77760.	93312.	108864.	124416.	139968.
	4	61011.	77038.	92891.	108651.	124355.	140023.	155666.	72104.	89188.	105752.	122022.	138110.	154077.	169961.
	6	71388.	88560.	105192.	121515.	137646.	153648.	169560.	86400.	106272.	124909.	142776.	160128.	177120.	193850.
	8	76763.	95013.	112440.	129371.	145980.	162367.	178596.	92151.	113989.	134277.	153526.	172043.	190027.	207609.
	10	76253.	94649.	112209.	129254.	145960.	162431.	178732.	90867.	112726.	133120.	152506.	171170.	189298.	207017.
	12	74995.	93312.	110838.	127872.	144576.	161049.	177355.	89115.	110592.	130752.	149990.	168558.	186624.	204302.
38	0	54872.	73162.	91453.	109744.	128034.	146325.	164616.	54872.	73162.	91453.	109744.	128034.	146325.	164616.
	4	71176.	89982.	108598.	127112.	145566.	163981.	182370.	83926.	103891.	123283.	142353.	161225.	179964.	198613.
	6	83320.	103399.	122877.	142013.	160936.	179718.	198401.	100924.	124056.	145784.	166641.	186921.	206799.	226385.
	8	91531.	113159.	133773.	153774.	173376.	192704.	211836.	110047.	136075.	160206.	183062.	205019.	226319.	247124.
	10	91207.	113071.	133891.	154066.	173814.	193267.	212505.	108801.	134964.	159315.	182414.	204613.	226142.	247162.
	12	89923.	111752.	132579.	152775.	172550.	192030.	211295.	106885.	132693.	156850.	179846.	201994.	223505.	244525.
40	0	64000.	85333.	106666.	128000.	149333.	170666.	192000.	64000.	85333.	106666.	128000.	149333.	170666.	192000.
	4	82381.	104269.	125952.	147525.	169033.	190500.	211938.	96911.	120064.	142589.	164763.	186722.	208539.	230257.
	6	96444.	119737.	142366.	164619.	186640.	208506.	230264.	116884.	143598.	168730.	192888.	216404.	239475.	262224.
	8	108106.	133510.	157680.	181104.	204041.	226645.	249009.	130152.	160881.	189319.	216212.	242015.	267020.	291423.
	10	108038.	133790.	158257.	181928.	205072.	227850.	250362.	129000.	160010.	188809.	216076.	242240.	267581.	292294.
	12	106770.	132549.	157076.	180814.	204023.	226861.	249428.	126944.	157647.	186313.	213541.	239717.	265099.	289867.
42	0	74088.	98784.	123480.	148176.	172872.	197568.	222264.	74088.	98784.	123480.	148176.	172872.	197568.	222264.
	4	94674.	119964.	145033.	169985.	194868.	219707.	244515.	111108.	137769.	163748.	189348.	214714.	239928.	265036.
	6	110808.	137638.	163740.	189432.	214870.	240140.	265294.	134332.	164966.	193831.	221616.	248691.	275277.	301510.
	8	124516.	153723.	181548.	208541.	234995.	261079.	286900.	150220.	185496.	218143.	249032.	278689.	307447.	335532.
	10	126861.	156944.	185468.	213022.	239935.	266400.	292541.	151602.	188035.	221803.	253722.	284304.	313889.	342711.
	12	125658.	155848.	184500.	212180.	239207.	265775.	292007.	149433.	185630.	219350.	251316.	281993.	311696.	340645.

TABLES 8/14–8/18 DISTRIBUTION OF MOMENTS TO COLUMN STRIPS IN A 2-WAY SLAB AND BEAM CONSTRUCTION, PER ACI-13.3.4

TABLE 8-14 VALUES OF $\alpha 1.\ell 2/\ell 1$ (FOR CHECKING ONLY) PAGE 1

BEAM h_b	b	h_s	$\ell 1 = 15$ γ=0.50	γ=0.75	γ=1.00	γ=1.50	γ=2.00	$\ell 1 = 18$ γ=0.50	γ=0.75	γ=1.00	γ=1.50	γ=2.00	$\ell 1 = 21$ γ=0.50	γ=0.75	γ=1.00	γ=1.50	γ=2.00
12	12	6	0.733	0.733	0.733	0.733	0.733	0.611	0.611	0.611	0.611	0.611	0.523	0.523	0.523	0.523	0.523
12	12	8	0.292	0.292	0.292	0.292	0.292	0.243	0.243	0.243	0.243	0.243	0.208	0.208	0.208	0.208	0.208
12	12	10	0.139	0.139	0.139	0.139	0.139	0.116	0.116	0.116	0.116	0.116	0.099	0.099	0.099	0.099	0.099
12	20	6	1.109	1.109	1.109	1.109	1.109	0.924	0.924	0.924	0.924	0.924	0.792	0.792	0.792	0.792	0.792
12	20	8	0.445	0.445	0.445	0.445	0.445	0.371	0.371	0.371	0.371	0.371	0.318	0.318	0.318	0.318	0.318
12	20	10	0.216	0.216	0.216	0.216	0.216	0.180	0.180	0.180	0.180	0.180	0.154	0.154	0.154	0.154	0.154
16	12	6	1.945	1.945	1.945	1.945	1.945	1.620	1.620	1.620	1.620	1.620	1.389	1.389	1.389	1.389	1.389
16	12	8	0.782	0.782	0.782	0.782	0.782	0.651	0.651	0.651	0.651	0.651	0.558	0.558	0.558	0.558	0.558
16	12	10	0.384	0.384	0.384	0.384	0.384	0.320	0.320	0.320	0.320	0.320	0.274	0.274	0.274	0.274	0.274
16	20	6	2.891	2.891	2.891	2.891	2.891	2.409	2.409	2.409	2.409	2.409	2.065	2.065	2.065	2.065	2.065
16	20	8	1.168	1.168	1.168	1.168	1.168	0.973	0.973	0.973	0.973	0.973	0.834	0.834	0.834	0.834	0.834
16	20	10	0.574	0.574	0.574	0.574	0.574	0.478	0.478	0.478	0.478	0.478	0.410	0.410	0.410	0.410	0.410
20	12	6	4.119	4.119	4.119	4.119	4.119	3.432	3.432	3.432	3.432	3.432	2.942	2.942	2.942	2.942	2.942
20	12	8	1.675	1.675	1.675	1.675	1.675	1.395	1.395	1.395	1.395	1.395	1.196	1.196	1.196	1.196	1.196
20	12	10	0.826	0.826	0.826	0.826	0.826	0.688	0.688	0.688	0.688	0.688	0.590	0.590	0.590	0.590	0.590
20	20	6	6.060	6.060	6.060	6.060	6.060	5.050	5.050	5.050	5.050	5.050	4.328	4.328	4.328	4.328	4.328
20	20	8	2.477	2.477	2.477	2.477	2.477	2.064	2.064	2.064	2.064	2.064	1.769	1.769	1.769	1.769	1.769
20	20	10	1.222	1.222	1.222	1.222	1.222	1.018	1.018	1.018	1.018	1.018	0.873	0.873	0.873	0.873	0.873

BEAM h_b	b	h_s	$\ell 1 = 24$ γ=0.50	γ=0.75	γ=1.00	γ=1.50	γ=2.00	$\ell 1 = 28$ γ=0.50	γ=0.75	γ=1.00	γ=1.50	γ=2.00	$\ell 1 = 32$ γ=0.50	γ=0.75	γ=1.00	γ=1.50	γ=2.00
16	18	8	0.670	0.670	0.670	0.670	0.670	0.575	0.575	0.575	0.575	0.575	0.503	0.503	0.503	0.503	0.503
16	18	10	0.329	0.329	0.329	0.329	0.329	0.282	0.282	0.282	0.282	0.282	0.247	0.247	0.247	0.247	0.247
16	18	12	0.182	0.182	0.182	0.182	0.182	0.156	0.156	0.156	0.156	0.156	0.137	0.137	0.137	0.137	0.137
16	30	8	1.020	1.020	1.020	1.020	1.020	0.874	0.874	0.874	0.874	0.874	0.765	0.765	0.765	0.765	0.765
16	30	10	0.504	0.504	0.504	0.504	0.504	0.432	0.432	0.432	0.432	0.432	0.378	0.378	0.378	0.378	0.378
16	30	12	0.282	0.282	0.282	0.282	0.282	0.242	0.242	0.242	0.242	0.242	0.211	0.211	0.211	0.211	0.211
22	18	8	1.967	1.967	1.967	1.967	1.967	1.686	1.686	1.686	1.686	1.686	1.475	1.475	1.475	1.475	1.475
22	18	10	0.972	0.972	0.972	0.972	0.972	0.833	0.833	0.833	0.833	0.833	0.729	0.729	0.729	0.729	0.729
22	18	12	0.544	0.544	0.544	0.544	0.544	0.466	0.466	0.466	0.466	0.466	0.408	0.408	0.408	0.408	0.408
22	30	8	2.930	2.930	2.930	2.930	2.930	2.511	2.511	2.511	2.511	2.511	2.197	2.197	2.197	2.197	2.197
22	30	10	1.456	1.456	1.456	1.456	1.456	1.248	1.248	1.248	1.248	1.248	1.092	1.092	1.092	1.092	1.092
22	30	12	0.817	0.817	0.817	0.817	0.817	0.700	0.700	0.700	0.700	0.700	0.613	0.613	0.613	0.613	0.613
28	18	8	4.411	4.411	4.411	4.411	4.411	3.781	3.781	3.781	3.781	3.781	3.308	3.308	3.308	3.308	3.308
28	18	10	2.205	2.205	2.205	2.205	2.205	1.890	1.890	1.890	1.890	1.890	1.654	1.654	1.654	1.654	1.654
28	18	12	1.241	1.241	1.241	1.241	1.241	1.064	1.064	1.064	1.064	1.064	0.931	0.931	0.931	0.931	0.931
28	30	8	6.490	6.490	6.490	6.490	6.490	5.563	5.563	5.563	5.563	5.563	4.868	4.868	4.868	4.868	4.868
28	30	10	3.262	3.262	3.262	3.262	3.262	2.796	2.796	2.796	2.796	2.796	2.446	2.446	2.446	2.446	2.446
28	30	12	1.841	1.841	1.841	1.841	1.841	1.578	1.578	1.578	1.578	1.578	1.380	1.380	1.380	1.380	1.380

TABLE 8-15 PERCENTAGE OF INT. NEG. MOMENTS IN A 2-WAY SLAB-BEAM CONSTR'N, ACI-13.3.4.1

BEAM h_b	b	h_s	ℓ1 = 15 γ=0.50	γ=0.75	γ=1.00	γ=1.50	γ=2.00	ℓ1 = 18 γ=0.50	γ=0.75	γ=1.00	γ=1.50	γ=2.00	ℓ1 = 21 γ=0.50	γ=0.75	γ=1.00	γ=1.50	γ=2.00
12	12	6	53.0	67.6	75.0	82.3	86.0	56.6	68.8	75.0	81.1	84.1	59.2	69.7	75.0	80.2	82.8
12	12	8	66.2	72.0	75.0	77.9	79.3	67.6	72.5	75.0	77.4	78.6	68.7	72.9	75.0	77.0	78.1
12	12	10	70.8	73.6	75.0	76.3	77.0	71.5	73.8	75.0	76.1	76.7	72.0	74.0	75.0	75.9	76.4
12	20	6	45.0	65.0	75.0	85.0	90.0	47.2	65.7	75.0	84.2	88.8	51.2	67.0	75.0	82.9	86.8
12	20	8	61.6	70.5	75.0	79.4	81.6	63.8	71.2	75.0	78.7	80.5	65.4	71.8	75.0	78.1	79.7
12	20	10	68.5	72.8	75.0	77.1	78.2	69.5	73.1	75.0	76.8	77.7	70.3	73.4	75.0	76.5	77.3
16	12	6	45.0	65.0	75.0	85.0	90.0	45.0	65.0	75.0	85.0	90.0	45.0	65.0	75.0	85.0	90.0
16	12	8	51.5	67.1	75.0	82.8	86.7	55.4	68.4	75.0	81.5	84.7	58.2	69.4	75.0	80.5	83.3
16	12	10	63.4	71.1	75.0	78.8	80.7	65.3	71.7	75.0	78.2	79.8	66.7	72.2	75.0	77.7	79.1
16	20	6	45.0	65.0	75.0	85.0	90.0	45.0	65.0	75.0	85.0	90.0	45.0	65.0	75.0	85.0	90.0
16	20	8	45.0	65.0	75.0	85.0	90.0	45.7	65.2	75.0	84.7	89.6	49.9	66.6	75.0	83.3	87.5
16	20	10	57.7	69.2	75.0	80.7	83.6	60.6	70.2	75.0	79.7	82.1	62.6	70.8	75.0	79.1	81.1
20	12	6	45.0	65.0	75.0	85.0	90.0	45.0	65.0	75.0	85.0	90.0	45.0	65.0	75.0	85.0	90.0
20	12	8	45.0	65.0	75.0	85.0	90.0	45.0	65.0	75.0	85.0	90.0	45.0	65.0	75.0	85.0	90.0
20	12	10	50.2	66.7	75.0	83.2	87.3	54.3	68.1	75.0	81.8	85.3	57.2	69.0	75.0	80.9	83.8
20	20	6	45.0	65.0	75.0	85.0	90.0	45.0	65.0	75.0	85.0	90.0	45.0	65.0	75.0	85.0	90.0
20	20	8	45.0	65.0	75.0	85.0	90.0	45.0	65.0	75.0	85.0	90.0	45.0	65.0	75.0	85.0	90.0
20	20	10	45.0	65.0	75.0	85.0	90.0	45.0	65.0	75.0	85.0	90.0	48.8	66.2	75.0	83.7	88.0

BEAM h_b	b	h_s	ℓ1 = 24 γ=0.50	γ=0.75	γ=1.00	γ=1.50	γ=2.00	ℓ1 = 28 γ=0.50	γ=0.75	γ=1.00	γ=1.50	γ=2.00	ℓ1 = 32 γ=0.50	γ=0.75	γ=1.00	γ=1.50	γ=2.00
16	18	8	54.8	68.2	75.0	81.7	85.0	57.7	69.2	75.0	80.7	83.6	59.9	69.9	75.0	80.0	82.5
16	18	10	65.1	71.7	75.0	78.2	79.9	66.5	72.1	75.0	77.8	79.2	67.5	72.5	75.0	77.4	78.7
16	18	12	69.5	73.1	75.0	76.8	77.7	70.2	73.4	75.0	76.5	77.3	70.8	73.6	75.0	76.3	77.0
16	30	8	45.0	65.0	75.0	85.0	90.0	48.7	66.2	75.0	83.7	88.1	52.0	67.3	75.0	82.6	86.4
16	30	10	59.8	69.9	75.0	80.0	82.5	62.0	70.6	75.0	79.3	81.4	63.6	71.2	75.0	78.7	80.6
16	30	12	66.5	72.1	75.0	77.8	79.2	67.7	72.5	75.0	77.4	78.6	68.6	72.8	75.0	77.1	78.1
22	18	8	45.0	65.0	75.0	85.0	90.0	45.0	65.0	75.0	85.0	90.0	45.0	65.0	75.0	85.0	90.0
22	18	10	45.8	65.2	75.0	84.7	89.5	49.9	66.6	75.0	83.3	87.5	53.1	67.7	75.0	82.2	85.9
22	18	12	58.6	69.5	75.0	80.4	83.1	60.9	70.3	75.0	79.6	82.0	62.7	70.9	75.0	79.0	81.1
22	30	8	45.0	65.0	75.0	85.0	90.0	45.0	65.0	75.0	85.0	90.0	45.0	65.0	75.0	85.0	90.0
22	30	10	45.0	65.0	75.0	85.0	90.0	45.0	65.0	75.0	85.0	90.0	45.0	65.0	75.0	85.0	90.0
22	30	12	50.4	66.8	75.0	83.1	87.2	53.9	67.9	75.0	82.0	85.5	56.6	68.8	75.0	81.1	84.1
28	18	8	45.0	65.0	75.0	85.0	90.0	45.0	65.0	75.0	85.0	90.0	45.0	65.0	75.0	85.0	90.0
28	18	10	45.0	65.0	75.0	85.0	90.0	45.0	65.0	75.0	85.0	90.0	45.0	65.0	75.0	85.0	90.0
28	18	12	45.0	65.0	75.0	85.0	90.0	45.0	65.0	75.0	85.0	90.0	47.0	65.6	75.0	84.3	88.9
28	30	8	45.0	65.0	75.0	85.0	90.0	45.0	65.0	75.0	85.0	90.0	45.0	65.0	75.0	85.0	90.0
28	30	10	45.0	65.0	75.0	85.0	90.0	45.0	65.0	75.0	85.0	90.0	45.0	65.0	75.0	85.0	90.0
28	30	12	45.0	65.0	75.0	85.0	90.0	45.0	65.0	75.0	85.0	90.0	45.0	65.0	75.0	85.0	90.0

$\%M_o$

$\%M_o$

TABLE 8-16 PERCENTAGE OF POSITIVE MOMENTS IN A 2-WAY SLAB-BEAM CONSTR'N, ACI-13.3.4.3

BEAM			ℓ1 = 15					ℓ1 = 18					ℓ1 = 21				
h_b	b	h_s	γ=0.50	γ=0.75	γ=1.00	γ=1.50	γ=2.00	γ=0.50	γ=0.75	γ=1.00	γ=1.50	γ=2.00	γ=0.50	γ=0.75	γ=1.00	γ=1.50	γ=2.00
12	12	6	49.0	63.6	71.0	78.3	82.0	50.8	63.0	69.1	75.2	78.3	52.1	62.6	67.8	73.0	75.7
12	12	8	55.6	61.4	64.3	67.3	68.7	56.3	61.2	63.6	66.0	67.3	56.8	61.0	63.1	65.2	66.2
12	12	10	57.9	60.6	62.0	63.4	64.1	58.2	60.5	61.7	62.9	63.4	58.5	60.4	61.4	62.4	62.9
12	20	6	45.0	65.0	75.0	85.0	90.0	46.1	64.6	73.8	83.1	87.7	48.1	63.9	71.8	79.8	83.7
12	20	8	53.3	62.2	66.6	71.1	73.3	54.4	61.8	65.5	69.2	71.1	55.2	61.5	64.7	67.9	69.5
12	20	10	56.7	61.0	63.2	65.4	66.4	57.2	60.9	62.7	64.5	65.4	57.6	60.7	62.3	63.8	64.6
16	12	6	45.0	65.0	75.0	85.0	90.0	45.0	65.0	75.0	85.0	90.0	45.0	65.0	75.0	85.0	90.0
16	12	8	48.2	63.9	71.7	79.5	83.4	50.2	63.2	69.7	76.2	79.5	51.6	62.7	68.3	73.9	76.7
16	12	10	54.2	61.9	65.7	69.6	71.5	55.1	61.6	64.8	68.0	69.6	55.8	61.3	64.1	66.8	68.2
16	20	6	45.0	65.0	75.0	85.0	90.0	45.0	65.0	75.0	85.0	90.0	45.0	65.0	75.0	85.0	90.0
16	20	8	45.0	65.0	75.0	85.0	90.0	45.3	64.8	74.6	84.3	89.2	47.4	64.1	72.5	80.8	85.0
16	20	10	51.3	62.8	68.6	74.3	77.2	52.8	62.3	67.1	71.9	74.3	53.8	62.0	66.1	70.2	72.3
20	12	6	45.0	65.0	75.0	85.0	90.0	45.0	65.0	75.0	85.0	90.0	45.0	65.0	75.0	85.0	90.0
20	12	8	45.0	65.0	75.0	85.0	90.0	45.0	65.0	75.0	85.0	90.0	45.0	65.0	75.0	85.0	90.0
20	12	10	47.6	64.1	72.3	80.6	84.7	49.6	63.4	70.3	77.2	80.6	51.1	62.9	68.8	74.7	77.7
20	20	6	45.0	65.0	75.0	85.0	90.0	45.0	65.0	75.0	85.0	90.0	45.0	65.0	75.0	85.0	90.0
20	20	8	45.0	65.0	75.0	85.0	90.0	45.0	65.0	75.0	85.0	90.0	45.0	65.0	75.0	85.0	90.0
20	20	10	45.0	65.0	75.0	85.0	90.0	45.0	65.0	75.0	85.0	90.0	46.9	64.3	73.0	81.8	86.1

BEAM			ℓ1 = 24					ℓ1 = 28					ℓ1 = 32				
h_b	b	h_s	γ=0.50	γ=0.75	γ=1.00	γ=1.50	γ=2.00	γ=0.50	γ=0.75	γ=1.00	γ=1.50	γ=2.00	γ=0.50	γ=0.75	γ=1.00	γ=1.50	γ=2.00
16	18	8	49.9	63.3	70.0	76.7	80.1	51.3	62.8	68.6	74.3	77.2	52.4	62.5	67.5	72.5	75.0
16	18	10	55.0	61.6	64.9	68.2	69.8	55.7	61.4	64.2	67.0	68.4	56.2	61.2	63.7	66.1	67.4
16	18	12	57.2	60.9	62.7	64.5	65.4	57.6	60.7	62.3	63.9	64.7	57.9	60.6	62.0	63.4	64.1
16	30	8	45.0	65.0	75.0	85.0	90.0	46.8	64.3	73.1	81.8	86.2	48.5	63.8	71.4	79.1	82.9
16	30	10	52.4	62.5	67.5	72.6	75.1	53.5	62.1	66.4	70.8	72.9	54.3	61.8	65.6	69.4	71.3
16	30	12	55.7	61.4	64.2	67.0	68.4	56.3	61.2	63.6	66.0	67.2	56.8	61.0	63.1	65.2	66.3
22	18	8	45.0	65.0	75.0	85.0	90.0	45.0	65.0	75.0	85.0	90.0	45.0	65.0	75.0	85.0	90.0
22	18	10	45.4	64.8	74.5	84.3	89.1	47.4	64.1	72.5	80.8	85.0	49.0	63.6	70.9	78.2	81.8
22	18	12	51.8	62.7	68.1	73.6	76.3	52.9	62.3	67.0	71.6	74.0	53.8	62.0	66.1	70.2	72.2
22	30	8	45.0	65.0	75.0	85.0	90.0	45.0	65.0	75.0	85.0	90.0	45.0	65.0	75.0	85.0	90.0
22	30	10	45.0	65.0	75.0	85.0	90.0	45.0	65.0	75.0	85.0	90.0	45.0	65.0	75.0	85.0	90.0
22	30	12	47.7	64.0	72.2	80.4	84.5	49.4	63.5	70.5	77.5	81.0	50.8	63.0	69.1	75.3	78.3
28	18	8	45.0	65.0	75.0	85.0	90.0	45.0	65.0	75.0	85.0	90.0	45.0	65.0	75.0	85.0	90.0
28	18	10	45.0	65.0	75.0	85.0	90.0	45.0	65.0	75.0	85.0	90.0	45.0	65.0	75.0	85.0	90.0
28	18	12	45.0	65.0	75.0	85.0	90.0	45.0	65.0	75.0	85.0	90.0	46.0	64.6	73.9	83.2	87.9
28	30	8	45.0	65.0	75.0	85.0	90.0	45.0	65.0	75.0	85.0	90.0	45.0	65.0	75.0	85.0	90.0
28	30	10	45.0	65.0	75.0	85.0	90.0	45.0	65.0	75.0	85.0	90.0	45.0	65.0	75.0	85.0	90.0
28	30	12	45.0	65.0	75.0	85.0	90.0	45.0	65.0	75.0	85.0	90.0	45.0	65.0	75.0	85.0	90.0

$\%M_o$

$\%M_o$

TABLE 8-17. PERCENTAGE OF EXT. NEG. MOMENTS IN A 2-WAY SLAB-BEAM CONSTR'N, ACI-13.3.4.2
THE SIZE OF EDGE BEAM IS THE SAME AS THE BEAM IN DIRECION OF ℓ1

PAGE 4

BEAM			ℓ1 = 15					ℓ1 = 18					ℓ1 = 21				
h_b	b	h_s	γ=0.50	γ=0.75	γ=1.00	γ=1.50	γ=2.00	γ=0.50	γ=0.75	γ=1.00	γ=1.50	γ=2.00	γ=0.50	γ=0.75	γ=1.00	γ=1.50	γ=2.00
12	12	6	96.0	95.9	95.8	95.5	95.3	96.9	96.7	96.5	96.0	95.5	97.5	97.2	97.0	96.4	95.8
12	12	8	98.8	98.5	98.2	97.6	97.1	99.0	98.8	98.5	98.0	97.5	99.2	98.9	98.7	98.2	97.8
12	12	10	99.4	99.3	99.1	98.7	98.4	99.5	99.4	99.2	98.9	98.6	99.6	99.5	99.3	99.1	98.8
12	20	6	87.5	88.1	88.6	89.8	90.9	90.0	90.3	90.5	91.0	91.6	92.1	92.0	91.9	91.7	91.5
12	20	8	96.3	95.8	95.2	94.1	93.0	97.1	96.5	96.0	94.9	93.8	97.6	97.1	96.6	95.5	94.5
12	20	10	98.4	98.0	97.6	96.7	95.8	98.7	98.3	98.0	97.2	96.4	98.9	98.6	98.2	97.5	96.8
16	12	6	90.9	91.3	91.8	92.6	93.4	92.4	92.8	93.1	93.8	94.5	93.5	93.8	94.1	94.7	95.3
16	12	8	96.6	96.5	96.5	96.3	96.2	97.4	97.2	97.0	96.7	96.4	97.9	97.7	97.5	97.0	96.6
16	12	10	98.7	98.4	98.2	97.7	97.2	98.9	98.7	98.5	98.0	97.6	99.1	98.9	98.7	98.3	97.8
16	20	6	76.2	77.3	78.4	85.0	90.0	80.2	81.1	82.0	85.0	90.0	83.0	83.8	84.5	86.1	90.0
16	20	8	89.9	90.3	90.8	91.7	92.6	91.7	92.0	92.3	93.0	93.6	93.4	93.4	93.4	93.4	93.4
16	20	10	96.0	95.6	95.3	94.6	93.8	96.9	96.5	96.1	95.2	94.4	97.5	97.0	96.6	95.8	94.9
20	12	6	86.5	87.1	87.8	89.0	90.2	88.8	89.3	89.8	90.8	91.8	90.4	90.8	91.2	92.1	93.0
20	12	8	94.0	94.2	94.5	95.1	95.6	95.0	95.2	95.4	95.9	96.3	95.7	95.9	96.1	96.5	96.8
20	12	10	97.2	97.2	97.2	97.1	97.1	97.8	97.7	97.6	97.4	97.2	98.2	98.1	98.0	97.7	97.4
20	20	6	65.2	66.8	75.0	85.0	90.0	71.0	72.3	75.0	85.0	90.0	75.1	76.3	77.4	85.0	90.0
20	20	8	85.0	85.6	86.3	87.7	90.0	87.5	88.0	88.6	89.7	90.9	89.2	89.7	90.2	91.2	92.2
20	20	10	92.3	92.6	93.0	93.7	94.4	93.5	93.8	94.1	94.7	95.3	94.8	94.9	95.0	95.1	95.2

BEAM			ℓ1 = 24					ℓ1 = 28					ℓ1 = 32				
h_b	b	h_s	γ=0.50	γ=0.75	γ=1.00	γ=1.50	γ=2.00	γ=0.50	γ=0.75	γ=1.00	γ=1.50	γ=2.00	γ=0.50	γ=0.75	γ=1.00	γ=1.50	γ=2.00
16	18	8	95.8	95.6	95.3	94.9	94.4	96.6	96.3	96.0	95.4	94.8	97.2	96.8	96.5	95.8	95.1
16	18	10	98.3	98.0	97.6	96.9	96.2	98.6	98.3	97.9	97.3	96.6	98.8	98.5	98.2	97.6	97.0
16	18	12	99.1	98.9	98.6	98.1	97.6	99.3	99.0	98.8	98.3	97.9	99.4	99.2	99.0	98.5	98.1
16	30	8	87.6	88.1	88.7	89.8	90.9	90.0	90.2	90.3	90.5	90.8	91.8	91.7	91.5	91.2	90.8
16	30	10	95.3	94.8	94.2	93.1	91.9	96.2	95.6	95.0	93.8	92.6	96.8	96.2	95.6	94.5	93.3
16	30	12	97.7	97.2	96.6	95.6	94.5	98.1	97.6	97.1	96.1	95.1	98.4	97.9	97.5	96.5	95.6
22	18	8	90.0	90.4	90.9	91.8	92.7	91.4	91.8	92.2	93.0	93.7	92.5	92.8	93.2	93.8	94.5
22	18	10	94.8	95.0	95.2	95.6	96.0	95.9	95.9	95.9	95.9	95.9	96.6	96.5	96.4	96.2	96.0
22	18	12	97.7	97.5	97.2	96.8	96.3	98.1	97.9	97.6	97.1	96.6	98.4	98.2	97.9	97.4	96.9
22	30	8	73.6	74.8	76.0	85.0	90.0	77.4	78.4	79.4	85.0	90.0	80.2	81.1	82.0	85.0	90.0
22	30	10	86.4	87.0	87.6	88.9	90.1	88.3	88.9	89.4	90.4	91.5	89.8	90.2	90.7	91.6	92.6
22	30	12	92.9	92.8	92.8	92.7	92.7	94.3	94.1	93.8	93.3	92.9	95.3	95.0	94.6	93.9	93.2
28	18	8	84.3	85.0	85.7	87.2	90.0	86.5	87.2	87.8	89.0	90.2	88.2	88.8	89.3	90.4	91.4
28	18	10	91.6	92.0	92.4	93.1	93.7	92.8	93.1	93.5	94.1	94.8	93.7	94.0	94.3	94.8	95.4
28	18	12	95.0	95.2	95.5	95.9	96.4	95.7	95.9	96.1	96.5	96.9	96.4	96.5	96.6	96.8	97.0
28	30	8	58.3	65.0	75.0	85.0	90.0	64.3	65.9	75.0	85.0	90.0	68.7	70.2	75.0	85.0	90.0
28	30	10	78.3	79.3	80.3	85.0	90.0	81.4	82.3	83.1	85.0	90.0	83.7	84.5	85.2	86.7	90.0
28	30	12	87.3	87.9	88.5	89.6	90.8	89.1	89.6	90.1	91.1	92.1	90.5	90.9	91.3	92.2	93.1

%Mo

%Mo

TABLE 8-18 PERCENTAGE OF EXT. NEG. MOMENTS IN A 2-WAY SLAB-BEAM CONSTR'N, ACI-13.3.4.2 PAGE 5
THE WIDTH OF EDGE BEAM EQUALS THE BEAM WIDTH IN DIRECTION OF ℓ1 MULTIPLIED BY ℓ2/ℓ1

BEAM			ℓ1 = 15					ℓ1 = 18					ℓ1 = 21				
h_b	b	h_s	γ=0.50	γ=0.75	γ=1.00	γ=1.50	γ=2.00	γ=0.50	γ=0.75	γ=1.00	γ=1.50	γ=2.00	γ=0.50	γ=0.75	γ=1.00	γ=1.50	γ=2.00
12	12	6	86.0	92.4	95.8	97.7	93.7	89.2	93.9	96.5	98.0	98.7	91.3	94.9	97.0	98.2	98.8
12	12	8	95.7	97.2	98.2	98.8	99.2	96.6	97.7	98.5	99.0	99.3	97.2	98.1	98.7	99.1	99.4
12	12	10	98.1	98.7	99.1	99.4	99.6	98.4	98.9	99.2	99.5	99.6	98.7	99.0	99.3	99.5	99.7
12	20	6	68.0	81.9	88.6	95.1	97.4	74.4	85.2	90.5	95.7	97.6	79.7	87.8	91.9	96.0	97.6
12	20	8	90.6	93.6	95.2	97.2	98.0	92.6	94.8	96.0	97.6	98.2	93.9	95.6	96.6	97.9	98.4
12	20	10	96.0	97.0	97.6	98.4	98.8	96.8	97.5	98.0	98.7	99.0	97.3	97.9	98.2	98.8	99.1
16	12	6	66.9	86.1	91.8	96.7	98.3	72.4	88.4	93.1	97.3	98.6	76.4	90.1	94.1	97.7	98.8
16	12	8	87.6	94.4	96.5	98.4	99.0	90.5	95.5	97.0	98.5	99.0	92.4	96.3	97.5	98.7	99.1
16	12	10	95.2	97.5	98.2	99.0	99.3	96.2	97.9	98.5	99.1	99.4	96.9	98.3	98.7	99.2	99.4
16	20	6	45.0	65.0	78.4	91.0	95.4	45.0	68.8	82.0	92.5	96.2	49.9	73.2	84.5	93.5	96.7
16	20	8	70.3	84.1	90.8	96.1	98.0	75.6	86.9	92.3	96.7	98.3	80.7	89.2	93.4	96.9	98.2
16	20	10	88.3	92.9	95.3	97.5	98.3	90.9	94.2	96.1	97.8	98.5	92.6	95.1	96.6	98.0	98.6
20	12	6	47.1	76.8	87.8	95.4	97.6	55.9	80.7	89.8	96.1	98.0	62.2	83.4	91.2	96.7	98.3
20	12	8	77.3	89.9	94.5	97.8	98.7	81.1	91.6	95.4	98.1	98.9	83.8	92.8	96.1	98.4	99.1
20	12	10	89.5	95.0	97.2	98.7	99.1	91.9	96.0	97.6	98.8	99.2	93.5	96.7	98.0	98.9	99.2
20	20	6	45.0	65.0	75.0	86.2	93.4	45.0	65.0	75.0	88.5	94.5	45.0	65.0	77.4	90.1	95.3
20	20	8	46.8	73.5	86.3	93.9	97.0	55.6	77.9	88.6	94.9	97.5	62.0	81.1	90.2	95.6	97.8
20	20	10	72.7	86.4	93.0	96.8	98.4	77.2	88.6	94.1	97.4	98.7	81.8	90.6	95.0	97.5	98.6

BEAM			ℓ1 = 24					ℓ1 = 28					ℓ1 = 32				
h_b	b	h_s	γ=0.50	γ=0.75	γ=1.00	γ=1.50	γ=2.00	γ=0.50	γ=0.75	γ=1.00	γ=1.50	γ=2.00	γ=0.50	γ=0.75	γ=1.00	γ=1.50	γ=2.00
16	18	8	86.8	92.4	95.3	97.6	98.5	89.4	93.7	96.0	97.8	98.6	91.2	94.6	96.5	98.0	98.7
16	18	10	94.7	96.5	97.6	98.5	99.0	95.7	97.1	97.9	98.7	99.1	96.3	97.4	98.2	98.8	99.2
16	18	12	97.3	98.1	98.6	99.1	99.4	97.8	98.4	98.8	99.2	99.4	98.1	98.6	99.0	99.3	99.5
16	30	8	69.2	82.3	88.7	94.8	97.4	75.4	85.3	90.3	95.2	97.3	79.9	87.6	91.5	95.5	97.4
16	30	10	88.5	92.2	94.2	96.5	97.7	90.6	93.5	95.0	96.9	97.9	92.2	94.4	95.6	97.2	98.1
16	30	12	94.4	95.8	96.6	97.7	98.4	95.4	96.5	97.1	98.0	98.6	96.1	96.9	97.5	98.2	98.8
22	18	8	64.1	83.9	90.9	96.3	98.2	69.2	86.2	92.2	96.9	98.4	73.1	87.9	93.2	97.2	98.6
22	18	10	81.8	91.7	95.2	98.0	98.9	85.6	93.2	95.9	98.1	98.9	88.2	94.2	96.4	98.3	98.9
22	18	12	91.9	95.8	97.2	98.5	99.0	93.5	96.5	97.6	98.7	99.1	94.5	97.0	97.9	98.8	99.1
22	30	8	45.0	65.0	76.0	90.2	94.9	45.0	65.4	79.4	91.6	95.6	45.0	69.7	82.0	92.6	96.1
22	30	10	62.0	79.2	87.6	94.9	97.3	67.4	82.2	89.4	95.6	97.7	71.5	84.4	90.7	96.2	97.9
22	30	12	80.2	88.6	92.8	96.7	98.0	84.2	90.6	93.8	96.9	98.0	86.9	92.0	94.6	97.2	98.1
28	18	8	45.0	73.5	85.7	94.7	97.4	47.4	77.3	87.8	95.5	97.7	54.0	80.1	89.3	96.0	98.0
28	18	10	68.3	86.1	92.4	97.0	98.4	72.8	88.1	93.5	97.4	98.6	76.2	89.6	94.3	97.7	98.8
28	18	12	81.5	91.8	95.5	98.1	99.0	84.1	93.0	96.1	98.4	99.1	86.6	94.0	96.6	98.5	99.1
28	30	8	45.0	65.0	75.0	85.0	92.3	45.0	65.0	75.0	86.3	93.4	45.0	65.0	75.0	88.0	94.2
28	30	10	45.0	65.0	80.3	91.5	95.8	45.0	68.5	83.1	92.7	96.4	46.0	72.5	85.2	93.6	96.9
28	30	12	58.2	78.6	88.5	95.0	97.5	64.2	81.7	90.1	95.7	97.8	68.6	84.0	91.3	96.2	98.1

%Mo

%Mo

THE EQUIVALENT FRAME METHOD

In cases when the limitations of the Direct Design Method are not met, an accurate analysis for moment calculations becomes a necessity. The structure is divided into equivalent frames on column lines both longitudinally and transversely. Each frame may be analyzed completely for lateral and vertical loads. However, for vertical loads only, each floor may be analyzed separately with its columns fixed at their far ends. For determining any support moment the slab beam may be considered fixed two panels beyond the span considered.

The moment distribution (or any other analytical or computer solution) provides the design moments due to flexure. All other steps described in detail for the Direct Design Method (shear, torsion, moment-transfer, etc.) are to be carried out after the flexural moments are established.

Step 1. Compute the moment of inertia of the slab beam.

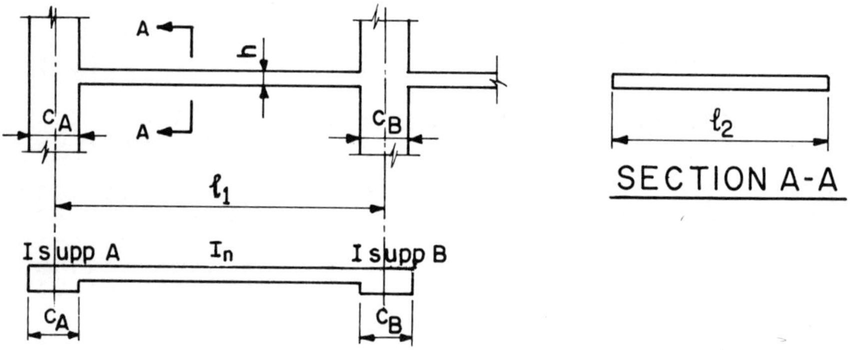

Fig. 8-4.

$$I_n = \text{(between supports)} = l_2 \times h^3/12$$

$$I_{supp} = \frac{I_n}{\left(1 - \frac{c_2}{l_2}\right)^2}$$

where c_2 is the length of transverse support

Compute $\frac{c_A}{l_1}$ and $\frac{c_B}{l_1}$.

Using Tables 13-1 and 13-2 of ACI Commentary 318-71, establish:

Stiffness factors: k_{AB} and k_{BA}

Carry-over factors: COF_{AB} and COF_{BA}

Fixed end moments: $M_{AB}\,(wl_2\,l_1^2)$ and $M_{BA}\,(wl_2\,l_1^2)$

Note: For $c/l \leqslant 0.10$ assume constant I.

Step 2. Compute the moment of inertia of column: Use the equivalent column for computing the stiffness (K_{ec}) of column for frame analysis (by moment distribution or by computer) as per (ACI-13.4.1.5) (see Table 8-7 for values of K_{ec}).

The equivalent column consists of:

(1) A column with $I = \infty$ at ends (K_c): (use Table 8-13).

Fig. 8-5.

(2) A torsional member, perpendicular to the direction of bending (K_t): (use Table 8-6). The largest of the following values are to be used:

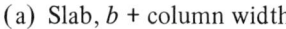

(a) Slab, b + column width

(b) Beam + slab

(c) Beam + slab (slab width = column width)

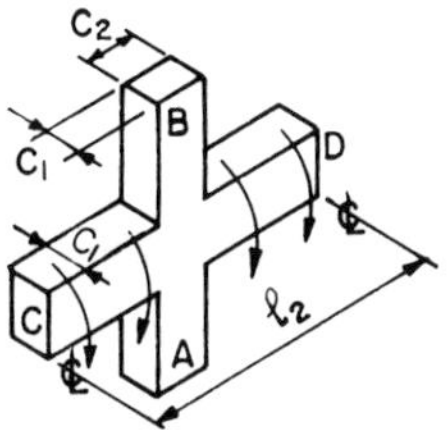

Fig. 8-6.

For cases (a) and (b) compute C (ACI-13-7): (use Table 8-5).

Compute the flexibility (reciprocal value of stiffness) of the torsional member as per equation (ACI-13-6):

$$\frac{1}{K_t} = \sum \frac{9EC}{l_2\left(1 - \frac{c_2}{l_2}\right)^3}$$

(3) The flexibility of the equivalent column is:

$$\frac{1}{K_{ec}} = \frac{1}{\Sigma K_c} + \frac{1}{\Sigma K_t} \qquad \text{(ACI-13-5; use Table 8-7 for } K_{ec}\text{)}$$

Step 3. Establish distribution factors from steps 1 and 2.

Step 4. Complete moment distribution procedure with alternate live loads as per attached Rogers' Method.

Step 5. Employ negative moment redistribution (10 percent +) as per 8.6 of Code.

Step 6. Establish maximum negative moments at faces of support, and maximum positive moments at about midspan.

Step 7. Add (if applicable) moments due to shear transfer (see Direct Design Method).

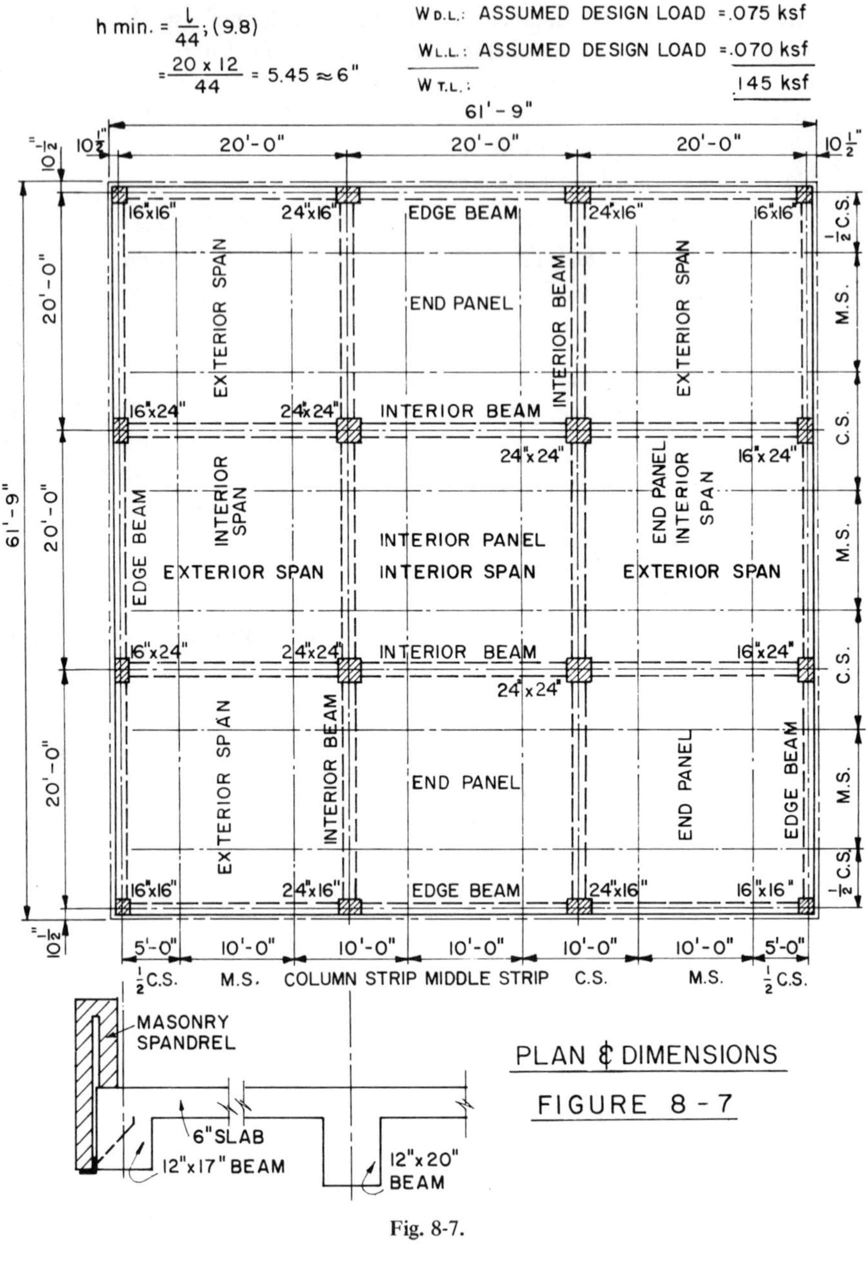
h min. = l/44; (9.8)
= 20 x 12 / 44 = 5.45 ≈ 6"
W D.L.: ASSUMED DESIGN LOAD = .075 ksf
W L.L.: ASSUMED DESIGN LOAD = .070 ksf
W T.L.: .145 ksf
61'-9"
10½"
20'-0"
20'-0"
20'-0"
10½"
16"x16"
24"x16"
EDGE BEAM
24"x16"
16"x16"
EXTERIOR SPAN
END PANEL
INTERIOR BEAM
EXTERIOR SPAN
16"x24"
24"x24"
INTERIOR BEAM
24"x24"
16"x24"
EDGE BEAM
INTERIOR SPAN
EXTERIOR SPAN
INTERIOR PANEL
INTERIOR SPAN
END PANEL INTERIOR SPAN
EXTERIOR SPAN
16"x24"
24"x24"
INTERIOR BEAM
16"x24"
24"x24"
EXTERIOR SPAN
INTERIOR BEAM
END PANEL
END PANEL
EDGE BEAM
16"x16"
24"x16"
EDGE BEAM
24"x16"
16"x16"
½ C.S.
M.S.
C.S.
M.S.
C.S.
M.S.
½ C.S.
5'-0"
10'-0"
10'-0"
10'-0"
10'-0"
10'-0"
5'-0"
½ C.S.
M.S.
COLUMN STRIP MIDDLE STRIP
C.S.
M.S.
½ C.S.
MASONRY SPANDREL
6" SLAB
12"x17" BEAM
12"x20" BEAM
PLAN ℄ DIMENSIONS
FIGURE 8-7

Fig. 8-7.

Step 8. Distribute maximum moments into column strips and middle strips (see Direct Design Method).

TWO-WAY REINFORCED CONCRETE SLABS SUPPORTED ON BEAMS
(ACI 318-1971 Standard Building Code)

The new Direct Design Method for slabs is intended to eliminate the differentiation between flat slabs, flat plates, and two-way slabs. While there are certain rules to be followed, and these vary for the above-mentioned different types of slabs, the basic system remains unchanged.

For illustration, a three-bay structure was selected. It is assumed that there are five stories and roof above the last floor. It is also assumed there are 70 psf as design live load; this is a realistic figure and represents residential live load and partition load. Figure 8-7 shows the plan view of the floor system with proper identifications and a partial section through the beams. Figure 8-8 provides the general information required and the computed coefficients as needed for the proposed design method.

Interior Span Moments; Interior Panel

Negative Moment: $0.65\, M_o = 0.65 \times 118 = 76.7$ ft kips
Positive Moment: $0.35\, M_o = 0.35 \times 118 = 41.3$ ft kips

Column strip carries 75 percent of the moments, and the beam carries 85 percent of the column strip moments. The beams also carry the full moments due to weight of the beam below the slab and moments due to wall load, if any. (Mostly applicable for spandrel beams.) The middle strip carries 25 percent of the moments.

Beam Moments, Column Strip:
M neg., beam = $0.75 \times 0.85 \times 76.7 = 48.8$ ft kips
M pos., beam = $0.75 \times 0.85 \times 41.3 = 26.3$ ft kips
Slab Moments, Column Strip:
M neg., slab = $0.75 \times 0.15 \times 76.7 = 8.6$ ft kips
M pos., slab = $0.75 \times 0.15 \times 41.3 = 4.7$ ft kips
Slab Moments, Middle Strip:
M neg., slab = $0.25 \times 76.7 = 19.2$ ft kips
M pos., slab = $0.25 \times 41.3 = 10.6$ ft kips
Interior Span Moments; Edge Panel
Middle strip is the same as for interior panel:
M neg., slab = $0.25 \times 76.7 = 19.2$ ft kips
M pos., slab = $0.25 \times 41.3 = 10.6$ ft kips

Edge Column Strip

Same intensity as interior column strip, but only half width:

M neg., slab = 8.6/2 = 4.3 ft kips

M pos., slab = 4.7/2 = 2.4 ft kips

Edge Beams

Half of interior beam moments. (Moments due to wall load and weight of beam to be added.)

M neg., beam = 48.8/2 = 24.4 ft kips

M pos., beam = 26.3/2 = 13.2 ft kips

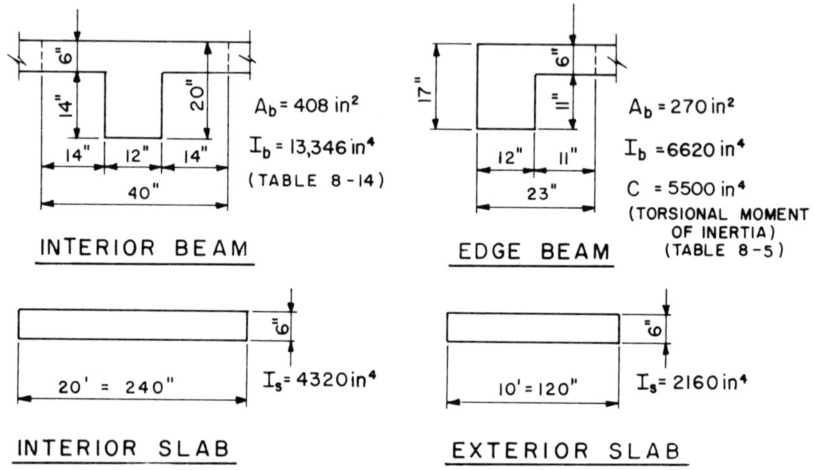

Assuming $E_b = E_s = E_c$ (modulus of elasticity)

α (int. beams) $= \dfrac{I_b}{I_s} = \dfrac{13346}{4320} = 3.1$; α (edge beams) $= \dfrac{I_b}{I_s} = \dfrac{6620}{2160} = 3.1$;

I_c (int. col's., int. panel) $= 24^4/12 = 27650 \text{ in}^4$; β_t (tors. rig.) $= \dfrac{c}{2I_s} = \dfrac{.5 \times 5500}{4320} = .637$;

I_c (int. col's., end panel) $= 16 \times 24^3/12 = 18430 \text{ in}^4$; STIFFNESS CONSTANTS:

I_c (ext. col's., int. panel) $= 24 \times 16^3/12 = 8190 \text{ in}^4$; $k_c = k_s = k_b = 4$;

I_c (ext. col's., end panel) $= 16^4/12 = 5460 \text{ in}^4$; FLEXURAL STIFFNESS OF END COL.:

c_1 (inter. col. size) $= 2.0'$; $c_1 = c_2$;

$\dfrac{c_1}{l_1} = \dfrac{2}{20} = .10$; $W = .145 \times 20^2 = 58^K$

α_{ec} end col. (top & bott) $= \dfrac{\Sigma(k_c E_c I_c/h)}{\Sigma[(k_s E_s I_s + k_b E_b I_b)/l]}$

$$= \frac{2 \times 8190/9 \times 12}{(4320 + 13346)/240} = 2.06$$

$$M_o \text{ (total static design moment)} = \frac{w l_2 l_n^2}{8} = \frac{W l_1}{8}\left(1 - \frac{c_1}{l_1}\right)^2 = \frac{58}{8} \times 20\,(1 - .10)^2 = 118^{IK}$$

Fig. 8-8 Basic information and constants.

Exterior Span Moments

Interior M neg. $= 0.75 - \dfrac{0.10}{1 + \dfrac{1}{\alpha_{ec}}} = 0.68\,M_0$

Positive design $M = 0.63 - \dfrac{0.28}{1 + \dfrac{1}{\alpha_{ec}}} = 0.44\,M_o$

Exterior M neg. $= \dfrac{0.65}{1 + \dfrac{1}{\alpha_{ec}}} = 0.44\,M_o$

By interpolation, (ACI-13.3.4.2) $M_{\text{neg.}} = 0.44\ (1 - 0.25)\,\dfrac{0.64}{2.5} = 0.41\ M_o$.

Beam Moments

0.85 × column strip M since α_1 is larger than unity

Interior M neg. = 0.85 × 0.75 × 0.68 × 118.0 = 51.2 ft kips
Positive M = 0.85 × 0.75 × 0.44 × 118.0 = 33.1 ft kips
Exterior M neg. = 0.85 × 0.41 × × 118.0 = 41.1 ft kips

Slab Moments, Column Strip

Interior M neg. = 0.15 × 0.75 × 0.68 × 118.0 = 9.0 ft kips
Positive M = 0.15 × 0.75 × 0.44 × 118.0 = 5.9 ft kips
Exterior M neg. = 0.15 × 0.41 × 118.0 = 7.3 ft kips

Slab Moments, Middle Strip

Interior M neg. = 0.25 × 0.68 × 118.0 = 20.0 ft kips
Positive M = 0.25 × 0.44 × 118.0 = 13.0 ft kips
Exterior M neg. = 0.03 × 118.0 = 3.6 ft kips

Corner Panel Moments

Beam Moments:

Interior M neg. = 51.2/2 = 25.6 ft kips
Positive M = 33.1/2 = 16.6 ft kips
Exterior M neg. = 41.1/2 = 20.6 ft kips

Slab Moments, Column Strip:

Interior M neg. = 9.0/2 = 4.5 ft kips
Positive M = 5.9/2 = 3.0 ft kips
Exterior M neg. = 7.3/2 = 3.7 ft kips

Moments due to additional loads:

Exterior beams: Wall load: 0.400 klf; Interior beams: *Weight of beam: 0.138 klf; Weight of beam: 0.175 klf; Total: 0.538 klf*

Exterior beams, l_n = 18.33 ft — Interior beams, l_n = 18.00 ft

Ext. M neg. $= \dfrac{18.33^2}{16} \times 0.538 = 11.3'^k$; — same $\times\ 0.175 = 3.7'^k$

Ext. M pos. $= \dfrac{18.33^2}{14} \times 0.538 = 12.9'^k$; — same $\times\ 0.175 = 4.2'^k$

$$\text{Int. } M \text{ neg.} = \frac{18.17^2}{10} \times 0.538 = 17.8'^k; \qquad \text{same} \times 0.175 = 5.8'^k$$

$$\text{Int. } M \text{ pos.} = \frac{18.00^2}{16} \times 0.538 = 10.9'^k; \qquad \text{same} \times 0.175 = 3.6'^k$$

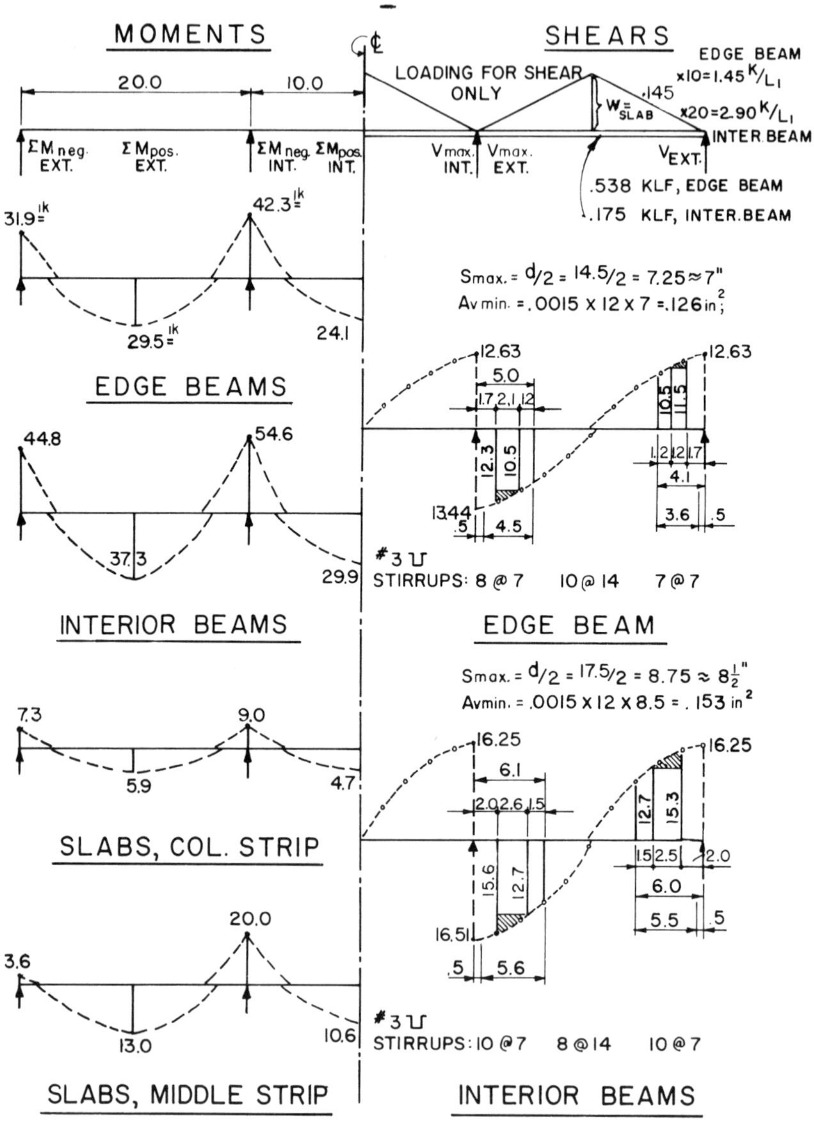

Fig. 8-9.

Combined Beam Moments

Edge Beams:	*Interior Beams:*
Ext. M neg. $= \dfrac{41.1}{2} + 11.3 = 31.9'^k$	$41.1 + 3.7 = 44.8'^k$
Ext. M pos. $= \dfrac{33.1}{2} + 12.9 = 29.5'^k$	$33.1 + 4.2 = 37.3'^k$
Int. M neg. $= \dfrac{48.8}{2} + 17.9 = 42.3'^k$	$48.8 + 5.8 = 54.6'^k$

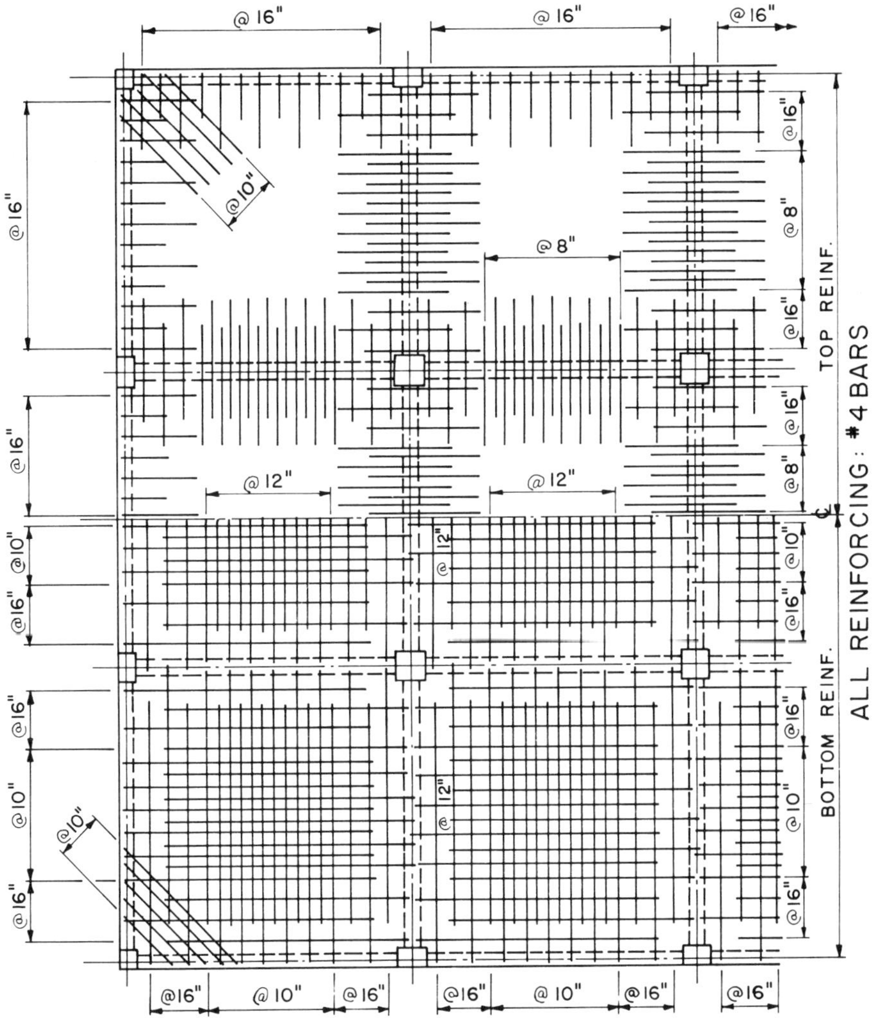

Fig. 8-10.

Int. M pos. $= \frac{26.3}{2} + 10.9 = 24.1'^k$ $\qquad$ $26.3 + 3.6 = 29.9'^k$

Figure 8-9 shows the moment and shear diagrams. The design of members to according to Strength Design Method. Figure 8-10 presents the slab reinforcement.

Note: The author does not believe that this method is an improvement over the previous Method 3 of the ACI Code 1963. On the contrary, this method is cumbersome, time-consuming, and does not provide any economies. Consequently, Method 3 of ACI 318-63 is recommended for the design of two-way slabs, unless the building official insists on the new design system.

METHOD 3 (ACI-318-63)

Notation

A = length of clear span in short direction

B = length of clear span in long direction

C = moment coefficients for two-way slabs as given in Tables 1, 2, and 3. Coefficients have identifying indexes, such as $C_{A\ \mathrm{neg}}$, $C_{B\ \mathrm{neg}}$, $C_{A\ \mathrm{DL}}$, $C_{B\ \mathrm{DL}}$, $C_{A\ \mathrm{LL}}$, $C_{B\ \mathrm{LL}}$.

m = ratio of short span to long span for two-way slabs

w = uniform load per sq ft. For negative moments and shears, w is the total dead load plus live load for use in Table 1. For positive moments, w is to be separated into dead and live loads for use in Tables 2 and 3.

(a) *Limitations*—A two-way slab shall be considered as consisting of strips in each direction as follows:

A middle strip one-half panel in width, symmetrical about panel center line and extending through the panel in the direction in which moments are considered.

A column strip one-half panel in width, occupying the two quarter-panel areas outside the middle strip.

Where the ratio of short to long span is less than 0.5, the slab shall be considered as a one-way slab and is to be designed in accordance with Chapter 9 except that negative reinforcement, as required for a ratio of 0.5, shall be provided along the short edge.

At discontinuous edges, a negative moment one-third (1/3) of the positive moment is to be used.

Critical sections for moment calculations are located as follows:

For negative moment along the edges of the panel at the faces of the supports.

For positive moment, along the center lines of the panels.

METHOD 3–TABLE 1–COEFFICIENTS FOR NEGATIVE MOMENTS IN SLABS*

$$\left.\begin{array}{l} M_{A\ neg} = C_{A\ neg} \times w \times A^2 \\ M_{B\ neg} = C_{B\ neg} \times w \times B^2 \end{array}\right\} \text{where } w = \text{uniform dead plus live load}$$

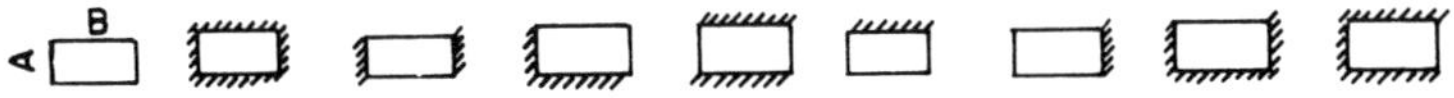

Ratio $m = \frac{A}{B}$		Case 1	Case 2	Case 3	Case 4	Case 5	Case 6	Case 7	Case 8	Case 9
1.00	$C_{A\ neg}$		0.045		0.050	0.075	0.071		0.033	0.061
	$C_{B\ neg}$		0.045	0.076	0.050			0.071	0.061	0.033
0.95	$C_{A\ neg}$		0.050		0.055	0.079	0.075		0.038	0.065
	$C_{B\ neg}$		0.041	0.072	0.045			0.067	0.056	0.029
0.90	$C_{A\ neg}$		0.055		0.060	0.080	0.079		0.043	0.068
	$C_{B\ neg}$		0.037	0.070	0.040			0.062	0.052	0.025
0.85	$C_{A\ neg}$		0.060		0.066	0.082	0.083		0.049	0.072
	$C_{B\ neg}$		0.031	0.065	0.034			0.057	0.046	0.021
0.80	$C_{A\ neg}$		0.065		0.071	0.083	0.086		0.055	0.075
	$C_{B\ neg}$		0.027	0.061	0.029			0.051	0.041	0.017
0.75	$C_{A\ neg}$		0.069		0.076	0.085	0.088		0.061	0.078
	$C_{B\ neg}$		0.022	0.056	0.024			0.044	0.036	0.014
0.70	$C_{A\ neg}$		0.074		0.081	0.086	0.091		0.068	0.081
	$C_{B\ neg}$		0.017	0.050	0.019			0.038	0.029	0.011
0.65	$C_{A\ neg}$		0.077		0.085	0.087	0.093		0.074	0.083
	$C_{B\ neg}$		0.014	0.043	0.015			0.031	0.024	0.008
0.60	$C_{A\ neg}$		0.081		0.089	0.088	0.095		0.080	0.085
	$C_{B\ neg}$		0.010	0.035	0.011			0.024	0.018	0.006
0.55	$C_{A\ neg}$		0.084		0.092	0.089	0.096		0.085	0.086
	$C_{B\ neg}$		0.007	0.028	0.008			0.019	0.014	0.005
0.50	$C_{A\ neg}$		0.086		0.094	0.090	0.097		0.089	0.088
	$C_{B\ neg}$		0.006	0.022	0.006			0.014	0.010	0.003

*A cross-hatched edge indicates that the slab continues across or is fixed at the support; an unmarked edge indicates a support at which torsional resistance is negligible.

METHOD 3–TABLE 2–COEFFICIENTS FOR DEAD LOAD POSITIVE MOMENTS IN SLABS*

$$\left.\begin{array}{l} M_{A\ \text{pos DL}} = C_{A\ \text{DL}} \times w \times A^2 \\ M_{B\ \text{pos DL}} = C_{B\ \text{DL}} \times w \times B^2 \end{array}\right\} \text{where } w = \text{uniform dead load}$$

Ratio $m = \frac{A}{B}$		*Case 1*	*Case 2*	*Case 3*	*Case 4*	*Case 5*	*Case 6*	*Case 7*	*Case 8*	*Case 9*
1.00	C_A DL	0.036	0.018	0.018	0.027	0.027	0.033	0.027	0.020	0.023
	C_B DL	0.036	0.018	0.027	0.027	0.018	0.027	0.033	0.023	0.020
0.95	C_A DL	0.040	0.020	0.021	0.030	0.028	0.036	0.031	0.022	0.024
	C_B DL	0.033	0.016	0.025	0.024	0.015	0.024	0.031	0.021	0.017
0.90	C_A DL	0.045	0.022	0.025	0.033	0.029	0.039	0.035	0.025	0.026
	C_B DL	0.029	0.014	0.024	0.022	0.013	0.021	0.028	0.019	0.015
0.85	C_A DL	0.050	0.024	0.029	0.036	0.031	0.042	0.040	0.029	0.028
	C_B DL	0.026	0.012	0.022	0.019	0.011	0.017	0.025	0.017	0.013
0.80	C_A DL	0.056	0.026	0.034	0.039	0.032	0.045	0.045	0.032	0.029
	C_B DL	0.023	0.011	0.020	0.016	0.009	0.015	0.022	0.015	0.010
0.75	C_A DL	0.061	0.028	0.040	0.043	0.033	0.048	0.051	0.036	0.031
	C_B DL	0.019	0.009	0.018	0.013	0.007	0.012	0.020	0.013	0.007
0.70	C_A DL	0.068	0.030	0.046	0.046	0.035	0.051	0.058	0.040	0.033
	C_B DL	0.016	0.007	0.016	0.011	0.005	0.009	0.017	0.011	0.006
0.65	C_A DL	0.074	0.032	0.054	0.050	0.036	0.054	0.065	0.044	0.034
	C_B DL	0.013	0.006	0.014	0.009	0.004	0.007	0.014	0.009	0.005
0.60	C_A DL	0.081	0.034	0.062	0.053	0.037	0.056	0.073	0.048	0.036
	C_B DL	0.010	0.004	0.011	0.007	0.003	0.006	0.012	0.007	0.004
0.55	C_A DL	0.088	0.035	0.071	0.056	0.038	0.058	0.081	0.052	0.037
	C_B DL	0.008	0.003	0.009	0.005	0.002	0.004	0.009	0.005	0.003
0.50	C_A DL	0.095	0.037	0.080	0.059	0.039	0.061	0.089	0.056	0.038
	C_B DL	0.006	0.002	0.007	0.004	0.001	0.003	0.007	0.004	0.002

*A cross-hatched edge indicates that the slab continues across or is fixed at the support; an unmarked edge indicates a support at which torsional resistance is negligible.

METHOD 3–TABLE 3–COEFFICIENTS FOR LIVE LOAD POSITIVE MOMENTS IN SLABS*

$$\left.\begin{aligned} M_{A\ \text{pos LL}} &= C_{A\ \text{LL}} \times w \times A^2 \\ M_{B\ \text{pos LL}} &= C_{B\ \text{LL}} \times w \times B^2 \end{aligned}\right\} \text{where } w = \text{uniform live load}$$

Ratio $m = \frac{A}{B}$		*Case 1*	*Case 2*	*Case 3*	*Case 4*	*Case 5*	*Case 6*	*Case 7*	*Case 8*	*Case 9*
		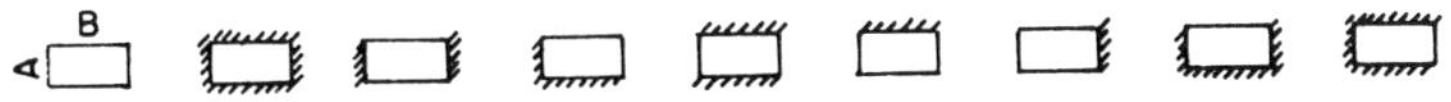								
1.00	C_A LL	0.036	0.027	0.027	0.032	0.032	0.035	0.032	0.028	0.030
	C_B LL	0.036	0.027	0.032	0.032	0.027	0.032	0.035	0.030	0.028
0.95	C_A LL	0.040	0.030	0.031	0.035	0.034	0.038	0.036	0.031	0.032
	C_B LL	0.033	0.025	0.029	0.029	0.024	0.029	0.032	0.027	0.025
0.90	C_A LL	0.045	0.034	0.035	0.039	0.037	0.042	0.040	0.035	0.036
	C_B LL	0.029	0.022	0.027	0.026	0.021	0.025	0.029	0.024	0.022
0.85	C_A LL	0.050	0.037	0.040	0.043	0.041	0.046	0.045	0.040	0.039
	C_B LL	0.026	0.019	0.024	0.023	0.019	0.022	0.026	0.022	0.020
0.80	C_A LL	0.056	0.041	0.045	0.048	0.044	0.051	0.051	0.044	0.042
	C_B LL	0.023	0.017	0.022	0.020	0.016	0.019	0.023	0.019	0.017
0.75	C_A LL	0.061	0.045	0.051	0.052	0.047	0.055	0.056	0.049	0.046
	C_B LL	0.019	0.014	0.019	0.016	0.013	0.016	0.020	0.016	0.013
0.70	C_A LL	0.068	0.049	0.057	0.057	0.051	0.060	0.063	0.054	0.050
	C_B LL	0.016	0.012	0.016	0.014	0.011	0.013	0.017	0.014	0.011
0.65	C_A LL	0.074	0.053	0.064	0.062	0.055	0.064	0.070	0.059	0.054
	C_B LL	0.013	0.010	0.014	0.011	0.009	0.010	0.014	0.011	0.009
0.60	C_A LL	0.081	0.058	0.071	0.067	0.059	0.068	0.077	0.065	0.059
	C_B LL	0.010	0.007	0.011	0.009	0.007	0.008	0.011	0.009	0.007
0.55	C_A LL	0.088	0.062	0.080	0.072	0.063	0.073	0.085	0.070	0.063
	C_B LL	0.008	0.006	0.009	0.007	0.005	0.006	0.009	0.007	0.006
0.50	C_A LL	0.095	0.066	0.088	0.077	0.067	0.078	0.092	0.076	0.067
	C_B LL	0.006	0.004	0.007	0.005	0.004	0.005	0.007	0.005	0.004

*A cross-hatched edge indicates that the slab continues across or is fixed at the support; an unmarked edge indicates a support at which torsional resistance is negligible.

(b) *Bending moments*—The bending moments for the middle strips shall be computed by the use of Tables 1, 2, and 3 from:

$$M_A = CwA^2 \quad \text{and} \quad M_B = CwB^2$$

The bending moments in the column strips shall be gradually reduced from the full value M_A and M_B from the edge of the middle strip to one-third (1/3) of these values at the edge of the panel.

Where the negative moment on one side of a support is less than 80 percent of that on the other side, the difference shall be distributed in proportion to the relative stiffnesses of the slabs.

(c) *Shear*—The shear stresses in the slab may be computed on the assumption that the load is distributed to the supports in accordance with 13.3.4.7 & 13-3.4.8 of ACI-Code 318-71.

(d) *Supporting beams*—The loads on the supporting beams for a two-way rectangular panel shall be computed using 13.3.4.7 & 13.3.4.8 of ACI-Code 318-71 for the percentages of loads in "*A*" and "*B*" directions. In no case shall the load on the beam along the short edge be less than that of an area bounded by the intersection of 45-deg lines from the corners. The equivalent uniformly distributed load per linear foot on this short beam is

$$\frac{wA}{3}$$

Sheridan & Devon, Chicago, Ill. Twenty-eight stories, flat plate, shear wall. Architects: Dubin, Dubin & Black. Structural Engineers: Rogers-Cohen-Barreto-Marchertas, Inc.

9

Development of Reinforcement and Splices in Reinforcement

According to the provisions of the previous ACI 318-63 Code, in order to prevent bond failure or splitting, the tension or compression in any bar had to be developed by proper embedment length, end anchorage, or by hooks (for tension only). It was limited to the maximum bond stress also, according to the location, size, and acting force of the bars.

The new ACI 318-71 Code simplifies the dealing with bond stress, using all new provisions in terms of minimum development lengths and minimum bar sizes. Thus the computation of bond stress, is considered unimportant, giving more meaning to the development-length concept.

In accordance with Section 9.2.2 of the new Code, the specified development lengths do not require a capacity-reduction factor (ϕ). The required lengths are the same for either the Strength Design Method or the Alternate Design Method.

For critical sections for development of reinforcement in a typical continuous beam see Fig. 9-1. Positive moment reinforcement of a member, which is part of the primary lateral load resisting system, must be anchored into the support.

At simple supports and points of inflection there is a limit to bar size for positive reinforcement, requiring that the development length l_d, computed for f_y does not exceed $(M_t/V_u + l_a)$ (see Figs. 9-2 and 9-3).

Figures 9-4, 9-5, and 9-6 illustrate the development length, l_d, for different negative reinforcement cases.

Section 12.4 of the new Code has provisions for end anchorage of tension bars in special members, such as: sloped, stepped, or tapered footings; brackets; deep beams; and others where f_s does not decrease linearly in proportion to a

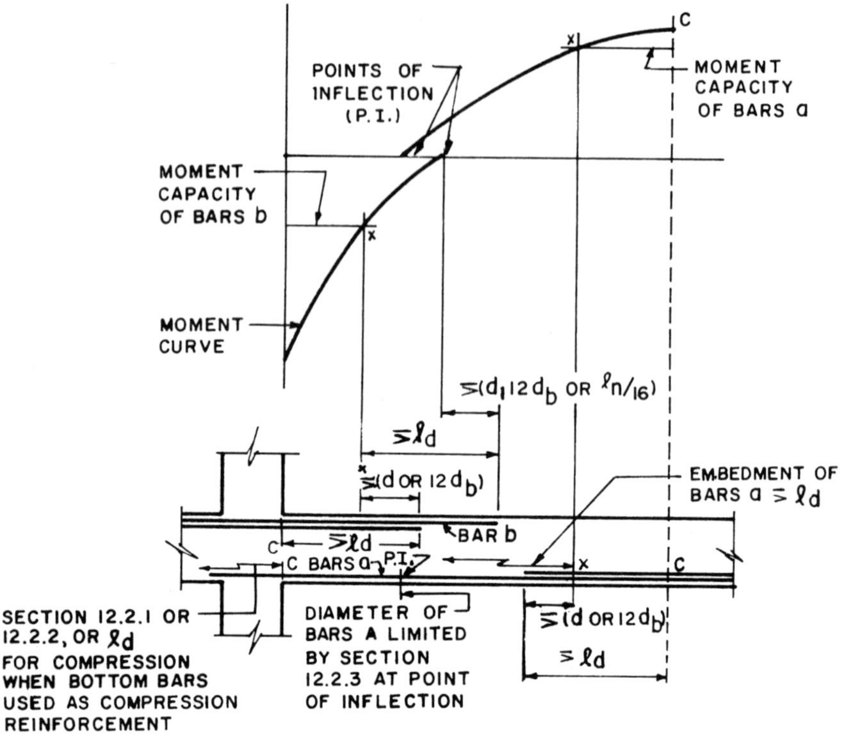
POINTS OF INFLECTION (P.I.)
MOMENT CAPACITY OF BARS a
MOMENT CAPACITY OF BARS b
MOMENT CURVE
≥(d, 12 d_b OR ℓn/16)
≥ℓd
≥(d OR 12 d_b)
EMBEDMENT OF BARS a ≥ ℓd
BAR b
≥ℓd
C BARS a
P.I.
SECTION 12.2.1 OR 12.2.2, OR ℓd FOR COMPRESSION WHEN BOTTOM BARS USED AS COMPRESSION REINFORCEMENT
DIAMETER OF BARS A LIMITED BY SECTION 12.2.3 AT POINT OF INFLECTION
≥(d OR 12 d_b)
≥ ℓd

Fig. 9-1.

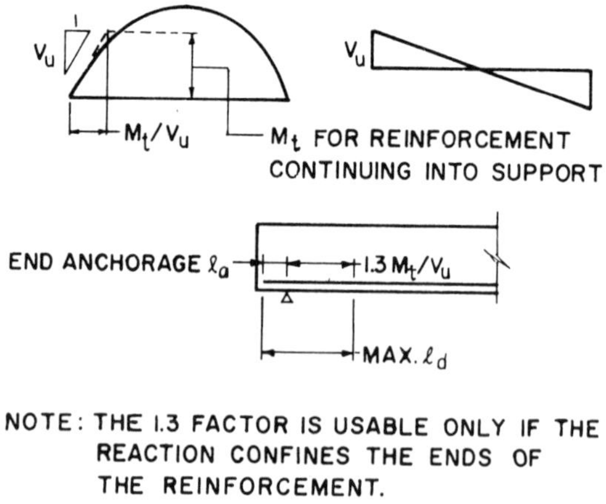
Vu
Mt/Vu
Mt FOR REINFORCEMENT CONTINUING INTO SUPPORT
Vu
END ANCHORAGE ℓa
1.3 Mt/Vu
MAX. ℓd
NOTE: THE 1.3 FACTOR IS USABLE ONLY IF THE REACTION CONFINES THE ENDS OF THE REINFORCEMENT.

Fig. 9-2.

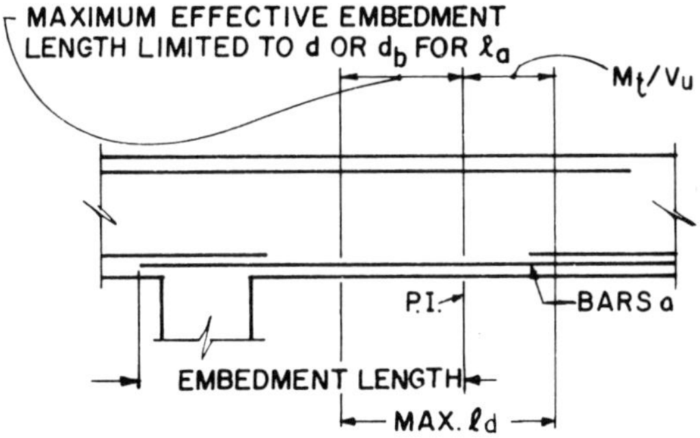

Fig. 9-3.

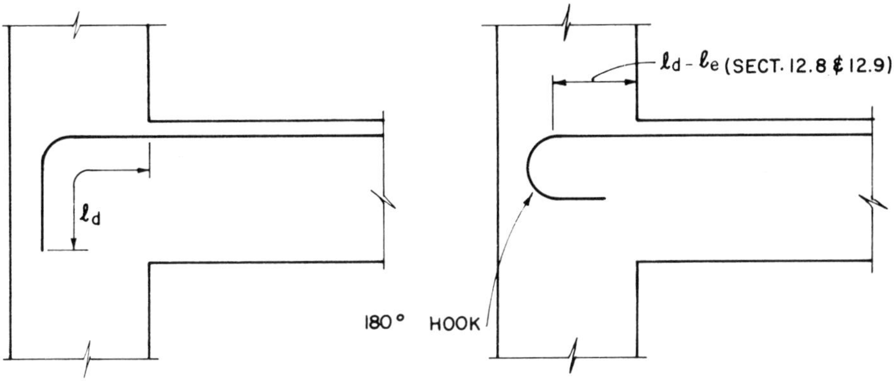

Fig. 9-4.

Fig. 9-5.

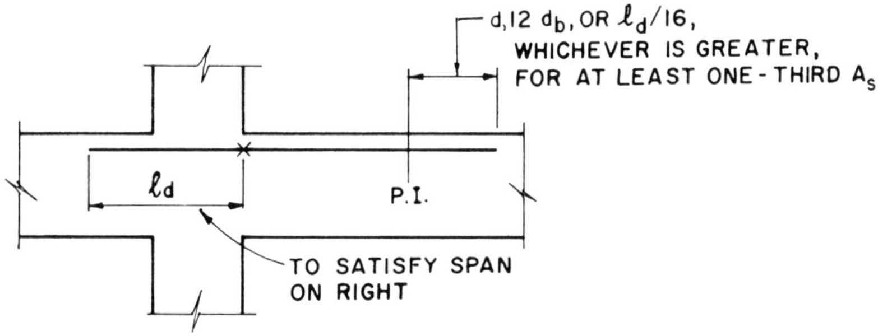

Fig. 9-6.

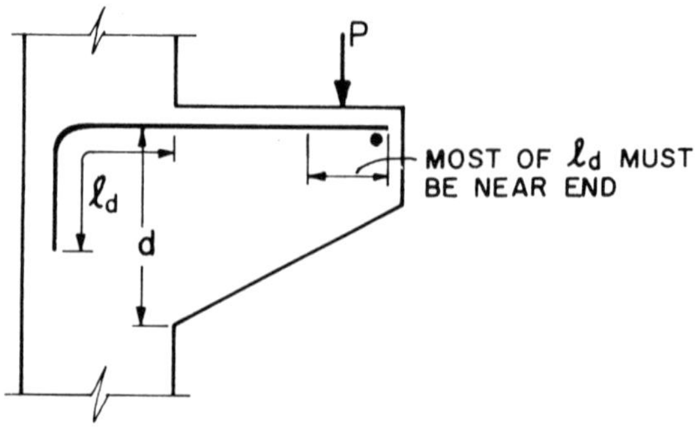

Fig. 9-7.

decreasing moment. Figure 9-7 shows the case of a bracket, where an l_d from the support is probably less critical than the required development length for a slightly smaller f_s existing near the load point.

Section 12.5 provides the l_d requirements to be used with various tension bar sizes and deformed wires, including modification factors for top bars, bars of yield stress greater than 60,000 psi, lightweight concrete, and other materials.

The basic development length shall be:

For #11 or smaller bars $0.04A_b(f_y)/\sqrt{f_c'}$ but not less than $0.0004d_bf_y$
For #14 bars $0.085(f_y)/\sqrt{f_c'}$
For #18 bars $0.11(f_y)/\sqrt{f_c'}$
For deformed wire $0.03d_b(f_y)/\sqrt{f_c'}$

The basic development length shall be multiplied by the applicable factor or factors:

Top reinforcement 1.4

Bars with f_y greater than 60,000 psi $2 - \dfrac{60{,}000}{f_y}$

When lightweight aggregate is used, the basic development lengths shall be multiplied by 1.33 for "all-lightweight" concrete and 1.18 for "sand-lightweight" concrete with linear interpolation when partial sand replacement is used; or the basic development length may be multiplied by $6.7\sqrt{f_c'}/f_{ct}$, but not less than 1.0, when f_{ct} is specified and the concrete is proportioned in accordance with Section 4.2 of the new Code.

Section 12.6 gives the development length of deformed bars in compression, with an absolute minimum length of 8 in. Increased development length for individual bars is required when three or four bars are bundled together. The increase shall be 20 and 30 percent respectively.

Table 9-1

ξ VALUES

BAR SIZE	fy = 60 ksi TOP BARS	OTHER BARS	fy = 40 ksi BARS ALL
#3 TO #5	540	540	360
#6	450	540	360
#7 TO #9	360	540	360
#10	360	480	360
#11	360	420	360
#14	330	330	330
#18	220	220	220

Table 9-1 gives the values of ξ, used in the formula,

$$f_h = \xi\sqrt{f_c'}$$

to consider the development length due to standard hook.

Section 12.13 has provisions for anchorage of web reinforcement, which remains the same as in the previous Code except for a new provision on length of the effective embedment of a stirrup leg, a new section on using welded plain wire fabric as stirrups, and a new requirement for *U*-stirrups.

The new increased anchorage condition is illustrated on Fig. 9-8, where the standard hook must have a development of $0.5(l_d)$ between $d/2$ of the beam and the point of tangency of the hook, instead of to the center of radius of the hook (ACI-12.13.1.1). This provision may require a reduction in size and thus spacing of web reinforcement, or increase the depth of the member.

The development length does not present a complicated design problem. In

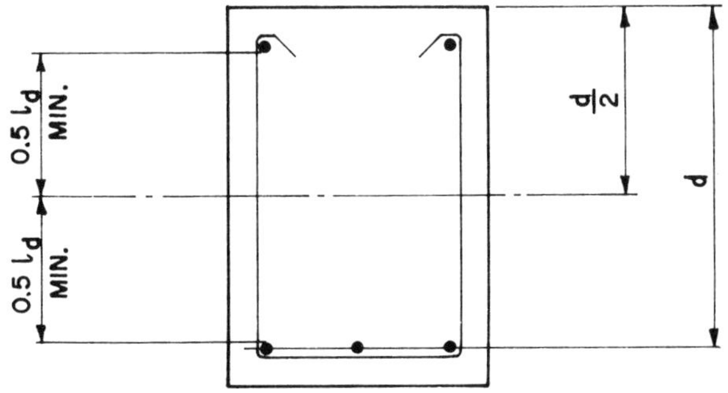

Fig. 9-8.

most cases of the design practice, the development length may be the governing factor for footing bar sizes, beyond-bar cutoff points, and for positive reinforcement of primary lateral load resisting members supported by small-size columns or walls.

Two examples of development calculations are presented as follows:

Example 1

An 8 ft by 8 ft footing carrying a 20-in.-square column is being designed for $f_c' = 3000$ psi and $f_y = 60{,}000$ psi. Determine the maximum bar size that may be used (see Figure 9-9). Three solutions are developed.

(1) Available development for straight bar:

$$0.5(96 - 20) - 2 \text{ in. (cover)} = 36 \text{ in.}$$

Assume #11 or smaller:

$$l_d = 0.04 A_b f_y / \sqrt{f_c'} \qquad \text{(ACI-12.5)}$$

A_b(required) $< 36\sqrt{3000}/0.04(60{,}000) = 0.821$
Use: #8 bars ($A_b = 0.79$)

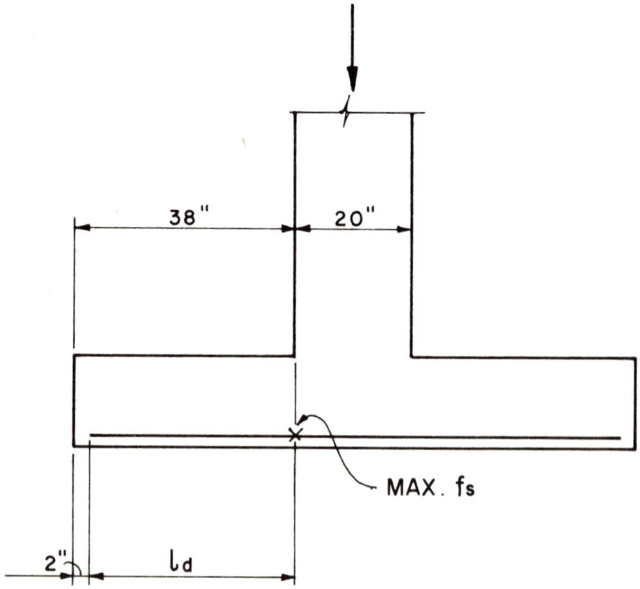

Fig. 9-9.

The lower limit:

$l_d = 0.004 d_b f_y$ governs only small bars, but check here for #8 bar:
min $l_d = 0.004(1.0)\,60{,}000 = 24$ in < 36, (O.K.)

(2) If bar spacing can be at least 6 in., Section 12.5(d), the new Code allows an 0.8 factor to be used for required l_d.

$$l_d = 0.8\left[0.04 A_b f_y / \sqrt{f_c'}\right] = 36 \text{ in.}$$

Therefore

$$A_b < 0.82/0.8 = 1.02 \text{ in.}^2$$

Use: #9 bars ($A_b = 1.00$).

(3) With a standard hook at the bar end, the available l_d to the point where the hook starts is reduced by $4d_b$. Hook values, by Code Section 12.8, are based on $f_h = \xi\sqrt{f_c'}$, with ξ for #3 to #9 bars = 540,

$$f_h = 540\sqrt{3000} = 29{,}600 \text{ psi}$$

Try: #11 bars at spacing 6 in.
l_d for remaining stress = 60,000 − 29,600 = 30,400 psi is needed.
$l_d = 0.8\langle 0.04 A_b (30{,}400)/\sqrt{f_c'}\rangle = 36 - 4d_b = 30.36$ in.
$A_b(\text{required}) < 30.36\sqrt{3000}/(0.8)\,0.04(30{,}400) < 1.71$
Use: #11 bars (A_b 1.56 in.^{2}) with standard hook if spacing is 6 in.

Example 2

Consider a continuous beam having two #9 bars running into the support and two #6 cut off within the point of inflection. f_y = 60,000 psi, and f_c' = 4000

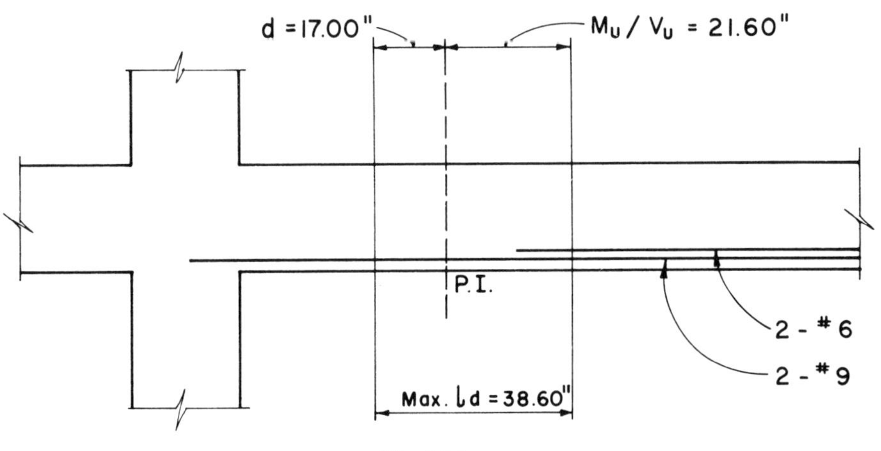

Fig. 9-10.

psi; design V_u at point of inflection 80.0 kips (see Figure 9-10). Check the #9 bars at the point of inflection for conformance with Section 12.2.3 of the Code.

Solution: Compute M_t, the moment capacity, with two #9 bars.

$$M_t = 0.9(2)\ 1.0(60{,}000)\ (16.0) = 1728 \text{ in.-kips.}$$

Available development length to right of point of inflection (PI)

$$= M_t/V_u = 1728/80 = 21.60 \text{ in.}$$

Available to left of P.I. $d = 17.00$ in.

Available for l_d 38.60 in. for #9 = 37.94 in.

Therefore, #9 bars are satisfactory.

Although splices in reinforcement are presented in Sections 7.5, 7.6, 7.7, 7.8, and 7.9 of Chapter 7 ("Details of Reinforcement") of the new Code, the author considers that, the length of splice generally being related to the development length, splices may be discussed in this chapter.

Due to lack of adequate researches on lap splices of #14 and #18, the use of such splices for these bars is prohibited, with one exception: the special case of a column-footing connection, as per Section 15.6.8.

The Code requires all welding of reinforcement to conform to AWS D 12, which demands that a chemical analysis of reinforcement be obtained and that the entire welding process, including method, material, amount of preheat, etc., be compatible with the chemical analysis.

Lap splices of bundled bars are based on the lap splice length required for individual bars of the same size because the bars spliced and such individual splices within the bundle shall not overlap each other. The length of lap, as specified in Sections 7.6 and 7.7 are to be increased by 20 and 33 percent for a three-bar and four-bar bundle respectively.

In order to prevent creation of an unreinforced section between widely spaced bars in a noncontact lap splice, the Code places a maximum distance between lap-spliced bars of one-fifth the required length of lap, or 6 in.

Section 7.5.5 provides for welded splices or positive connections. A full welded splice must develop 125 percent of the specified yield strength of the bar to avoid splice failure before yielding of the bar. The same is true with full positive connections (Cadweld, Yee splice, etc.).

Section 7.6.1 classifies the tension lap splices, and the required lengths are multiples of the development lengths of reinforcement given in Section 12.5 of the Code. This means that all items affecting development length must be considered in calculating the tension lap splice length. Four classes of splices are defined, with corresponding increase of development length required for each class in order to obtain maximum economy consistent with uniform structural capacity.

The most severe category is the Class D splice (ACI-7.6.1.4), which is intended

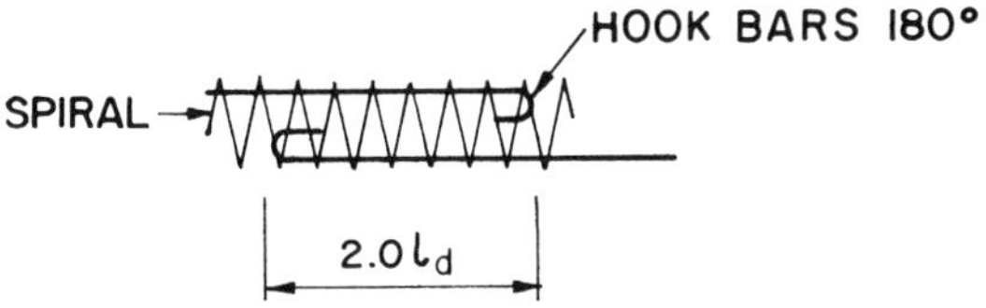

Fig. 9-11.

to be used as a tension-tie member such as a tie-beam between the bases of a rigid frame (see Fig. 9-11 for Class D splice).

In conventional concrete members (columns, beams, slabs), tension splices must be of Class A, B, or C, depending on the tensile stress in the bar and the percentage of bars spliced within a splice length.

In regions of flexure members with maximum moment or high computed stresses greater than 0.5 f_y, Class B or C splices are required.

The various tension lap splice conditions are summarized in Table 9-2.

Class C splices [1.7(l_d)] are required where more than one half of the bars are lap spliced within a required lap length. A Class B splice [1.3(l_d)] is allowed if no more than one half of the bars are lap spliced.

In regions of low computed stresses, less than 0.5f_y, Class A or B splices are satisfactory. Class A splices [1.0(l_d)] are used where no more than three-quarters of the bars are lap-spliced within a required lap length. Class B splices [1.3(l_d)] are allowed where more than three-quarters of the bars are lap-spliced within a required lap length. A reduction in length is allowed if Class A, B, and C splices are enclosed within spirals.

The computed "tensile stress" referred to in the Code is the design stress using the basic strength method (that is, using load factors).

Section 7.7 of the new Code deals with the splices in compression. The minimum length of a lap splice in compression should be the development length in compression l_d (Section 12.6 of ACI 318-71), but not less in inches than $0.0005 f_y d_b$ for f_y of 60,000 psi or less, nor $(0.0009 f_y - 24) d_b$ for f_y greater than 60,000 psi, nor 12 in.

When the specified concrete strength f_c' is less than 3000 psi, the lap is to be increased by one-third.

The basic compression lap length may be reduced in both tied or spiral columns, reflecting the fact that appropriate ties or spirals restrain against splitting of the concrete.

In bars required for compression only, the compressive stress may be transmitted by bearing of square cut ends held in concentric contact by a suitable device (G-Loc, etc.).

Table 9-2
Tension Lap Splices

Member	*Maximum stress*	*Percent spliced*	*Lap*	*Section*	*Notes*
Flexure, with or without axial	$>0.5f_y$ tension	>50	$1.7l_d$ (Class C)	7.6.3.1.1	Avoid if possible
		≤50	$1.3l_d$ (Class B)	7.6.3.1.1	
	$\leq 0.5f_y$ tension	>75	$1.3l_d$ (Class B)	7.6.3.2.1	Preferred
		≤75	$1.0l_d$ (Class A)	7.6.3.2.1	
Tension tie (welded or positive connection preferred)		Stagger if possible	$2.0l_d$ (Class D)	7.6.1 7.6.2	Avoid if possible. Spiral Required

Compression Lap Splice Lengths Per ACI 318-63

	*Minimum lap length**		*Calculated lap length required to satisfy development bond for full f_y*	
f_y	$f_c' \geq 3000$	$f_c' < 3000$	$f_c' = 2300$	$f_c' = 1300$
40	$20d_b$	$26.7d_b$	$16d_b$	$21.4d_b$
50	$20d_b$	$26.7d_b$	$20d_b$	$26.7d_b$
60	$24d_b$	$32\ d_b$	$24d_b$	$32.0d_b$
75	$30d_b$	$40\ d_b$	$30d_b$	$40.0d_b$

d_b = bar diameters

*Note that the minimum lap lengths will control length of splice in all practical conditions. The minimum lap lengths for $f_c' \geq 3000$ psi are based on the calculated lap lengths for 2300 psi concrete, while those for $f_c' < 3000$ psi are based on 1300 psi concrete. Only if concrete with $f_c' < 1300$ psi were used, or if splices were to be stressed before concrete strength reached 1300 psi, would the calculated development length control.

Comparison of Compression Lap Splice Requirements–1963 Versus 1971 Code in Bar Diameters

	Minimum lap splice lengths $f_c' \geq 3000$				*Calculated lap required by bond for full f_y with $f_c' = 2300$*	
	1963 Code	*1971 Code*				
f_y	*All bars*	*Spiral column*	*Tied column*	*Loose*	*1963 Code*	*1971 Code**
40	20	15.0	16.6	20	16	16.7
50	20	18.75	20.75	25	20	20.85
60	24	22.5	24.9	30	24	25.0
75	30	32.6	36.2	43.5	30	31.2
80	–	36.0	39.9	48.0	–	33.3

*For f_c' = 2300 psi for splices of loose bars or bars in tied columns.

For deformed wire fabric, the Code contains equations for both a minimum splice length and the minimum amount that the outermost cross wires must lap.

One of the most efficient and economical full positive bar splices is the Alfred A. Yee developed ductile iron splice sleeve with double-tapered frustrums (see Fig. 9-12). This ductile iron sleeve is filled with high-strength nonshrink cement grout (Embeco, etc.) to effect joinery between separate bars. The action of the sleeve is to contain the filler grout, which tends to move outward when the deformed bars are under tension or compression. The grout in the wedge-shaped section of the double-tapered frustrums also contributes mechanical resistance to tension action in the bars. For compressive action between spliced bars, the

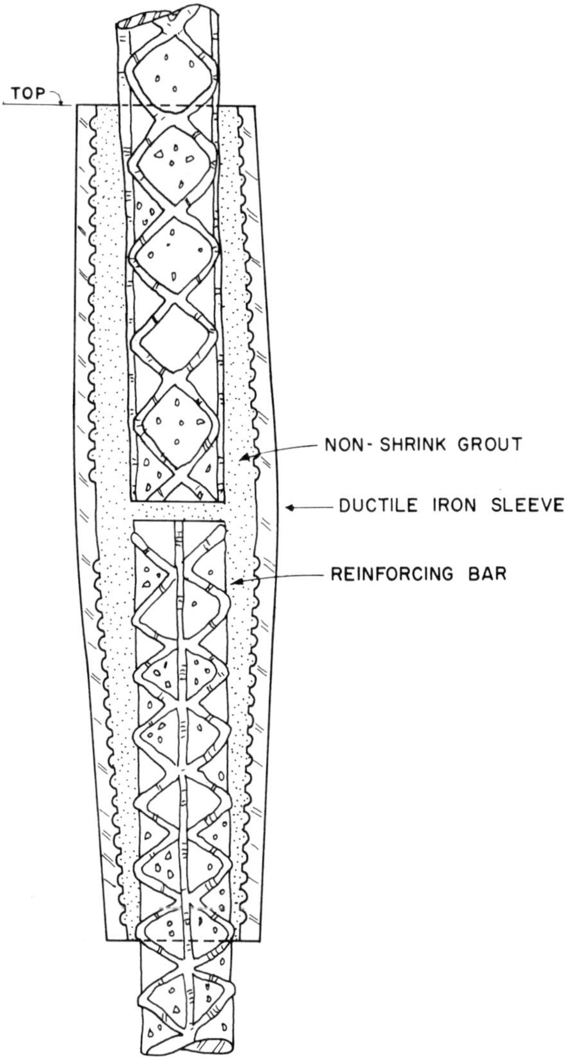

Fig. 9-12 Ductile iron splice sleeve.

containment effect of the entire grout-filled sleeve will prevent any one bar from moving against the other.

Laboratory test results for isolated ductile iron sleeves have indicated that the effective reinforcing steel stresses attained are approximately 60 to 80,000 psi in tension and 100,000 psi in compression when the nonshrink filler grout is 4 days old. The filler grout strength average is approximately 10,000 psi at 28 days. Figure 9-12 describes the Yee sleeve.

Century City Medical Center, Los Angeles, Calif. Daniel, Mann, Johnson & Mendenhall, Architects. Paul Rogers, Chief Structural Engineer.

10

Footings

The only important change in design criteria for reinforced concrete footings in accordance with the ACI 318-71 Code is the concept of reinforcement-development. Thus, after checking shear and flexural stresses and calculating the required reinforcement, the development length of the selected bars should be investigated according to the criteria described in Chapter 9.

The following examples are step-by-step calculation methods for footing design. The data of Table 10-1 are for instant selection of footings for reinforced concrete columns of given size, ultimate loading capacity and properties (f_c', f_y), and for given factored soil pressure. The given reinforcement area is for a 12 in. width of footing slab in one direction. In computing the effective slab thickness, 4.7 in. distance was assumed for the sum of the half-diameter of reinforcement and the average cover.

The capacity of columns are computed for concentric loads, and for 0.05, 0.10, 0.167, and 0.20 eccentricity ratios, where the eccentricity ratio is expressed as a factor of the value of the eccentricity (based on the moment to column load ratio) to the total length of footing. The given ultimate capacity is for column loads, which is the total capacity of the footing, reduced by the increased weight of footing by load factor of 1.4. The weight of footing reduces slightly the eccentricity, whereby the minimum soil pressure is not exactly zero. No tension in the soil is allowed for 0.2 eccentricity ratio.

Example 1

A column 23 in. square supports a total load of 1450 kips. The safe soil pressure is 6.0 ksf. Assuming 1.6 load factor, the design load will be P_u =

Table 10-1

Square footings					*f'c = 3.00 ksi*				*fy = 60.0 ksi*
Soil fact'd (ksf)	*—Footing— Size (ft)*	*Thick (in.)*	A_s *sq in. /ft*	*Col. size (in.)*	*Ult. capacity (kip) ecc/ftg size 0.00*	*0.05*	*0.10*	*0.167*	*0.20*
4.0	4	12	0.148	10	61	47	38	30	25
	5	12	0.259	10	95	73	59	47	40
	6	14	0.309	10	135	104	84	68	56
	7	16	0.359	10	182	140	114	91	74
	8	18	0.408	10	236	181	147	118	95
	9	20	0.455	10	296	227	185	148	117
	10	22	0.502	10	362	278	226	181	142
	11	24	0.548	10	433	333	271	216	167
	12	26	0.593	10	510	393	319	255	194
	13	28	0.628	11	593	456	371	296	221
	14	29	0.694	12	685	527	428	342	253
	15	31	0.727	13	778	598	486	389	283
	16	33	0.759	14	876	674	548	438	313
7.0	4	13	0.234	10	108	83	68	54	47
	5	15	0.326	10	168	130	105	84	72
	6	18	0.384	10	241	185	150	120	102
	7	21	0.442	10	325	250	203	162	136
	8	23	0.529	10	422	325	264	211	176
	9	26	0.573	11	530	408	331	265	218
	10	28	0.646	12	651	501	407	325	266
	11	30	0.719	13	783	603	490	391	318
	12	33	0.760	14	925	711	578	462	370
	13	35	0.831	15	1079	830	675	539	429
	14	37	0.901	16	1245	958	778	622	490
	15	39	0.959	18	1421	1093	888	710	555
	16	41	1.028	19	1608	1237	1005	803	623
10.0	4	14	0.302	10	156	120	98	78	68
	5	18	0.361	10	242	186	151	121	105
	6	21	0.450	10	347	267	217	173	149
	7	24	0.539	10	469	361	293	234	200
	8	27	0.612	11	610	469	381	305	258
	9	30	0.669	13	767	590	480	383	322
	10	32	0.771	14	944	726	590	472	394
	11	35	0.826	16	1136	874	710	567	470
	12	38	0.896	17	1344	1034	840	671	552
	13	40	0.981	19	1572	1209	982	785	642
	14	43	1.048	20	1813	1394	1133	905	735
	15	45	1.147	21	2073	1594	1296	1035	835
	16	48	1.198	23	2345	1804	1466	1171	937
14.0	4	16	0.349	10	220	169	137	110	96
	5	20	0.443	10	341	263	213	170	149
	6	24	0.536	10	489	376	306	244	212
	7	27	0.624	12	663	510	414	331	286
	8	30	0.710	14	862	663	539	431	370
	9	33	0.814	15	1087	836	680	543	464
	10	36	0.900	17	1337	1028	836	668	567
	11	39	0.984	19	1611	1240	1007	805	680
	12	42	1.068	21	1910	1469	1194	954	801

Table 10-1 (Continued)

Square footings					*f'c = 3.00 ksi*				*fy = 60.0 ksi*
Soil fact'd (ksf)	*—Footing— Size (ft)*	*Thick (in.)*	*A_s sq in. /ft*	*Col. size (in.)*	*Ult. capacity (kip) ecc/ftg size 0.00*	*0.05*	*0.10*	*0.167*	*0.20*
	13	46	1.137	22	2230	1715	1394	1114	929
	14	48	1.251	24	2579	1984	1612	1288	1070
	15	51	1.333	26	2949	2269	1843	1473	1217
	16	55	1.399	27	3338	2567	2086	1667	1366
18.0	4	18	0.381	10	283	218	177	141	125
	5	22	0.506	10	440	339	275	220	193
	6	26	0.588	12	632	486	395	315	275
	7	29	0.700	14	857	659	536	428	372
	8	33	0.781	16	1115	858	697	557	481
	9	36	0.892	18	1407	1082	879	703	605
	10	40	0.992	19	1730	1331	1081	864	740
	11	43	1.101	21	2087	1605	1304	1042	889
	12	47	1.179	23	2474	1903	1546	1236	1048
	13	50	1.287	25	2894	2226	1809	1446	1221
	14	53	1.395	27	3346	2574	2091	1671	1406
	15	56	1.502	29	3830	2946	2393	1913	1602
	16	60	1.577	31	4339	3338	2712	2167	1805
22.0	4	19	0.435	10	347	267	217	173	153
	5	24	0.532	11	540	415	337	269	237
	6	28	0.637	13	774	596	484	387	339
	7	31	0.771	15	1051	809	657	525	459
	8	35	0.875	17	1369	1053	856	684	594
	9	39	0.977	19	1727	1328	1079	862	747
	10	42	1.087	22	2127	1636	1329	1062	917
	11	46	1.188	24	2565	1973	1603	1281	1100
	12	49	1.320	26	3045	2342	1903	1521	1302
	13	53	1.420	28	3561	2739	2226	1779	1516
	14	57	1.518	30	4116	3167	2573	2056	1745
	15	61	1.616	32	4710	3623	2944	2353	1987
	16	64	1.723	35	5345	4112	3341	2670	2248

Square footings					*f'c = 4.00 ksi*				*fy = 60.0 ksi*
Soil fact'd (ksf)	*—Footing— Size (ft)*	*Thick (in.)*	*A_s sq in. /ft*	*Col. size (in.)*	*Ult. capacity (kip) ecc/ftg size 0.00*	*0.05*	*0.10*	*0.167*	*0.20*
4.0	4	12	0.147	10	61	47	38	30	25
	5	12	0.257	10	95	73	59	47	40
	6	13	0.348	10	136	104	85	68	57
	7	15	0.394	10	183	141	114	91	75
	8	17	0.441	10	237	182	148	118	96
	9	19	0.487	10	297	229	186	148	119
	10	21	0.533	10	363	279	227	181	143
	11	23	0.578	10	435	335	272	217	169
	12	25	0.622	10	513	395	321	256	196
	13	26	0.692	11	599	461	374	299	227
	14	28	0.724	12	688	529	430	344	257

Table 10-1 (Continued)

Square footings					$f'c$ = 4.00 ksi				fy = 60.0 ksi
Soil fact'd (ksf)	—Footing— Size (ft)	Thick (in.)	A_s sq in. /ft	Col. size (in.)	Ult. capacity (kip) ecc/ftg size 0.00	0.05	0.10	0.167	0.20
	15	29	0.792	13	786	604	491	393	291
	16	30	0.861	14	890	684	556	444	326
7.0	4	12	0.266	10	109	84	68	54	47
	5	14	0.361	10	169	130	106	84	73
	6	17	0.414	10	241	186	151	121	103
	7	20	0.470	10	326	251	204	163	137
	8	22	0.559	10	423	326	265	211	177
	9	24	0.634	11	533	410	333	266	221
	10	26	0.709	12	655	503	409	327	270
	11	29	0.747	13	786	604	491	392	320
	12	31	0.819	14	930	715	581	464	375
	13	33	0.891	15	1085	835	678	542	435
	14	34	0.986	17	1255	966	785	627	501
	15	36	1.057	18	1433	1103	896	716	567
	16	39	1.089	19	1617	1244	1011	808	632
10.0	4	14	0.299	10	156	120	98	78	68
	5	17	0.389	10	243	187	152	121	105
	6	20	0.479	10	347	267	217	174	149
	7	23	0.567	10	470	362	294	235	201
	8	26	0.639	11	611	470	382	305	259
	9	28	0.727	13	770	593	481	385	325
	10	30	0.832	14	948	729	592	473	398
	11	33	0.885	16	1140	877	713	569	475
	12	35	0.988	17	1352	1040	845	675	560
	13	37	1.075	19	1581	1216	988	789	651
	14	40	1.140	20	1823	1402	1139	910	745
	15	42	1.242	21	2085	1604	1303	1041	847
	16	45	1.290	23	2358	1814	1474	1178	950
14.0	4	16	0.346	10	220	169	137	110	96
	5	19	0.472	10	342	263	214	171	149
	6	23	0.564	10	490	377	306	245	212
	7	26	0.651	12	664	511	415	332	286
	8	28	0.771	14	865	665	540	432	372
	9	31	0.875	15	1090	839	681	544	466
	10	34	0.959	17	1341	1031	838	670	571
	11	37	1.043	19	1616	1243	1010	807	684
	12	40	1.126	21	1915	1473	1197	957	806
	13	43	1.227	22	2239	1722	1399	1118	938
	14	45	1.344	24	2590	1992	1619	1294	1080
	15	48	1.425	26	2961	2278	1851	1479	1229
	16	51	1.524	27	3356	2581	2097	1676	1384
18.0	4	17	0.411	10	283	218	177	141	125
	5	21	0.535	10	441	339	276	220	193
	6	24	0.649	12	633	487	396	316	276
	7	28	0.727	14	858	660	536	429	373
	8	31	0.839	16	1117	859	698	558	484
	9	34	0.951	18	1410	1084	881	704	608

Table 10-1 (Continued)

Square footings					*f′c = 4.00 ksi*				*fy = 60.0 ksi*
Soil fact'd (ksf)	*—Footing— Size (ft)*	*Thick (in.)*	*A_s sq in. /ft*	*Col. size (in.)*	*Ult. capacity (kip) ecc/ftg size 0.00*	*0.05*	*0.10*	*0.167*	*0.20*
	10	37	1.062	20	1735	1335	1085	867	745
	11	41	1.159	21	2091	1609	1307	1045	893
	12	44	1.268	23	2481	1909	1551	1239	1056
	13	47	1.377	25	2903	2233	1814	1450	1230
	14	50	1.486	27	3357	2582	2098	1677	1416
	15	53	1.593	29	3841	2955	2401	1919	1614
	16	56	1.700	31	4357	3352	2723	2176	1823
22.0	4	18	0.466	10	347	267	217	173	153
	5	22	0.594	11	540	416	338	270	238
	6	26	0.696	13	776	597	485	387	340
	7	30	0.798	15	1052	809	658	526	459
	8	33	0.934	17	1371	1055	857	685	597
	9	37	1.034	19	1730	1330	1081	864	749
	10	40	1.145	22	2130	1638	1331	1064	920
	11	43	1.280	24	2571	1978	1607	1284	1107
	12	46	1.414	26	3052	2348	1908	1525	1310
	13	50	1.511	28	3570	2746	2231	1783	1525
	14	53	1.644	30	4130	3177	2581	2063	1759
	15	56	1.752	33	4730	3638	2956	2362	2007
	16	60	1.847	35	5363	4126	3352	2679	2266

1.6(1450) = 2320 kips and the design soil pressure = 1.6(6) = 9.6 ksf. Assuming the weight of the footing as 6 percent of the column load, or 140 kips, the required bearing area is 2,460/9.6 = 256.2 sq ft. A base 16 ft-0 in. square is selected, furnishing 256.0 sq ft. The net upward pressure due to column load only is 2,320/256 = 9.062 ksf.

The depth of footings is mostly governed by shear stress, but in order to locate the critical section for shear, the depth must be assumed. An approximate value is obtained by first computing the depth required for bending.

$$M_u = 9.062(1.0)\,\frac{(7.042)^2}{2} = 224.6 \text{ ft-kips}/1'\text{-}0''$$

Assume ratio $M_{u_{\text{required}}}/\overline{M}_{\text{concr.}} = 0.2$

$$\overline{M}_{u_{\text{concr.}}} = 224.6/0.2 = 1{,}123 \text{ ft-kips}$$

From Table 3-7, $d = 40$ in.

Since the depth required for shear is usually greater, $d = 43$ in. will be assumed for a trial. The critical section for beam shear is 43 in. from the face of the column, and its width is

$$23 + 2(43) = 109 \text{ in.} = 9.08 \text{ ft}$$

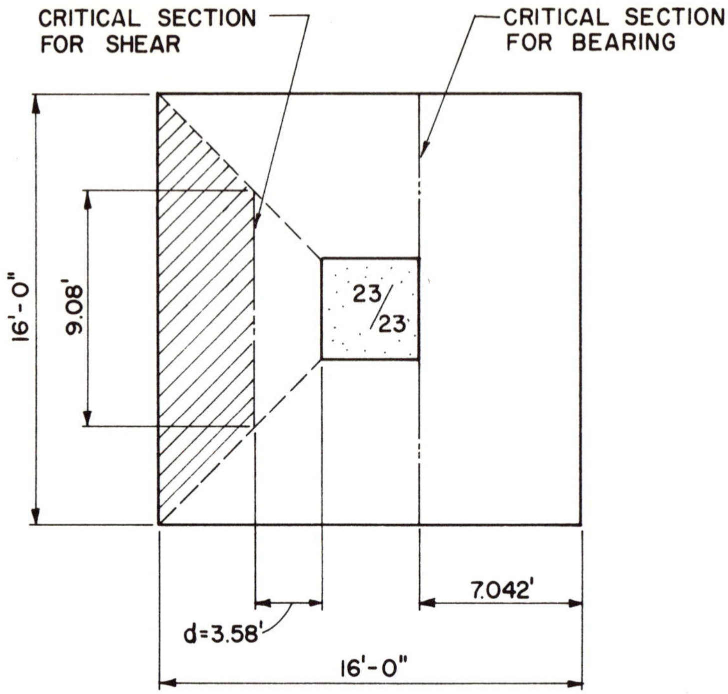

Fig. 10-1.

The shear on this section is

$$V_u = \frac{9.08 + 16.0}{2}(7.042 - 3.58)(9.062) = 393.4 \text{ kips}$$

$$v_u = \frac{V_u}{b_w d\phi} = \frac{393{,}400}{109(43)0.85} = 98.7 \text{ psi}$$

Shear stress carried by the concrete is:

$$v_c = 2\sqrt{f_c'} = 2(63.25) = 126.50 \text{ psi}$$

$$v_c = v_u (\text{O.K.})$$

The critical section for two-way action is at the periphery of the reaction area at $d/2$ distance (see Fig. 10-2).

b_o(periphery) = 4(23 + 43) = 264 in.

$$V_u = (16.0^2 - 5.5^2)(9.062) = 2045.75 \text{ kips}$$

$$v_u = \frac{V_u}{\phi b_o d} = \frac{2{,}045{,}750}{0.85(264)(43)} = 212 \text{ psi}$$

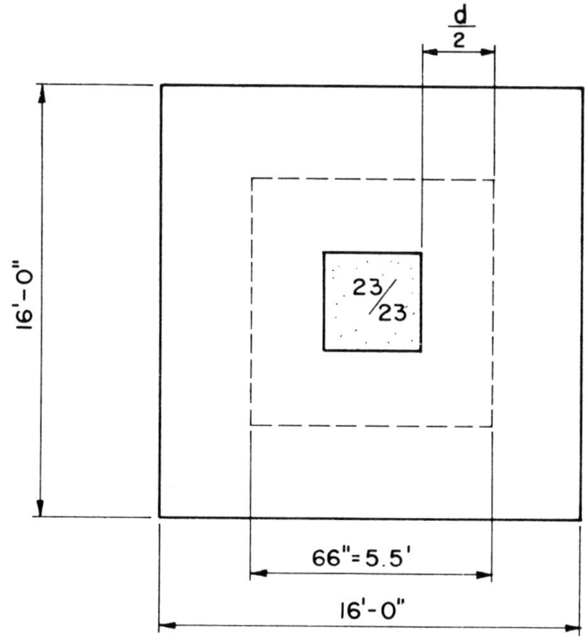

Fig. 10-2.

Shear stress carried by the concrete:

$$v_c = 4\sqrt{f_c'} = 4(63.25) = 253 \text{ psi}$$

$$v_c > v_u \text{ O.K.}$$

Check required reinforcement for bending:

$$M_{u_{\text{required}}} = 224.6 \text{ ft-kips per } 1.0 \text{ ft}$$

Using $f_c' = 3$ ksi, $f_y = 60$ ksi, $b = 12$ in., $d = 43$ in. From Table 3-7:

$$\overline{M}_{u_{\text{concr.}}} = 1{,}300.0 \text{ ft-kips}$$

$$\text{Ratio: } \frac{M_{u_{\text{required}}}}{\overline{M}_{u_{\text{concr.}}}} = 224.6/1{,}300.0 = 0.172$$

From Fig. 3-8, $a_u = 4.37$

$$A_s = \frac{M_{u_{\text{required}}}}{a_u d} = \frac{224.6}{4.37(43)} = 1.195 \text{ sq in.}$$

Check column design load capacity from Table 10-1. Using: $f_c' = 3$ ksi, $f_y = 60.0$ ksi, soil = 10.0 ksf, column size, 23 by 23 in. Capacity is 2345.0 kips; required reinforcement, 1.198 sq in. per ft in both directions.

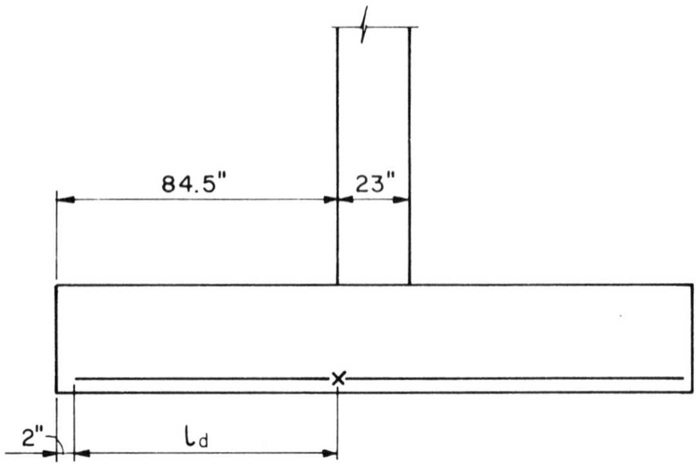

Fig. 10-3.

Check maximum reinforcing-bar size that may be used (see Fig. 10-3). Available development length for straight bar = 84.5 in. - 2 in. = 82.5 in. Assume #11 or smaller bars,

$$l_d = 0.04A_b f_y/\sqrt{f_c'} = 84.5 \text{ in.}$$

Required $A_b < 84.5\sqrt{3000}/(0.04(60{,}000)) < 2.26$ sq in. Use: #11 bar (A_b = 1.56 sq in.) at 15 in. o.c. (=) 1.25 sq in.

Example 2

Check the same footing as described in Example 1, but here the column is of 1800 kips design load acting with an eccentricity of 0.80 ft (see Fig. 10-4).

P_u = 1,800 kips, e = 0.80 ft, $M_u = P_u(e)$ = 1,800(0.8) = 1,440 ft-kips

$$S = \frac{16.0^3}{6} = 682.6 \text{ ft}^3$$

$$f_1 = \frac{1800.0}{256.0} + \frac{1{,}440.0}{682.6} = 7.03 + 2.11 = 9.14 \text{ ksf}$$

$$f_2 = 7.03 - 2.11 = 4.92 \text{ ksf}$$

Maximum soil pressure: 9.14 + 1.4(4.0) 0.15 = 10.00 ksf (O.K.)

The net average upward pressure due to column load is:

$$\frac{9.14 + 4.92}{2} = 7.03 \text{ ksf}$$

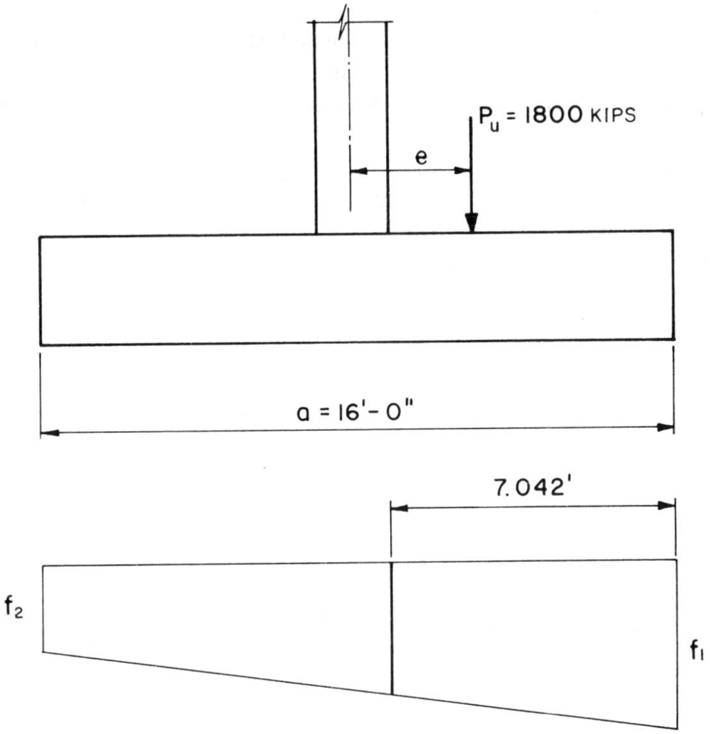

Fig. 10-4.

Since the average upward pressure is less than the value in Example 1, shear and bending check are neglected. By Tables 10-1, maximum design load capacity of column for

$$\beta - \frac{e}{a} - 0.80/16.0 - 0.05 \qquad\qquad P_u = 1{,}804 \text{ kips}$$

COMBINED FOOTINGS

The single-column footings are generally the simplest and most economical type of building column foundations. Their use under exterior columns meets with difficulties if property rights prevent the use of footings projecting beyond the exterior walls. In this case combined footings or strap footings are used, which enables one to design a kind which will not project beyond the wall column.

Combined footings under two or more columns are also used under closely spaced, heavily loaded interior columns where single footings, if they were provided, would completely or nearly merge.

The design of combined footings by means of subgrade reactions has been restricted from general application to singular cases for two principal reasons: (1) the uncertainties in connection with the selection of the correct subgrade reaction; (2) the difficulties of the theory and its application to practical cases.

A considerable amount of research work was undertaken to explore the problems in connection with the subgrade reaction produced by various soils; in 1956 Karl Terzaghi presented an article, "Evaluation of Coefficients of Subgrade Reactions," published by the Institution of Civil Engineers, London. This paper provides sufficient information to determine, at least for design purposes, the magnitude of the subgrade reaction from data obtained from a standard soil exploration (see Table 10-2). Since the calculation of moments, contact pres-

Table 10-2*

Size Factor δ for Footings on Sand

$$\rho = \frac{1}{4}\left(1 + \frac{1}{b}\right)^2$$

b (ft)	5′	10′	15′	20′	Large
ρ	0.36	0.30	0.28	0.275	0.25

Side-Ratio Factor ρ for Footings on Clay

$$\rho = \frac{n + 0.5}{1.5n}$$

$n\bar{b}$

BASE AREA $\bar{b}$

For strips $\rho = 0.67$

Coefficient of Subgrade Reaction *ksi*

	SAND			*CLAY*		
	Loose	*Medium*	*Dense*	*Stiff*	*Very Stiff*	*Hard*
N	4–10	10–30	30–50	8–15	15–30	> 30
ksi	20–60	60–300	300–1000	50–100	100–200	> 200

$$\lambda = \sqrt[4]{\frac{k}{4E\bar{I}}} = \sqrt[4]{\frac{(2k_{si}\,\rho)\,\bar{b}}{4\left(\frac{1000 f_c'\,144}{1000}\right)\left(\frac{bt^3}{12 \times 1728}\right)}} = \boxed{2.9\sqrt[4]{\frac{k_{si}\rho}{f_c' t^3}} = \lambda}\left(\frac{1}{ft}\right)$$

$k = 2_{si}\,\delta$

*The following formulas and tables are based on "Evaluation of Coefficients of Subgrade Reaction" by K. Terzaghi, published by the Institution of Civil Engineers, London, 1956, and *Soil Mechanics in Engineering Practice*, by K. Terzaghi and R. B. Peck, Wiley, New York.

sures, etc., by means of the exact theory, are not very sensitive to the selection of the k-value, such an estimate is usually sufficient. If necessary, it can be checked later by actual testing.

Much work was done also to develop the basic idea theoretically. Since its original concept (Winkler, 1867), engineers have always been intrigued by the mathematical potentialities of this method. Realizing its difficulties, tables have been worked out (Hayashi, Hetenyi, etc.) and methods have been developed (Malter, Gazis, Ray, Popov, etc., to mention only a few) to facilitate the calculations. Most of the papers produced quite ingenious new methods of approach to the exact solution, but none appeared to be simple enough to be used for general application. Consequently, most of the economical advantage of the theory remained untapped.

As practical engineers, we must realize that the uncertainty of our assumptions often does not warrant exactness of the result. We know, for example, that soil in general does not satisfy the basic assumption of elastic support; that the k-factor is by no means a constant; that the homogeneity of the soil is certainly questionable; and that the rigidity of the superstructure is hard to evaluate—and this is only part of a list of deficiencies that can be enumerated. We must keep in mind, therefore, that the great importance of this method lies primarily in its value as a "tool," not in the extreme exactness of its results.

The following derivations will show that, if certain requirements are fulfilled, a simplified approximate method can be developed for the calculation of moments, shears, and contact pressures for a combined footing, which will furnish sufficiently accurate results.*

Although the method described is for a continuous beam foundation, other combined footings, such as grid foundations and mat foundations, can be calculated on the same basis by dividing them into strips. Each strip, then, has to be considered as an independent unit and has to be calculated by using the full column loads in each direction. In the case of a mat foundation, each strip will have a width extending from center to center of adjacent bays and will consist of a column strip bordered by two half middle strips. The moments, shears, and contact pressures obtained by this method are for the entire strip. They will have to be subdivided and apportioned to the column and to the middle strip.

In order to ascertain the characteristics of a continuous beam, each footing strip must have not less than three bays or four columns in a row.

The ratio between adjacent spans and adjacent column loads may differ by a maximum of 20 percent of the greater value. It will be shown that even such liberal deviations from the basic case of equal loads and equal spacings can be tolerated without excessive loss in the accuracy of the end results.

All formulae and derivations are based on the assumption that the continuous

*See also: Simplified Design of Combined Footings, by F. Kramrisch & P. Rogers; ASCE Prod. Vol. 87, No. SM 5, Oct. 1961.

beam footings or strips are of prismatic cross section. Small variations in the width of the footing, making the narrowest part not less than three-fourths of the widest, will be tolerated without difficulty. In a continuous beam footing, the column spacings must stay between certain limits to keep the footing responsive to the reactions of the subgrade. Column spacings which are too small will make the foundation act as infinitely stiff; whereas column spacings which are too big will dissolve the continuous footing into single units with wasted interconnecting pieces. The evaluation of the contact pressures, positive moments, and shears was simplified by the assumption of a straight-line distribution of the contact pressure with a maximum value under the columns and a minimum value at the center of each bay.

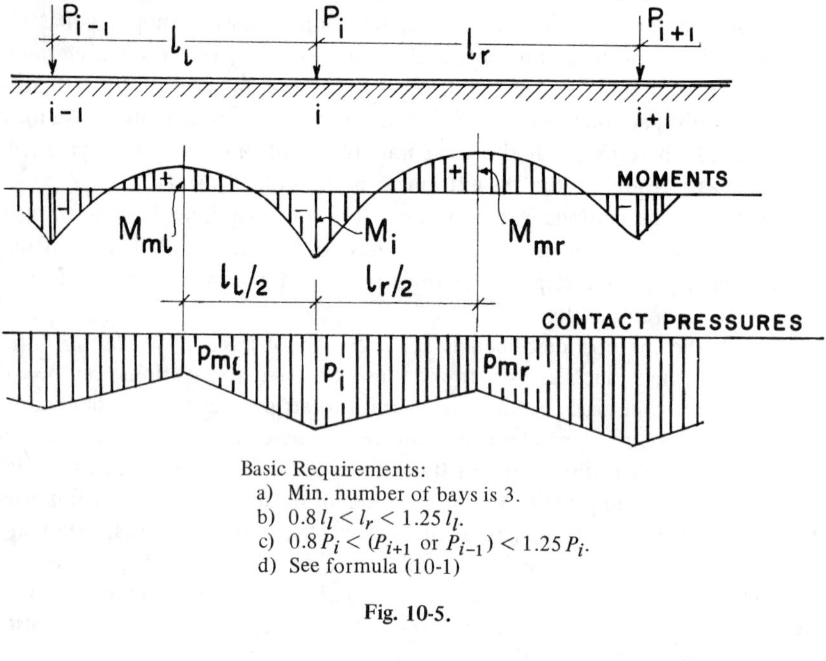

Basic Requirements:

a) Min. number of bays is 3.
b) $0.8\,l_l < l_r < 1.25\,l_l$.
c) $0.8\,P_i < (P_{i+1}$ or $P_{i-1}) < 1.25\,P_i$.
d) See formula (10-1)

Fig. 10-5.

Figure 10-5 schematically shows the moments and the simplified distribution of the contact pressures under a continuous beam footing. But it has to be kept in mind that the moments shown in that Figure are produced by the column loads only and have to be superposed, if necessary, with moments caused by the settlement deformation (dishing) of the combined footing as a whole.

In order not to expand the scope of this treatise unnecessarily, we presume that the classical theory of a beam supported by subgrade reactions (or, as previously

NOTATION

a	(f)	length of end projection of combined footing, measured from exterior column to tip of end projection.
$\bar{b}$	(f)	average width of combined footing (beam or strip), averaged between centers of bays or between center of first bay and tip of end projection.
E	(ksf)	modulus of elasticity of concrete used for combined footing.
$\bar{I}$	(f^4)	average moment of inertia of combined footing, based on average width $\bar{b}$.
k	(ksf)	coefficient of vertical subgrade reaction for the width $\bar{b}$.
k_{sl}	(tcf)	basic coefficient of vertical subgrade reaction for a square area 1 foot wide.
l	(f)	distance between two adjacent columns.
$\bar{l}$	(f)	average of two adjacent column distances.
M_e	(fk)	moment at exterior column for entire width of combined footing.
M_i	(fk)	moment at interior column, for entire width of combined footing.
M_m	(fk)	moment at center of bay, for entire width of combined footing.
M_o	(fk)	simple span beam moment (for the entire width of the combined footing) for a beam supported by two adjacent columns acted upon by contact pressure due to column loads only.
N	....	number of blows in standard penetration soil test.
P_e	(k)	load of exterior column, consisting of weight of superstructure (D.L. + L.L.) excluding weight of footing and surcharge.
P_i	(k)	load of interior column, consisting of weight of superstructure (D.L. + L.L.) excluding weight of footing and surcharge.
p_a	(k/f)	contact pressure at tip of end projection due to column loads only, for entire width of combined footing.
p_e	(k/f)	contact pressure at exterior column due to column loads only, for entire width of combined footing.
p_i	(k/f)	contact pressure at interior column due to column loads only, for entire width of combined footing.
p_m	(k/f)	contact pressure at center of bay, due to column loads only, for entire width of combined footing.
$\bar{p}_m$	(k/f)	average between the p_m values at the centers of two adjacent bays.
$\bar{\bar{p}}_m$	(k/f)	average between the p_m values at the center of same bay as determined from adjacent bays
p'	(ksf)	total unit contact pressure, due to column loads plus footing weight plus footing weight plus surcharge.
ρ	....	size- or side-ratio factor of subgrade reaction.
t	(in.)	thickness of combined footing.

Fig. 10-6.

called, "beam on elastic foundation") is known, and refer to the standard articles and textbooks on the subject.*

Figure 10-6 illustrates the notations used in this combined footing design method.

MOMENT UNDER INTERIOR COLUMNS

For an infinitely long beam of prismatic cross section, supported by subgrade reaction, the exact theory furnishes the following formula for the bending

*K. Hayashi, *Theorie des Traegers auf elastisher Unterlage*, Springer, Berlin, 1921. M. Hetenyi, *Beams on Elastic Foundation*, Univ. of Michigan Press, Ann Arbor, 1946. E. Winkler, *Die Lehre von Elastizitaet und Festigkeit*, Prag, 1867.

moment, M_x, due to the load P:

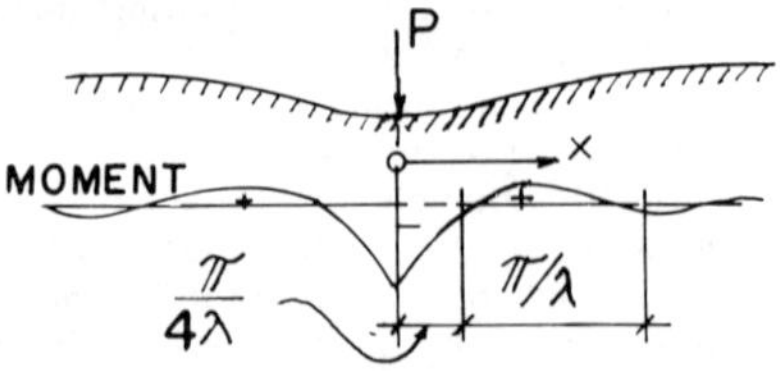

Fig. 10-7.

$$M_x = \frac{P}{4\lambda} C_{\lambda x}$$

In this formula the "characteristic"

$$\lambda = \sqrt[4]{\frac{k}{4EI}}$$

k is the coefficient of subgrade reaction and

$$C_{\lambda x} = e^{-\lambda x} (\cos \lambda x - \sin \lambda x)$$

is a trigonometric function for which tables have been prepared for various values of λ_x. If several loads are acting on the beam, the total moment at point c will be the sum of all moment influences produced by the loads so acting.

$$M_c = \frac{P_1}{4\lambda} C_{\lambda x_1} + \frac{P_2}{4\lambda} C_{\lambda x_2} + \cdots \frac{P_n}{4\lambda} C_{\lambda x_n} = \sum \frac{P_{1,2,\ldots n}}{4\lambda} C_{\lambda x\,1,2,\ldots n}$$

$x_1, x_2, \cdots x_n$ are distances of loads $P_1, P_2, \cdots P_n$ from point c.

If we restrict, arbitrarily for the time being, the spacing of the loads to values between $1.75/\lambda$ and $3.50/\lambda$, we can replace the trigonometric function for $C_{\lambda x}$, closely enough, by a straight-line relation. The total moment under an interior column load at point i can then be written as

$$\boxed{M_i = -\frac{P_i}{4\lambda} (0.24\lambda \bar{l} + 0.16)} \qquad (10\text{-}1)$$

Basic requirement d) $\dfrac{1.75}{\lambda} < \bar{l} < \dfrac{3.50}{\lambda}$

$$\bar{l} = \frac{l_e + l_r}{2}$$

This moment is the sum of the moment due to the column load P_i at point i and the influence of *one* adjacent column load at each side, P_{i-1} and P_{i+1}, located at $\bar{l} = \dfrac{l_e + l_r}{2}$ (average) from point i.

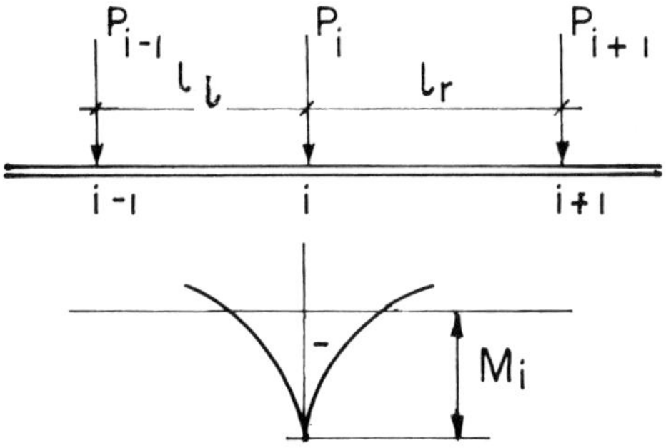

Fig. 10-8.

The following will explain the selection of the limiting values for the average spacing of the column loads: If concentrated loads are placed very close together, the foundation will act like an infinitely stiff beam and the distribution of the contact pressure will follow a straight line. In this case, the maximum negative moment at the column will approach the value of the negative moment of an inverted continuous beam carrying an equal distributed load (contact pressure).

$$-\frac{P_i}{4\lambda}(0.24\lambda\bar{l}_{\text{min}} + 0.16) = -\frac{P_i\bar{l}_{\text{min}}}{12}$$

From this equation, the lower limit $\bar{l}_{\text{min}}$ can be evaluated as $1.78/\lambda$. For reasons of simplicity it was assumed as $1.75/\lambda$. For column spacings closer than this value, equation (10-1) will furnish moments which are greater than those produced by a straight-line distribution of the contact pressure. Such moments are statically produced only by loadings, or contact pressures, which are greater at the center of the span than at the support; but such a condition is in actual practice not feasible. In addition to the selection of the minimum column spacing with $1.75/\lambda$, the maximum negative moment must be checked not to exceed $Pl/12$; otherwise, the foundation has to be assumed infinitely stiff and the distribution of the contact pressure will follow a straight line.

From the evaluation of the function for $C_{\lambda x}$ (see also Fig. 10-7), it can be seen that the moment influence from column loads located at a distance greater than $\sim\pi/\lambda$ can be disregarded. The upper limit was selected with $3.50/\lambda$ to satisfy this requirement. Its magnitude eliminates, at the same time, the need for adding the influence of the second adjacent column in either direction, even for the smallest admissible column spacing ($3.5 = 2 \times 1.75$).

Equation (10-1) is an approximate formula; but within the stipulated limitations it furnishes quite close results, even considering the permissible variations

(Fig. 10-5) of column loads and column spacings up to 20 percent of the greater value. The following equations, based on Table 10-3 and shown in Figure 10-9 give the relationship between the approximate equation (10-1) and the exact results for the most unfavorable combinations of variations in loads and spacings.

TABLE OF MOMENT FACTORS F_0 TO F_6

$\bar{l}\lambda$	F_0	F_1	F_2	F_3	F_4	F_5	F_6
1.75	0.5800	0.5960	0.6768	0.4950	0.6062	0.5914	0.6013
2.00	.6400	.6414	.7131	.5517	.6478	.6282	.6497
2.25	.7000	.7036	.7629	.6295	.7058	.6840	.7130
2.50	.7600	.7702	.8162	.7127	.7679	.7458	.7783
2.75	.8200	.8330	.8664	.7912	.8257	.8053	.8374
3.00	.8800	.8874	.9099	.8587	.8783	.8607	.8898
3.25	.9400	.9314	.9451	.9142	.9205	.9060	.9310
3.50	1.0000	.9646	.9717	.9557	.9529	.9416	.9618

Table 10-3

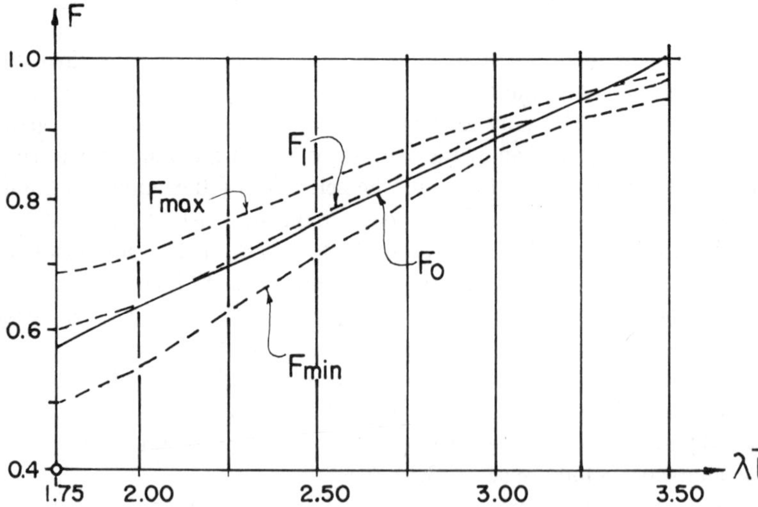

Fig. 10-9 Deviations of moment factors F_1 to F_6.

From formula (10-1)

$$M_i = -\frac{P}{4\lambda}(0.24\lambda\bar{l} + 0.16) = -\frac{P}{4\lambda}F_0$$

Condition No. 1:

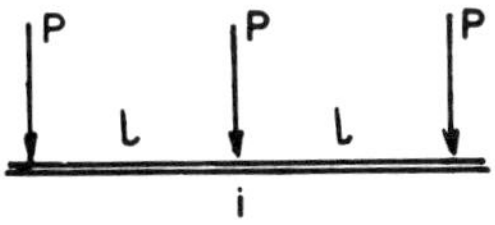

$$M_{i(1)} = -\frac{P}{4\lambda}(C_{\lambda l} + C_{\lambda 0} + C_{\lambda l}) = -\frac{P}{4\lambda}F_1$$

Condition No. 2:

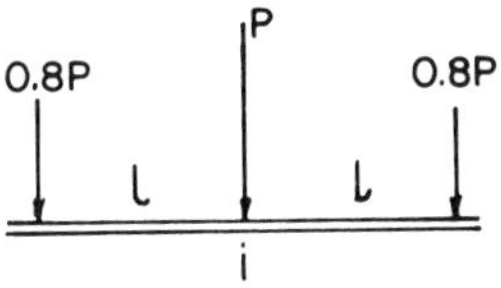

$$M_{i(2)} = -\frac{P}{4\lambda}(0.8C_{\lambda l} + C_{\lambda 0} + 0.8C_{\lambda l}) = -\frac{P}{4\lambda}F_2$$

Condition No. 3:

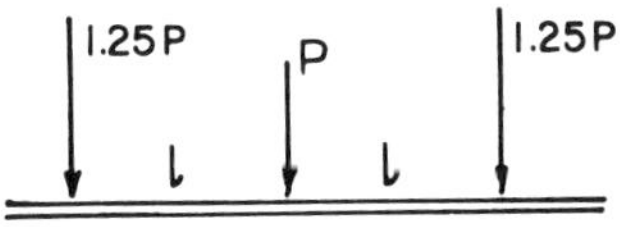

$$M_{i(3)} = -\frac{P}{4\lambda}(1.25C_{\lambda l} + C_{\lambda 0} + 1.25C_{\lambda l}) = -\frac{P}{4\lambda}F_3$$

Condition No. 4:

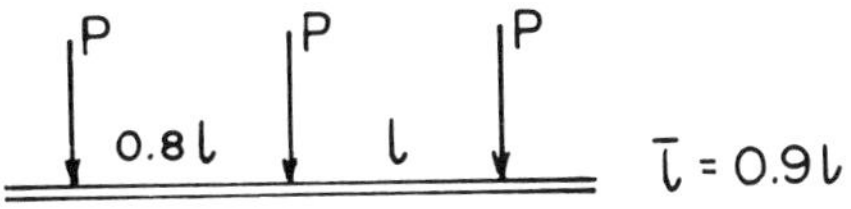

$$M_{i(4)} = -\frac{P}{4\lambda}(C_{\lambda 0.89\bar{l}}) + C_{\lambda 0} + C_{\lambda 1.11\bar{l}}) = -\frac{P}{4\lambda}F_4$$

Condition No. 5:

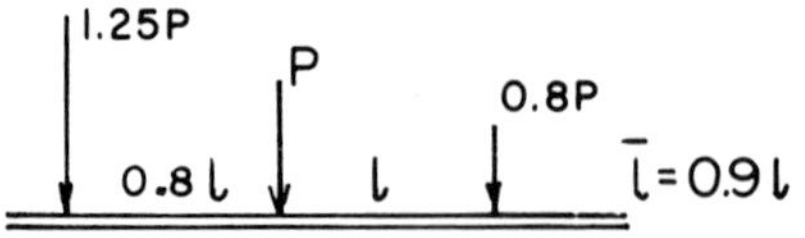

$$M_{i(5)} = -\frac{P}{4\lambda}(1.25C_{\lambda 0.89\bar{l}} + C_{\lambda 0} + 0.8C_{\lambda 1.11\bar{l}}) = -\frac{P}{4\lambda}F_5$$

Condition No. 6:

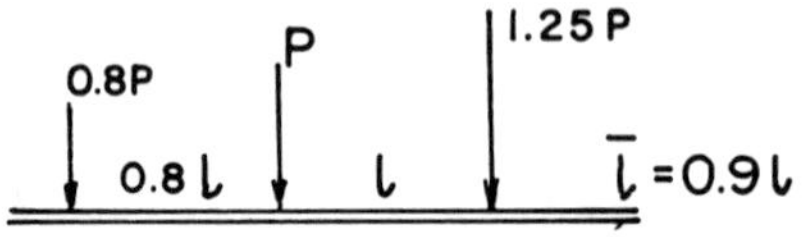

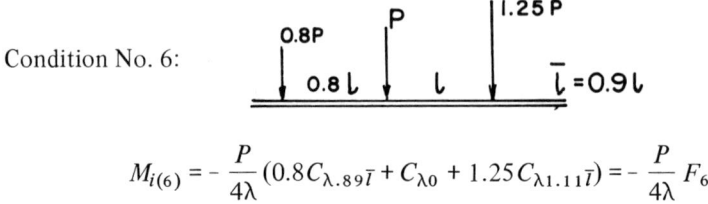

$$M_{i(6)} = -\frac{P}{4\lambda}(0.8C_{\lambda .89\bar{l}} + C_{\lambda 0} + 1.25C_{\lambda 1.11\bar{l}}) = -\frac{P}{4\lambda}F_6$$

CONTACT PRESSURE UNDER INTERIOR COLUMNS

The magnitude and distribution of the contact pressure under an interior column load is not determined directly from the exact theory, but from the moment obtained by equation (10-1). In order to simplify the procedure, the following two assumptions were made:

Assumption no. 1. The resultant of all subgrade reactions *acting on each interior foundation part* shall be equal, concentric, and of opposite direction to the column load acting on this part. (An interior foundation part is that portion of a combined footing which primarily supports one interior column load. It is assumed to extend from center to center of adjacent bays.)

Assumption no. 1 satisfies the fundamental requirement that the resultant of all subgrade reactions acting on a combined footing shall be equal, concentric and of opposite direction to the resultant of all column loads acting on the combined footing, but due to its restricted form, it simplifies the analysis and facilitates the calculation.

Assumption no. 2. The contact pressures under an interior foundation part are assumed to have a maximum value under the column, a minimum value at the center of the bay, and shall follow a straight-line distribution in between (see Fig. 10-10).

Since the negative moment at a column is affected but little by variations in the loads of adjacent columns, the maximum contact pressure under a column can be approximated from a symmetrical loading condition over the average span $\bar{l}$ (see Fig. 10-11).

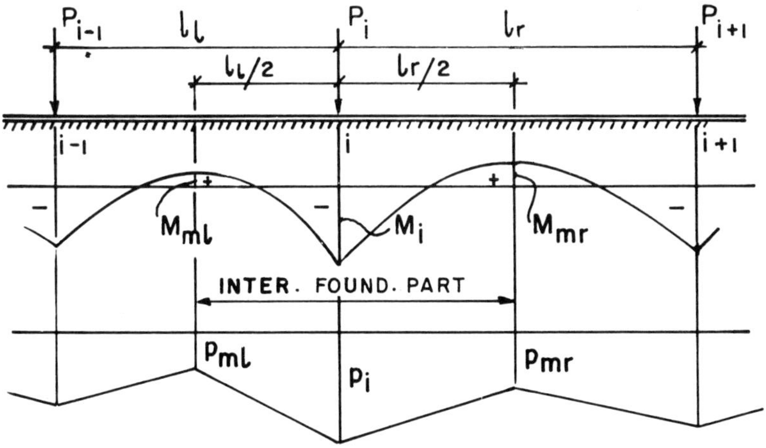

Fig. 10-10.

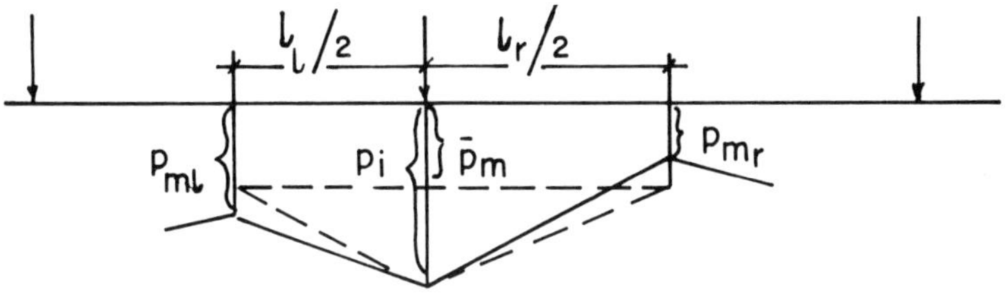

$$\bar{p}_m = \frac{p_{ml} + p_{mr}}{2} \qquad \bar{l} = \frac{l_l + l_r}{2}$$

Fig. 10-11.

$$-M_i = \frac{p_m \bar{l}^2}{12} - (p_i - \bar{p}_m)\frac{\bar{l}^2}{32} = -\frac{\bar{l}^2}{96}(3p_i + 5\bar{p}_m)$$

$$p_i = \frac{(\bar{p}_m + p_i}{2}\bar{l} \qquad \bar{p}_m = \frac{2P_i}{\bar{l}} - p_i$$

By introducing $\bar{p}_m$ into equation for M_i, we obtain p_i, the approximate maximum contact pressure for the entire width of the beam, as

$$\boxed{p_i = \frac{5P_i}{\bar{l}} + \frac{48M_i}{\bar{l}^2}} \tag{10-2}$$

The minimum contact pressures at the middle of the adjacent spans can then be determined as shown below:

$$P_i = \frac{(p_{ml} + p_i)}{2}\frac{l_l}{2} + \frac{(p_i + p_{mr})}{2}\frac{l_r}{2}$$

$$\frac{(p_{ml} + p_i)}{2}\frac{l_l}{2}\frac{l_l}{6}\frac{(2p_{ml} + p_i)}{(p_{ml} + p_i)} = \frac{(p_i + p_{mr})}{2}\frac{l_r}{2}\frac{l_r}{6}\frac{(p_i + 2p_{mr})}{(p_i + p_{mr})}$$

$$\boxed{\begin{aligned} p_{ml} &= 2P_i \frac{l_r}{l_l \bar{l}} - p_i \frac{\bar{l}}{l_l} \\ p_{mr} &= 2P_i \frac{l_l}{l_r \bar{l}} - p_i \frac{\bar{l}}{l_r} \end{aligned}} \qquad (10\text{-}3)$$

For equal spans and equal loads,

$$\boxed{p_m = \frac{2P_i}{l} - p_i = -\frac{48M_i}{l^2} - \frac{3P_i}{l}} \qquad (10\text{-}3a)$$

MOMENT BETWEEN COLUMNS

By considering *each part* of a combined footing, extending from center to center of adjacent bays, *separately*, we obtain at the middle of each bay two minimum contact pressures. These contact pressures, $p_{mr(l)}$ belonging to the left part, and $p_{ml(r)}$ belonging to the right part are not necessarily alike because each was computed on a different basis. For the calculation of the positive

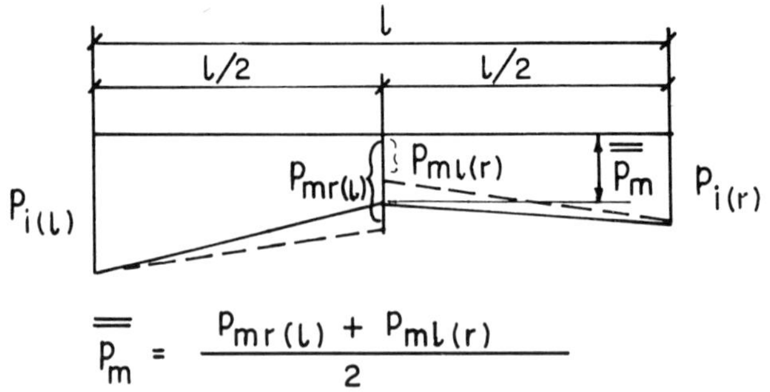

Fig. 10-12.

moment, they can be replaced by $\bar{\bar{p}}_m$. The moment M_o of the simply supported beam then becomes

$$M_o = \frac{l^2}{48}(p_{i(l)} - \overline{\overline{p_m}}) + \frac{l^2}{8}\overline{\overline{p_m}} + \frac{l^2}{48}(p_{i(r)} - \overline{\overline{p_m}})$$

$$\boxed{M_o = \frac{l^2}{48}(p_{i(l)} + 4\overline{\overline{p_m}} + p_m + p_{i(r)})} \qquad (10\text{-}4)$$

and the positive moment results as

$$M_m = M_o + \overline{M_i}$$

where $\overline{M_i}$ is the average of the negative moments under the columns at each end of the bay.

$$\overline{M_i} = \tfrac{1}{2}(M_{i(l)} + M_{i(r)})$$

From the final distribution of the contact pressures, the shear can be determined for any location.

MOMENT UNDER EXTERIOR COLUMNS

The moment under an exterior column consists of the moment due to the exterior column load itself, the influence of the next interior columns and the influence of the free end. The first two parts can be treated similarly as for an interior column.

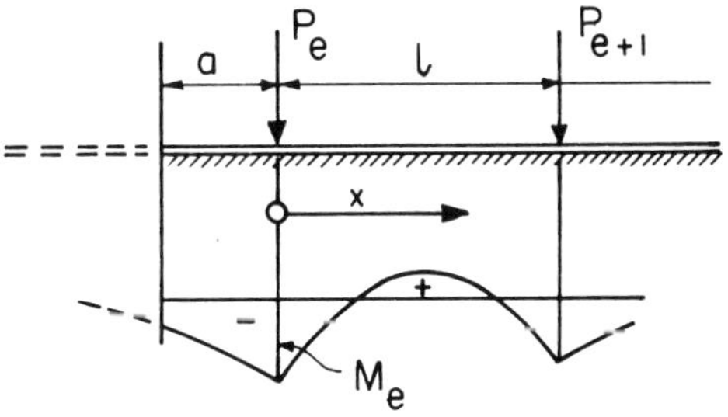

Fig. 10-13.

$$M'_e = -\frac{1}{4\lambda}(P_e C_{\lambda 0} + P_{e+1} C_{\lambda l 1} + \cdots)$$

Using the same limitations as for the interior column spacing, a straight-line expression can be substituted and the equation becomes

$$\boxed{M'_e = -\frac{P_e}{4\lambda}(0.13\lambda l_1 + 0.56)} \qquad (10\text{-}5a)$$

The following equations, based on Table 10-4 and shown in Fig. 10-14, give the relationship between the exact results for the various conditions of loadings and

Table 10-4. Moment factors F_7 to F_{10}

$l_1\lambda$	F_7	F_8	F_0	F_{10}
1.75	0.7875	0.7980	0.8384	0.7475
2.00	.8200	.8207	.8565	.7758
2.25	.8525	.8518	.8814	.8147
2.50	.8850	.8851	.9081	.8563
2.75	.9175	.9165	.9332	.8956
3.00	.9500	.9437	.9549	.9293
3.25	.9825	.9657	.9725	.9571
3.50	1.0150	.9823	.9858	.9778

those of the approximate equation (10-5a):
From formula (10-5a)

$$M'_{e(7)} = -\frac{P}{4\lambda}(0.13\lambda l_1 + 0.56) = -\frac{P}{4\lambda}F_7$$

condition no. 8:

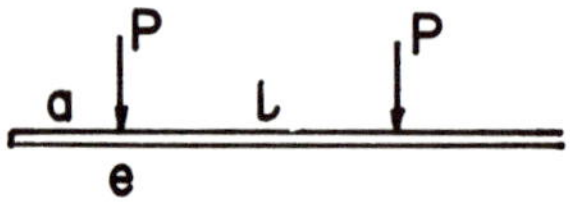

$$M'_{e(8)} = \frac{P}{4\lambda}(C_{\lambda 0} + C_{\lambda l_1}) = -\frac{P}{4\lambda}F_8$$

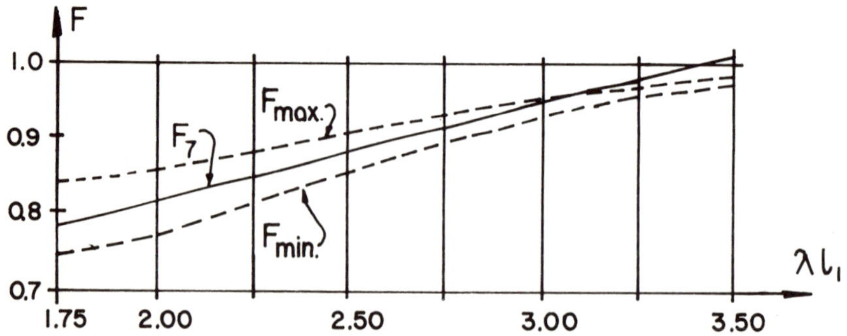

Fig. 10-14 Deviations of moment factors F_8 to F_{10}.

condition no. 9:

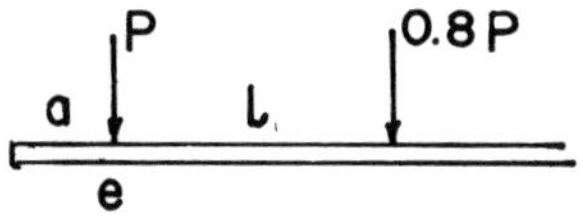

$$M'_{e(9)} = -\frac{P}{4\lambda}(C_{\lambda 0} + 0.8C_{\lambda l_1}) = -\frac{P}{4\lambda}F_9$$

condition no. 10:

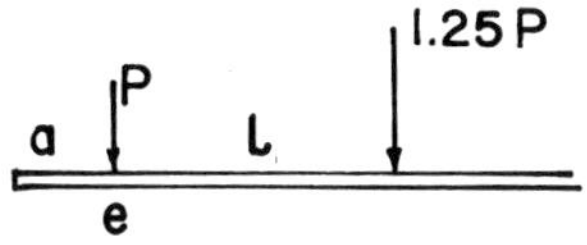

$$M'_{e(10)} = -\frac{P}{4\lambda}(C_{\lambda 0} + 1.25C_{\lambda l_1}) = -\frac{P}{4\lambda}F_{10}$$

The influence of a free end of a semi-infinite long beam supported by subgrade reaction on the moment under the first column can be found through the application of a moment and a shear force at the free end. These two forces, called by Hetenyi "end-conditioning forces," make the actual moment and shear of the infinite long beam vanish to simulate the condition of a free end.

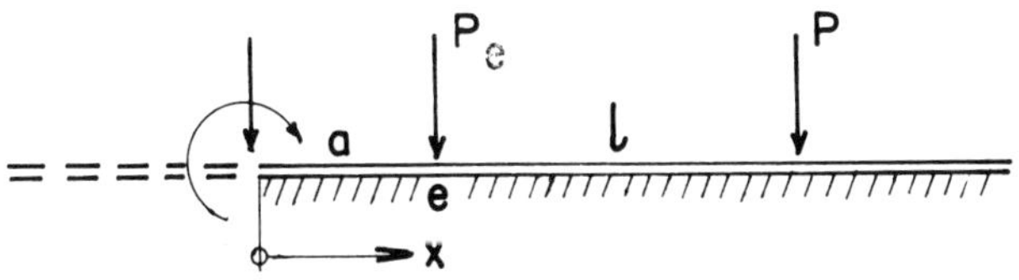

Fig. 10-15.

The influence of the end-conditioning forces on the moment M_e under the exterior column is

$$M''_e = -\frac{P_e}{4\lambda}[(C_{\lambda_a} + 2D_{\lambda_a})C_{\lambda_a} - 2(C_{\lambda_a} + D_{\lambda_a})D_{\lambda_a}] =$$

$$-\frac{P_e}{4\lambda}(C^2_{\lambda_a} - 2D^2_{\lambda_a}) = -\frac{P_e}{4\lambda}F_{11}$$

Evaluating this equation with the help of tables given in Hetenyi's *Beams on Elastic Foundation*, we obtain the curve for F_{11} shown in Fig. 10-16.

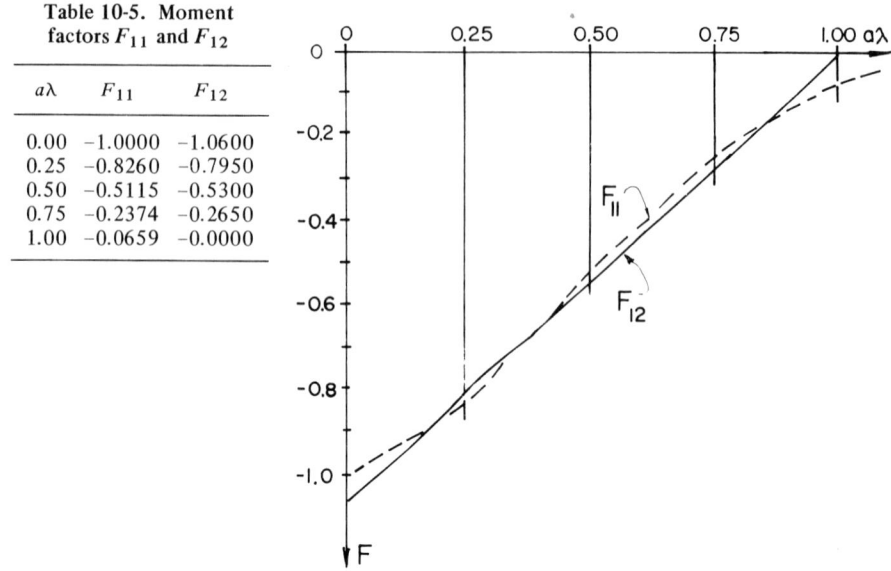

Table 10-5. Moment factors F_{11} and F_{12}

$a\lambda$	F_{11}	F_{12}
0.00	−1.0000	−1.0600
0.25	−0.8260	−0.7950
0.50	−0.5115	−0.5300
0.75	−0.2374	−0.2650
1.00	−0.0659	−0.0000

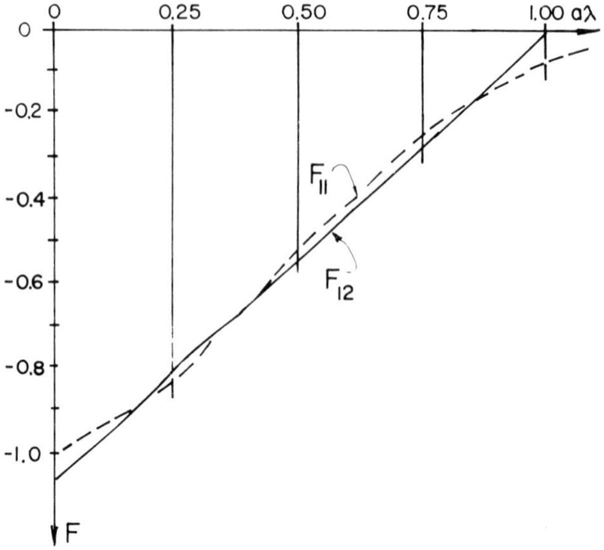

Fig. 10-16.

Moment factor F_{11} crosses the zero line at a distance $\lambda a = 3/8\pi$. Stipulating $1/\lambda$ as the maximum length for the end projection a, we can substitute a straight-line relation F_{12} for the trigonometric expression of F_{11}.

$$M_e'' = -\frac{P_e}{4\lambda}(1.06_{\lambda_a} - 1.06) = -\frac{P_e}{4\lambda}F_{12} \qquad (10\text{-}5b)$$

The final moment M_e consists of the sum of M_e' and M_e''.

$$M_e = M_e' + M_e'' = -\frac{P_e}{4\lambda}(0.13_{\lambda_{l_1}} + 0.56 + 1.06_{\lambda_a} - 1.06) =$$

$$M_e = -\frac{P_e}{4\lambda}(0.13_{\lambda_{l_1}} + 1.06_{\lambda_a} - 0.50) \qquad (10\text{-}5)$$

Equation (10-5) furnishes, especially for small end projections, moments which require high subgrade reactions on the projecting part of the footing. Subsequent settlement will reduce the moment at the exterior column and shift the resultant of the subgrade reactions for the exterior part toward the interior of the beam. This self-balancing action led to the assumption that for small projections the maximum contact pressure at the end of the projection (p_a) should not exceed the contact pressure under the exterior column. This is shown in Fig. 10-17, which indicates also the usually slight offset of the resultant of the contact pressures for the exterior footing part from the location of the exterior column.

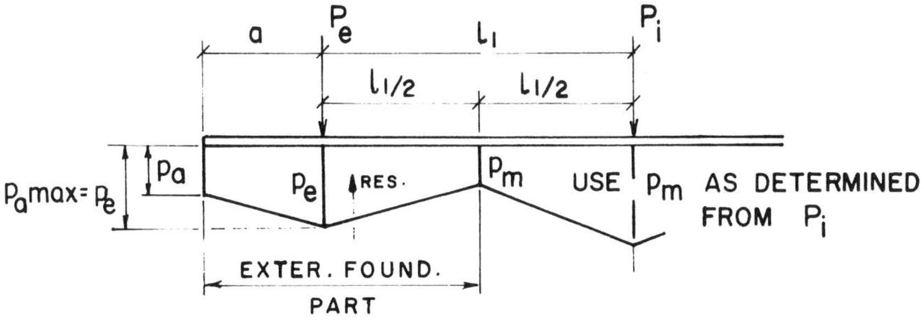

Fig. 10-17.

To account for these adjustments, assumptions nos. 1 and 2 have been changed to nos. 1a and 2a, as given below, if used in connection with moments and contact pressures under exterior columns.

Assumption no. 1a. Under exterior columns, the resultant of all subgrade reactions acting on the exterior part of a combined beam footing must be equal and opposite to the load of the exterior column. (An exterior foundation part is that portion of a combined footing which primarily supports an exterior column load. It is assumed to extend from the center of the first bay to the tip of the projection.)

Assumption no. 2a. The contact pressures under an exterior foundation part are assumed to have a maximum value under the column, a minimum value at the tip of the projection and the center of the first bay, and must follow a straight-line distribution in between. The contact pressure at the tip of the projection must never exceed the contact pressure under the exterior column.

The limitation for the contact pressure at the tip of the end projection leads to a second formula for the negative moment under an exterior column.

$$P_e = p_e{}^{a} \left(\frac{p_e + p_m}{2}\right)\frac{l_1}{2}, \quad p_e = \frac{4P_e - p_m l_1}{4a + l_1}$$

$$\boxed{M_e = -p_e \frac{a^2}{2} = -\left(\frac{4P_e - p_m l_1}{4a + l_1}\right)\frac{a^2}{2}} \qquad (10\text{-}6)$$

To assure the limitation, the smaller one equations (10-5) or (10-6) shall be used.

CONTACT PRESSURE UNDER EXTERIOR COLUMNS

Based on assumptions nos. 1a and 2a, the contact pressures under an exterior column can be determined as follows:

If equation (10-5) governs:

$$P_e = p_e\left(\frac{a}{2} + \frac{l_1}{4}\right) + p_a\frac{a}{2} + p_m\frac{l_1}{4}$$

$$M_e = -p_e\frac{a^2}{6} - p_a\frac{a^2}{3}; \quad p_a = -\frac{3M_e}{a^2} - \frac{p_e}{2}$$

introducing p_a into the equation for P_e we obtain:

$$p_e = \frac{4P_e + \dfrac{6M_e}{a} - p_m l_1}{a + l_1} \qquad p_a = -\frac{3M_e}{a^2} - \frac{p_e}{2} \tag{10-7}$$

if equation (10-6) governs:

$$p_e = p_a = \frac{4P_e - p_m l_1}{4a + l_1} \tag{10-8}$$

[for derivation see equation (10-6)].

MUTUAL INFLUENCE OF END CONDITIONS

Since the foundation shall be continuous over at least three bays, the minimum length of a beam, between exterior columns, is $5.25/\lambda$. Such beams can be considered as "long"—that is, they may be considered as semi-infinite beams and with end conditions which do not influence each other.

SUMMARY

Approximate formulae have been developed for the moments under interior and exterior columns supported by a continuous beam footing; these permit calculation without the use of tables. If certain requirements are met, they will furnish sufficiently accurate results. By making simplifying assumptions about the distribution of the contact pressures under the footing, the magnitude of the contact pressures can be found directly from the moments. Liberal deviations from equal column loads and equal column spacings can be tolerated and permit wide application to practical design. Grid and mat foundations can be treated as cases of intersecting beam foundations and can be solved by the same method. Examples illustrate the practical application of the method.

Additional moments caused by uneven settlements (dishing) of the combined footings are not included in the formulas.

Examples

The following examples illustrate the practical application of the method. For reasons of clarity, any influence due to uneven settlements of the combined footing has been omitted.

EXAMPLE ON MAT FOUNDATION:

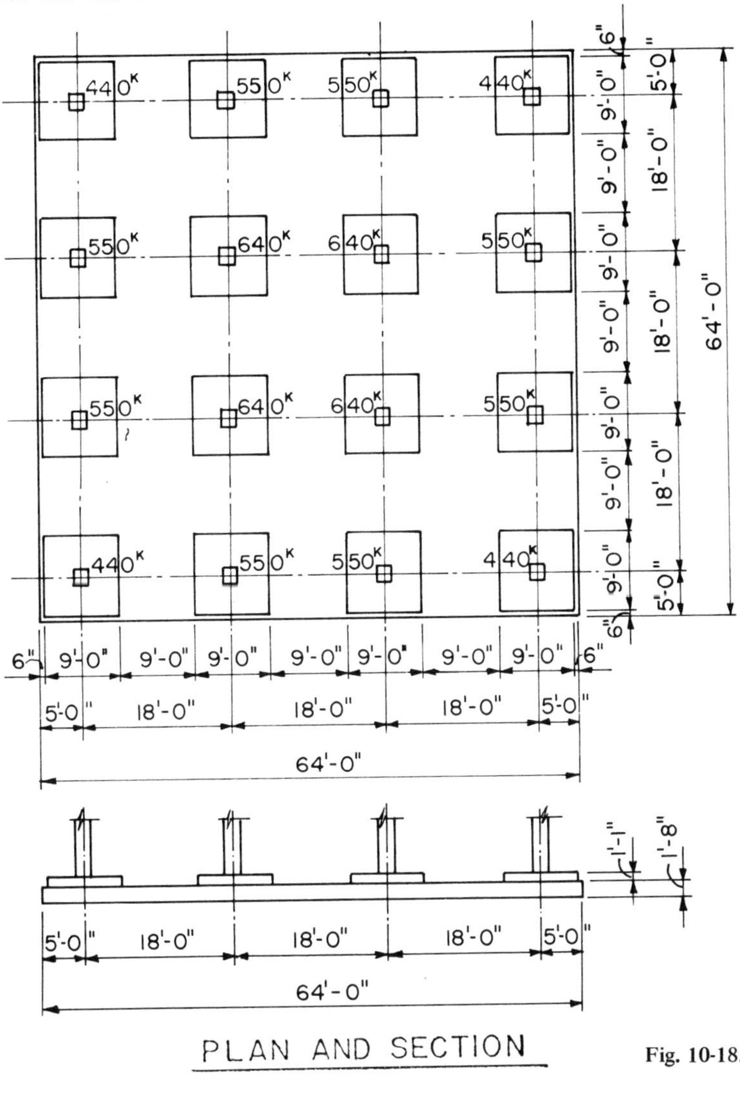

Fig. 10-18.

Assume average 18 in. slab thickness (t), 0.225^K/sq ft
Assume soil to be stiff clay, allowable soil pressure: 3000 psf, $k_{si} = 100t/\text{ft}^3$

$$P_i = 640^K, P_e = 550^K, l = 18'\text{-}0'', b = 18'\text{-}0'', a = 5'\text{-}0'', k = 2\rho k_{si}$$

$$\rho = \frac{1 + 0.5}{1.5} = 1.0, f'_c = 3000 \text{ psi}$$

$$E_c = 432 \times 10^{3K}/\text{sq ft}, I = 18 \times 1.5^3 \div 12 = 5.06 \text{ ft}^4$$

$$\lambda = 2.9 \sqrt[4]{\frac{k_{si}\rho}{f'_c t^3}} = 2.9 \sqrt[4]{\frac{100}{3000 \times 18^3}} = 2.9 \frac{1}{\sqrt[4]{30 \times 18^3}} = 0.139$$

$$l \times \lambda = 18 \times 0.139 = 2.5 \begin{matrix} > 1.75 \\ < 3.50 \end{matrix}$$

Interior Columns and Bay

$$M_i = -\frac{P_i}{4\lambda}(0.24\lambda \bar{l} + 0.16) = \frac{640}{4 \times 0.139}(0.24 \times 0.139 \times 18 + 0.16)$$

$$= -\mathit{870} \text{ ft-kips, or}$$

$$M_i = \frac{P_i l}{12} = -\frac{640 \times 18}{12} = -960 \text{ ft-kips}, 870 < 960$$

$$p_i = \frac{5P_i}{\bar{l}} + \frac{48M_i}{l^2} = \frac{5 \times 640}{18} - \frac{48 \times 870}{18^2} = 49^K/\text{ft}$$

$$p_i^1 = \frac{49}{18} + 0.225 = \mathit{2.95}^K/\text{sq ft}$$

$$p_m = -\frac{48M_i}{l^2} - \frac{3P}{l} = \frac{48 \times 870}{18^2} - \frac{3 \times 640}{18} = 22^K/\text{ft}$$

$$p_m^1 = \frac{22}{18} + 0.225 = \mathit{1.45}^K/\text{sq ft}$$

$$M_o = \frac{l^2}{48}(p_{i_{\text{left}}} + 4p_m + p_{i_{\text{right}}}) = \frac{18^2}{48}(49 + 4 \times 22 + 49) = \mathit{1250} \text{ ft-kips}$$

$$M_m = M_o + \overline{M}_i = 1250 - 870 = +\mathit{380} \text{ ft-kips}$$

Exterior Columns and Bay (Perpendicular to Edge)

$$M_e = -\frac{P_e}{4\lambda}(0.13\lambda l_1 + 1.06\lambda a - 0.50) =$$

$$= -\frac{550}{4 \times 0.139}(0.13 \times 0.139 \times 18 + 1.06 \times 0.139 \times 5 - 0.50)$$

$$= -\mathit{560} \text{ ft-kips, or}$$

$$M_e = -\frac{a^2}{2}\left(\frac{4P_e - p_m l_1}{4a + l_1}\right) = -\frac{5^2}{2}\left(\frac{4 \times 550 - 22 \times 18}{4 \times 5 + 18}\right) = -595 \text{ ft-kips, } 560 < 595$$

$$p_e = \frac{4P_e + \dfrac{6M_e}{a} - p_m l_1}{a + l} = \frac{4 \times 550 - \dfrac{6 \times 560}{5} - 22 \times 18}{5 + 18} = 50^K/\text{ft}$$

$$p_e^1 = \frac{50}{18} + 0.225 = \mathit{3.00}^K/\text{sq ft}$$

$$p_a = -\frac{3M_e}{a^2} - \frac{p_e}{2} = \frac{3 \times 560}{5^2} - 25 = 4.2^K/\text{ft}$$

$$p_a^1 = \frac{42}{18} + 0.225 = \mathit{2.55}^K/\text{sq. ft.}$$

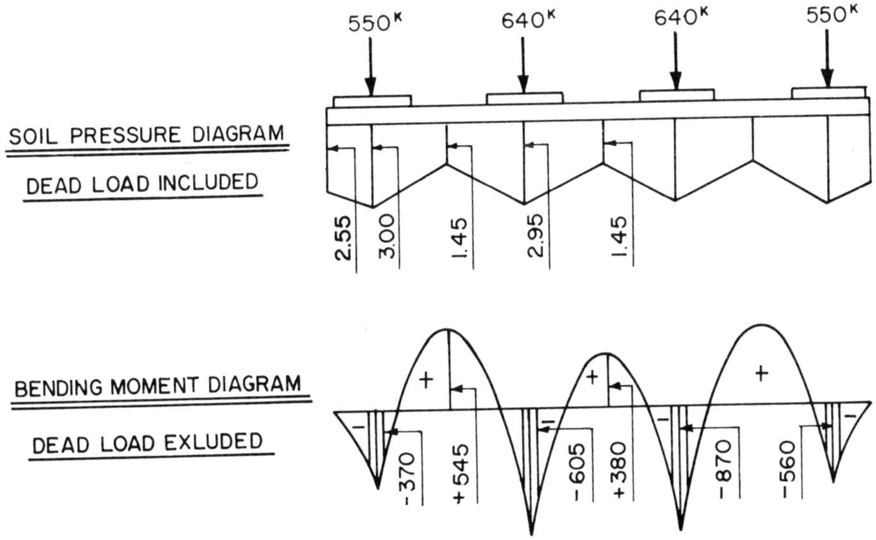

Fig. 10-19.

EXAMPLE ON STRAP FOUNDATION

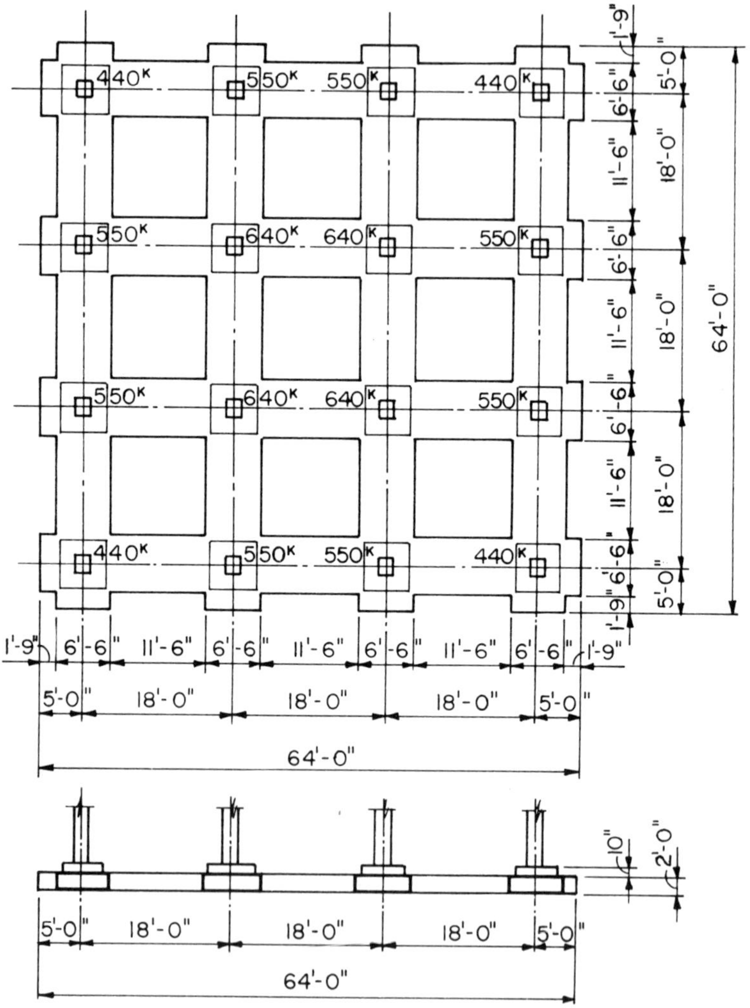

PLAN AND SECTION

Fig. 10-20.

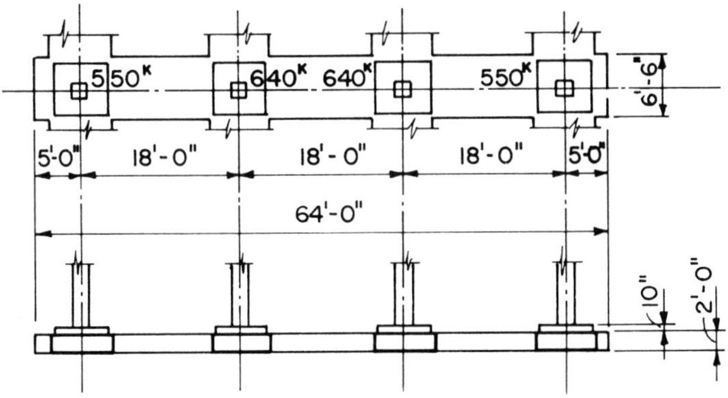

Fig. 10-21 Typical designed strap-plan and section.

$P_i = 640^K, P_e = 550^K, l = 18'\text{-}0'', a = 5'\text{-}0'', b = 6'\text{-}6'', t = 24$ in.

$\rho = 0.67$, allowable soil pressure = 8000 psf, $k_{si} = 250$

$$\lambda = 2.9\sqrt[4]{\frac{250 \times 0.67}{3000 \times 24^3}} = 2.9\sqrt[4]{\frac{16.67}{4{,}147{,}200}} = 29\sqrt[4]{\frac{1}{249{,}000}} = \frac{2.9}{22.34} = 0.13$$

$$\frac{1.75}{0.13} = 13.45 \text{ ft} \quad \frac{3.5}{0.13} = 26.9 \text{ ft}$$

$$M_i = -\frac{640}{4 \times 0.13}(0.24 \times 0.13 \times 18 + 0.16) = -886 \text{ ft-kips, or}$$

$$M_i = -\frac{640 \times 18}{12} = -960 \text{ ft kips, } 886 < 960$$

$$p_i = \frac{5 \times 640}{18} - \frac{48 \times 886}{18^2} = 47^K/\text{ft}$$

$$p_i^1 = \frac{47}{6.5} + 0.3 = 7.55^K/\text{sq ft}$$

$$p_m = \frac{48 \times 886}{18^2} - \frac{3 \times 640}{18} = 24^K/\text{ft}$$

$$p_m^1 = \frac{24}{6.5} + 0.3 = 4.00^K/\text{sq ft}$$

$$M_o = \frac{18^2}{48}(2 \times 47 + 4 \times 24) = 1280 \text{ ft-kips}$$

$$M_m = 1280 - 886 = \mathit{394} \text{ ft-kips}$$

$$M_e = \frac{550}{4 \times 0.13}(0.13^2 \times 18 + 1.05 \times 0.13 \times 5 - 0.50) = \mathit{513} \text{ ft-kips, or}$$

$$M_e = -\frac{5^2}{2}\left(\frac{4 \times 550 - 24 \times 18}{4 \times 5 + 18}\right) = 583 \text{ ft-kips, } 513 < 583$$

$$p_e = \frac{4 \times 550 - \dfrac{6 \times 513}{5} - 24 \times 18}{5 + 18} = 50.2^K/\text{ft.}$$

$$p_e^1 = \frac{50.2}{6.5} + 0.3 = \mathit{8.00}^K/\text{sq ft}$$

$$p_a = \frac{3 \times 513}{25} - 25 = 36.5^K/\text{ft}$$

$$p_a^1 = \frac{36.5}{6.5} + 0.3 = \mathit{5.9}^K/\text{sq ft}$$

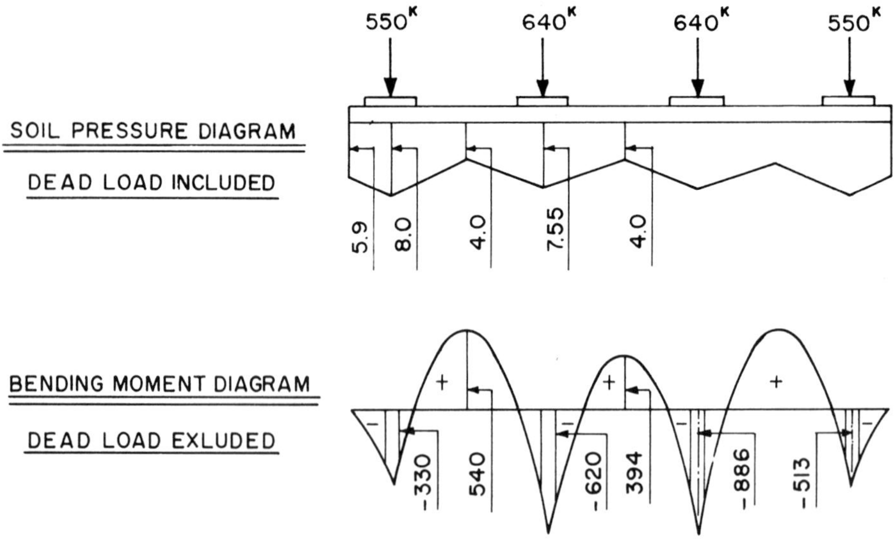

Fig. 10-22.

2000 South Michigan Avenue, Chicago, Ill. Dubin, Dubin and Black, Architects. Rogers-Cohen-Barreto-Marchertas, Structural Engineers.

11

Cantilevered Retaining Walls

Among the various types of freestanding retaining walls, only the design of the most common type, the cantilevered, will be outlined here.

The reinforced concrete contilever wall (Fig. 11-1) employs a vertical stem; which retains the earth and is held in position by the footing or base slab. In this case the weight of the fill on top of the heel, in addition to the weight of the wall, contributes to the stability of the structure. Since the stem represents a vertical cantilever, its required thickness increases rapidly with increasing height.

For the calculation of the lateral earth pressure, generally the value of equivalent fluid pressure is given, simplifying the design process. In case of surcharge, this will be transformed for additional earth-fill height.

The retaining wall may fail in two different ways: (1) its individual parts may not be strong enough to resist the acting forces, and (2) the wall as a whole may be bodily displaced by the acting earth pressure, without breaking up internally. Proper design of stem and base slab, to resist the moments and shears, provides safety against the first possibility. Adequate external stability and controlled bearing pressure for the soil offers safety against the second possibility.

The importance of drainage behind the wall cannot be overemphasized. Drainage can be provided by weep holes, consisting of 4 in. pipes embedded in the wall, with a horizontal spacing of 15 to 20 ft, or by using a continuous back drain, consisting of a layer of gravel or crushed stone covering the entire rear face of the wall, with discharge at the ends of the wall or at a few intermediate points. It is desirable to provide weakened plane contraction joints at intervals of about 25 ft, and expansion joints 100 ft apart.

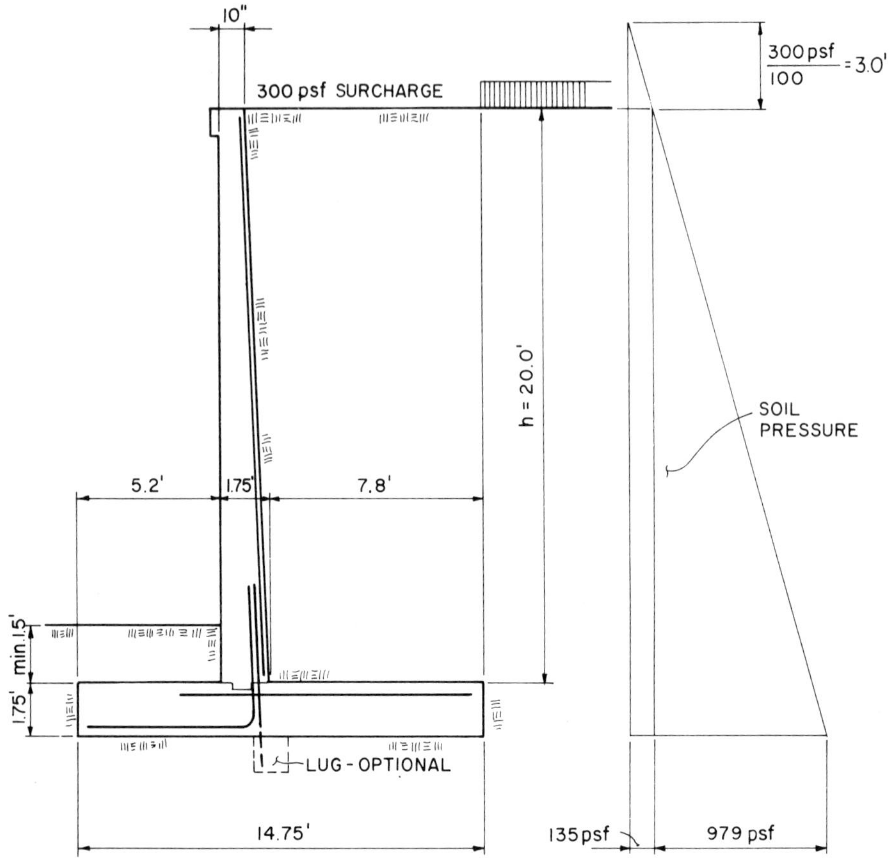

Fig. 11-1.

The data in Table 11-1 give instant information for reinforced concrete cantilevered retaining walls for heights from 5 to 20 ft. Above 20 ft the use of counterforts is recommended from economical point of view. While the Table covers a wide range, it may be necessary to compute values of different wall configurations and to show how the tabulated values were worked out. The following example is therefore included.

Example

A cantilever wall is to be designed to retain a bank 20 ft-0 in. high whose horizontal surface is subject to a live-load surcharge of 300 psf. From experiences, reasonable dimensions are assumed and earth pressure with 45.0 lb/

TABLE 11-1 CANTILEVERED RETAINING WALLS F'C = 3.00 KSI FY = 40.0 KSI
REMARKS -- M = MIN. STEEL

HORIZONTAL BACKFILL

H	WALL	HEEL	-----EQUIV. FLUID PRESSURE 30.0 LBS/CU.FT-----						-----EQUIV. FLUID PRESSURE 45.0 LBS/CU.FT-----					
FT	TOP IN.	WIDTH FT.	WALL BOT.	FTG. WIDTH	AS SQ.IN	P(1) KSF	P(2) KSF	SAFETY FACTOR	WALL BOT.	FTG. WIDTH	AS SQ.IN	P(1) KSF	P(2) KSF	SAFETY FACTOR
5	8	1.25	10.00	2.75	0.450M	1.13	0.01	2.46	10.00	3.50	0.450M	0.84	0.11	2.46
6	8	1.58	11.00	3.50	0.510M	1.11	0.14	2.74	11.00	4.00	0.510M	1.09	0.05	2.28
7	8	1.91	11.00	3.75	0.510M	1.47	0.00	2.40	11.00	4.50	0.510M	1.26	0.01	2.17
8	8	2.25	12.00	4.50	0.570M	1.45	0.14	2.62	12.00	5.25	0.570M	1.34	0.06	2.24
9	8	2.58	12.00	5.00	0.570M	1.58	0.15	2.58	12.00	5.75	0.570M	1.52	0.02	2.16
10	8	2.91	13.00	5.50	0.630M	1.79	0.14	2.54	13.00	6.50	0.630M	1.60	0.08	2.22
11	8	3.25	13.00	6.00	0.630M	1.91	0.15	2.51	13.00	7.00	0.630M	1.77	0.04	2.15
12	8	3.58	14.00	6.50	0.690M	2.13	0.14	2.49	14.00	7.75	0.690M	1.86	0.10	2.20
13	10	3.91	14.00	6.75	0.690M	2.50	0.00	2.33	14.00	8.25	0.853	2.03	0.07	2.16
14	10	4.25	15.00	7.50	0.750M	2.47	0.14	2.46	15.00	8.75	0.984	2.28	0.01	2.11
15	10	4.58	15.00	8.00	0.798	2.59	0.16	2.45	15.00	9.50	1.226	2.29	0.08	2.15
16	10	4.91	16.00	8.50	0.899	2.81	0.14	2.43	16.00	10.00	1.382	2.54	0.03	2.11
17	10	5.25	16.00	9.00	1.088	2.93	0.16	2.42	16.00	10.75	1.684	2.54	0.10	2.15
18	10	5.58	17.00	9.50	1.205	3.14	0.14	2.41	17.00	11.25	1.866	2.79	0.05	2.11
19	10	5.91	17.00	9.75	1.433	3.50	0.00	2.30	17.00	11.75	2.236	2.96	0.01	2.07
20	10	6.25	18.00	10.50	1.566	3.47	0.14	2.39	18.00	12.50	2.446	3.04	0.06	2.11

HORIZONTAL BACKFILL WITH 300 PSF SURCHARGE

H	WALL	HEEL	-----EQUIV. FLUID PRESSURE 30.0 LBS/CU.FT-----						-----EQUIV. FLUID PRESSURE 45.0 LBS/CU.FT-----					
FT	TOP IN.	WIDTH FT.	WALL BOT.	FTG. WIDTH	AS SQ.IN	P(1) KSF	P(2) KSF	SAFETY FACTOR	WALL BOT.	FTG. WIDTH	AS SQ.IN	P(1) KSF	P(2) KSF	SAFETY FACTOR
5	8	1.75	10.00	4.50	0.450M	0.83	0.07	2.31	10.00	5.50	0.450M	0.75	0.03	2.09
6	8	2.00	11.00	5.00	0.510M	1.01	0.05	2.23	11.00	6.25	0.510M	0.86	0.04	2.09
7	8	2.33	11.00	5.50	0.510M	1.15	0.05	2.19	11.00	6.75	0.510M	1.01	0.01	1.99
8	8	2.58	12.00	6.00	0.570M	1.33	0.03	2.15	12.00	7.50	0.570M	1.12	0.03	2.01
9	8	2.91	12.00	6.50	0.570M	1.46	0.03	2.13	12.00	8.25	0.684	1.18	0.06	2.03
10	8	3.16	13.00	7.00	0.630M	1.64	0.02	2.11	13.00	8.75	0.810	1.37	0.01	1.97
11	8	3.41	13.00	7.50	0.682	1.75	0.02	2.09	13.00	9.50	1.048	1.42	0.04	1.98
12	8	3.75	14.00	8.00	0.780	1.95	0.00	2.09	14.00	10.00	1.199	1.63	0.00	1.95
13	10	4.00	14.00	8.50	0.969	2.06	0.01	2.08	14.00	10.75	1.502	1.67	0.04	1.97
14	10	4.33	15.00	9.00	1.083	2.26	0.00	2.09	15.00	11.25	1.680	1.88	0.01	1.95
15	10	4.58	15.00	9.50	1.314	2.37	0.01	2.08	15.00	12.00	2.057	1.91	0.04	1.96
16	10	4.83	16.00	10.00	1.444	2.54	0.00	2.07	16.00	12.50	2.263	2.11	0.00	1.92
17	10	5.16	16.00	10.50	1.720	2.67	0.00	2.07	16.00	13.25	2.725	2.16	0.04	1.95
18	10	5.41	17.00	11.25	1.866	2.67	0.11	2.14	18.00	13.75	2.708	2.36	0.00	1.92
19	10	5.75	17.00	11.50	2.192	2.97	0.00	2.07	19.00	14.50	2.948	2.40	0.04	1.95
20	10	6.00	18.00	12.25	2.354	2.97	0.11	2.13	21.00	15.00	2.978	2.60	0.00	1.92

TABLE 11-1 CANTILEVERED RETAINING WALLS   F'C = 4.00 KSI   FY = 40.0 KSI
REMARKS -- M = MIN. STEEL

HORIZONTAL BACKFILL

H	WALL	HEEL	-----EQUIV. FLUID PRESSURE 30.0 LBS/CU.FT-----						-----EQUIV. FLUID PRESSURE 45.0 LBS/CU.FT-----					
FT	TOP IN.	WIDTH FT.	WALL BOT.	FTG. WIDTH	AS SQ.IN	P(1) KSF	P(2) KSF	SAFETY FACTOR	WALL BOT.	FTG. WIDTH	AS SQ.IN	P(1) KSF	P(2) KSF	SAFETY FACTOR
5	8	1.25	10.00	2.75	0.450M	1.13	0.01	2.46	10.00	3.50	0.450M	0.84	0.11	2.46
6	8	1.58	11.00	3.50	0.510M	1.11	0.14	2.74	11.00	4.00	0.510M	1.09	0.05	2.28
7	8	1.91	11.00	3.75	0.510M	1.47	0.00	2.40	11.00	4.50	0.510M	1.26	0.01	2.17
8	8	2.25	12.00	4.50	0.570M	1.45	0.14	2.62	12.00	5.25	0.570M	1.34	0.06	2.24
9	8	2.58	12.00	5.00	0.570M	1.58	0.15	2.58	12.00	5.75	0.570M	1.52	0.02	2.16
10	8	2.91	13.00	5.50	0.630M	1.79	0.14	2.54	13.00	6.50	0.630M	1.60	0.08	2.22
11	8	3.25	13.00	6.00	0.630M	1.91	0.15	2.51	13.00	7.00	0.630M	1.77	0.04	2.15
12	8	3.58	14.00	6.50	0.690M	2.13	0.14	2.49	14.00	7.75	0.690M	1.86	0.10	2.20
13	10	3.91	14.00	6.75	0.690M	2.50	0.00	2.33	14.00	8.25	0.842	2.03	0.07	2.16
14	10	4.25	15.00	7.50	0.750M	2.47	0.14	2.46	15.00	8.75	0.970	2.28	0.01	2.11
15	10	4.58	15.00	8.00	0.789	2.59	0.16	2.45	15.00	9.50	1.204	2.29	0.08	2.15
16	10	4.91	16.00	8.50	0.888	2.81	0.14	2.43	16.00	10.00	1.356	2.54	0.03	2.11
17	10	5.25	16.00	9.00	1.073	2.93	0.16	2.42	16.00	10.75	1.645	2.54	0.10	2.15
18	10	5.58	17.00	9.50	1.187	3.14	0.14	2.41	17.00	11.25	1.822	2.79	0.05	2.11
19	10	5.91	17.00	9.75	1.407	3.50	0.00	2.30	17.00	11.75	2.170	2.96	0.01	2.07
20	10	6.25	18.00	10.50	1.537	3.47	0.14	2.39	18.00	12.50	2.371	3.04	0.06	2.11

HORIZONTAL BACKFILL WITH 300 PSF SURCHARGE

H	WALL	HEEL	-----EQUIV. FLUID PRESSURE 30.0 LBS/CU.FT-----						-----EQUIV. FLUID PRESSURE 45.0 LBS/CU.FT-----					
FT	TOP IN.	WIDTH FT.	WALL BOT.	FTG. WIDTH	AS SQ.IN	P(1) KSF	P(2) KSF	SAFETY FACTOR	WALL BOT.	FTG. WIDTH	AS SQ.IN	P(1) KSF	P(2) KSF	SAFETY FACTOR
5	8	1.75	10.00	4.50	0.450M	0.83	0.07	2.31	10.00	5.50	0.450M	0.75	0.03	2.09
6	8	2.00	11.00	5.00	0.510M	1.01	0.05	2.23	11.00	6.25	0.510M	0.86	0.04	2.09
7	8	2.33	11.00	5.50	0.510M	1.15	0.05	2.19	11.00	6.75	0.510M	1.01	0.01	1.99
8	8	2.58	12.00	6.00	0.570M	1.33	0.03	2.15	12.00	7.50	0.570M	1.12	0.03	2.01
9	8	2.91	12.00	6.50	0.570M	1.46	0.03	2.13	12.00	8.25	0.676	1.18	0.06	2.03
10	8	3.16	13.00	7.00	0.630M	1.64	0.02	2.11	13.00	8.75	0.799	1.37	0.01	1.97
11	8	3.41	13.00	7.50	0.674	1.75	0.02	2.09	13.00	9.50	1.029	1.42	0.04	1.98
12	8	3.75	14.00	8.00	0.770	1.95	0.00	2.09	14.00	10.00	1.177	1.63	0.00	1.95
13	10	4.00	14.00	8.50	0.955	2.06	0.01	2.08	14.00	10.75	1.466	1.67	0.04	1.97
14	10	4.33	15.00	9.00	1.066	2.26	0.00	2.09	15.00	11.25	1.638	1.88	0.01	1.95
15	10	4.58	15.00	9.50	1.289	2.37	0.01	2.08	15.00	12.00	1.992	1.91	0.04	1.96
16	10	4.83	16.00	10.00	1.416	2.54	0.00	2.07	16.00	12.50	2.189	2.11	0.00	1.92
17	10	5.16	16.00	10.50	1.679	2.67	0.00	2.07	16.00	13.25	2.614	2.16	0.04	1.95
18	10	5.41	17.00	11.25	1.822	2.67	0.11	2.14	17.00	13.75	2.836	2.36	0.00	1.92
19	10	5.75	17.00	11.50	2.128	2.97	0.00	2.07	17.00	14.50	3.340	2.40	0.04	1.95
20	10	6.00	18.00	12.25	2.286	2.97	0.11	2.13	18.00	15.00	3.588	2.60	0.00	1.92

TABLE 11-1 CANTILEVERED RETAINING WALLS   F'C = 3.00 KSI   FY = 60.0 KSI
REMARKS -- M = MIN. STEEL

HORIZONTAL BACKFILL

H FT	WALL TOP IN.	HEEL WIDTH FT.	EQUIV. FLUID PRESSURE 30.0 LBS/CU.FT WALL BOT.	FTG. WIDTH	AS SQ.IN	P(1) KSF	P(2) KSF	SAFETY FACTOR	EQUIV. FLUID PRESSURE 45.0 LBS/CU.FT WALL BOT.	FTG. WIDTH	AS SQ.IN	P(1) KSF	P(2) KSF	SAFETY FACTOR
5	8	1.25	10.00	2.75	0.300M	1.13	0.01	2.46	10.00	3.50	0.300M	0.84	0.11	2.46
6	8	1.58	11.00	3.50	0.340M	1.11	0.14	2.74	11.00	4.00	0.340M	1.09	0.05	2.28
7	8	1.91	11.00	3.75	0.340M	1.47	0.00	2.40	11.00	4.50	0.340M	1.26	0.01	2.17
8	8	2.25	12.00	4.50	0.380M	1.45	0.14	2.62	12.00	5.25	0.380M	1.34	0.06	2.24
9	8	2.58	12.00	5.00	0.380M	1.58	0.15	2.58	12.00	5.75	0.380M	1.52	0.02	2.16
10	8	2.91	13.00	5.50	0.420M	1.79	0.14	2.54	13.00	6.50	0.420M	1.60	0.08	2.22
11	8	3.25	13.00	6.00	0.420M	1.91	0.15	2.51	13.00	7.00	0.420M	1.77	0.04	2.15
12	8	3.58	14.00	6.50	0.460M	2.13	0.14	2.49	14.00	7.75	0.460M	1.86	0.10	2.20
13	10	3.91	14.00	6.75	0.460M	2.50	0.00	2.33	14.00	8.25	0.569	2.03	0.07	2.16
14	10	4.25	15.00	7.50	0.500M	2.47	0.14	2.46	15.00	8.75	0.656	2.28	0.01	2.11
15	10	4.58	15.00	8.00	0.532	2.59	0.16	2.45	15.00	9.50	0.817	2.29	0.08	2.15
16	10	4.91	16.00	8.50	0.599	2.81	0.14	2.43	16.00	10.00	0.921	2.54	0.03	2.11
17	10	5.25	16.00	9.00	0.726	2.93	0.16	2.42	16.00	10.75	1.123	2.54	0.10	2.15
18	10	5.58	17.00	9.50	0.803	3.14	0.14	2.41	17.00	11.25	1.244	2.79	0.05	2.11
19	10	5.91	17.00	9.75	0.955	3.50	0.00	2.30	17.00	11.75	1.490	2.96	0.01	2.07
20	10	6.25	18.00	10.50	1.044	3.47	0.14	2.39	18.00	12.50	1.631	3.04	0.06	2.11

HORIZONTAL BACKFILL WITH 300 PSF SURCHARGE

H FT	WALL TOP IN.	HEEL WIDTH FT.	EQUIV. FLUID PRESSURE 30.0 LBS/CU.FT WALL BOT.	FTG. WIDTH	AS SQ.IN	P(1) KSF	P(2) KSF	SAFETY FACTOR	EQUIV. FLUID PRESSURE 45.0 LBS/CU.FT WALL BOT.	FTG. WIDTH	AS SQ.IN	P(1) KSF	P(2) KSF	SAFETY FACTOR
5	8	1.75	10.00	4.50	0.300M	0.83	0.07	2.31	10.00	5.50	0.300M	0.75	0.03	2.09
6	8	2.00	11.00	5.00	0.340M	1.01	0.05	2.23	11.00	6.25	0.340M	0.86	0.04	2.09
7	8	2.33	11.00	5.50	0.340M	1.15	0.05	2.19	11.00	6.75	0.340M	1.01	0.01	1.99
8	8	2.58	12.00	6.00	0.380M	1.33	0.03	2.15	12.00	7.50	0.380M	1.12	0.03	2.01
9	8	2.91	12.00	6.50	0.380M	1.46	0.03	2.13	12.00	8.25	0.456	1.18	0.06	2.03
10	8	3.16	13.00	7.00	0.420M	1.64	0.02	2.11	13.00	8.75	0.540	1.37	0.01	1.97
11	8	3.41	13.00	7.50	0.455	1.75	0.02	2.09	13.00	9.50	0.699	1.42	0.04	1.98
12	8	3.75	14.00	8.00	0.520	1.95	0.00	2.09	14.00	10.00	0.800	1.63	0.00	1.95
13	10	4.00	14.00	8.50	0.646	2.06	0.01	2.08	14.00	10.75	1.002	1.67	0.04	1.97
14	10	4.33	15.00	9.00	0.722	2.26	0.00	2.09	15.00	11.25	1.120	1.88	0.01	1.95
15	10	4.58	15.00	9.50	0.876	2.37	0.01	2.08	15.00	12.00	1.372	1.91	0.04	1.96
16	10	4.83	16.00	10.00	0.963	2.54	0.00	2.07	16.00	12.50	1.508	2.11	0.00	1.92
17	10	5.16	16.00	10.50	1.147	2.67	0.00	2.07	16.00	13.25	1.817	2.16	0.04	1.95
18	10	5.41	17.00	11.25	1.244	2.67	0.11	2.14	18.00	13.75	1.805	2.36	0.00	1.92
19	10	5.75	17.00	11.50	1.461	2.97	0.00	2.07	19.00	14.50	1.965	2.40	0.04	1.95
20	10	6.00	18.00	12.25	1.569	2.97	0.11	2.13	21.00	15.00	1.985	2.60	0.00	1.92

TABLE 11-1 CANTILEVERED RETAINING WALLS   F'C = 4.00 KSI   FY = 60.0 KSI

REMARKS -- M = MIN. STEEL

HORIZONTAL BACKFILL

H FT	WALL TOP IN.	HEEL WIDTH FT.	EQUIV. FLUID PRESSURE 30.0 LBS/CU.FT WALL BOT.	FTG. WIDTH	AS SQ.IN	P(1) KSF	P(2) KSF	SAFETY FACTOR	EQUIV. FLUID PRESSURE 45.0 LBS/CU.FT WALL BOT.	FTG. WIDTH	AS SQ.IN	P(1) KSF	P(2) KSF	SAFETY FACTOR
5	8	1.25	10.00	2.75	0.300M	1.13	0.01	2.46	10.00	3.50	0.300M	0.84	0.11	2.46
6	8	1.58	11.00	3.50	0.340M	1.11	0.14	2.74	11.00	4.00	0.340M	1.09	0.05	2.28
7	8	1.91	11.00	3.75	0.340M	1.47	0.00	2.40	11.00	4.50	0.340M	1.26	0.01	2.17
8	8	2.25	12.00	4.50	0.380M	1.45	0.14	2.62	12.00	5.25	0.380M	1.34	0.06	2.24
9	8	2.58	12.00	5.00	0.380M	1.58	0.15	2.58	12.00	5.75	0.380M	1.52	0.02	2.16
10	8	2.91	13.00	5.50	0.420M	1.79	0.14	2.54	13.00	6.50	0.420M	1.60	0.08	2.22
11	8	3.25	13.00	6.00	0.420M	1.91	0.15	2.51	13.00	7.00	0.420M	1.77	0.04	2.15
12	8	3.58	14.00	6.50	0.460M	2.13	0.14	2.49	14.00	7.75	0.460M	1.86	0.10	2.20
13	10	3.91	14.00	6.75	0.460M	2.50	0.00	2.33	14.00	8.25	0.561	2.03	0.07	2.16
14	10	4.25	15.00	7.50	0.500M	2.47	0.14	2.46	15.00	8.75	0.647	2.28	0.01	2.11
15	10	4.58	15.00	8.00	0.526	2.59	0.16	2.45	15.00	9.50	0.803	2.29	0.08	2.15
16	10	4.91	16.00	8.50	0.592	2.81	0.14	2.43	16.00	10.00	0.904	2.54	0.03	2.11
17	10	5.25	16.00	9.00	0.715	2.93	0.16	2.42	16.00	10.75	1.097	2.54	0.10	2.15
18	10	5.58	17.00	9.50	0.791	3.14	0.14	2.41	17.00	11.25	1.214	2.79	0.05	2.11
19	10	5.91	17.00	9.75	0.938	3.50	0.00	2.30	17.00	11.75	1.446	2.96	0.01	2.07
20	10	6.25	18.00	10.50	1.025	3.47	0.14	2.39	18.00	12.50	1.581	3.04	0.06	2.11

HORIZONTAL BACKFILL WITH 300 PSF SURCHARGE

H FT	WALL TOP IN.	HEEL WIDTH FT.	EQUIV. FLUID PRESSURE 30.0 LBS/CU.FT WALL BOT.	FTG. WIDTH	AS SQ.IN	P(1) KSF	P(2) KSF	SAFETY FACTOR	EQUIV. FLUID PRESSURE 45.0 LBS/CU.FT WALL BOT.	FTG. WIDTH	AS SQ.IN	P(1) KSF	P(2) KSF	SAFETY FACTOR
5	8	1.75	10.00	4.50	0.300M	0.83	0.07	2.31	10.00	5.50	0.300M	0.75	0.03	2.09
6	8	2.00	11.00	5.00	0.340M	1.01	0.05	2.23	11.00	6.25	0.340M	0.86	0.04	2.09
7	8	2.33	11.00	5.50	0.340M	1.15	0.05	2.19	11.00	6.75	0.340M	1.01	0.01	1.99
8	8	2.58	12.00	6.00	0.380M	1.33	0.03	2.15	12.00	7.50	0.380M	1.12	0.03	2.01
9	8	2.91	12.00	6.50	0.380M	1.46	0.03	2.13	12.00	8.25	0.451	1.18	0.06	2.03
10	8	3.16	13.00	7.00	0.420M	1.64	0.02	2.11	13.00	8.75	0.533	1.37	0.01	1.97
11	8	3.41	13.00	7.50	0.449	1.75	0.02	2.09	13.00	9.50	0.686	1.42	0.04	1.98
12	8	3.75	14.00	8.00	0.514	1.95	0.00	2.09	14.00	10.00	0.784	1.63	0.00	1.95
13	10	4.00	14.00	8.50	0.637	2.06	0.01	2.08	14.00	10.75	0.977	1.67	0.04	1.97
14	10	4.33	15.00	9.00	0.711	2.26	0.00	2.09	15.00	11.25	1.092	1.88	0.01	1.95
15	10	4.58	15.00	9.50	0.859	2.37	0.01	2.08	15.00	12.00	1.328	1.91	0.04	1.96
16	10	4.83	16.00	10.00	0.944	2.54	0.00	2.07	16.00	12.50	1.459	2.11	0.00	1.92
17	10	5.16	16.00	10.50	1.120	2.67	0.00	2.07	16.00	13.25	1.742	2.16	0.04	1.95
18	10	5.41	17.00	11.25	1.214	2.67	0.11	2.14	17.00	13.75	1.891	2.36	0.00	1.92
19	10	5.75	17.00	11.50	1.419	2.97	0.00	2.07	17.00	14.50	2.227	2.40	0.04	1.95
20	10	6.00	18.00	12.25	1.524	2.97	0.11	2.13	18.00	15.00	2.392	2.60	0.00	1.92

RET. WALLS

TABLE 11-1 CANTILEVERED RETAINING WALLS F'C = 3.00 KSI FY = 60.0 KSI
REMARKS -- M = MIN. STEEL

HORIZONTAL BACKFILL

H FT	WALL TOP IN.	HEEL WIDTH FT.	30.0 LBS/CU.FT WALL BOT.	30.0 FTG. WIDTH	30.0 AS SQ.IN	30.0 P(1) KSF	30.0 P(2) KSF	30.0 SAFETY FACTOR	45.0 LBS/CU.FT WALL BOT.	45.0 FTG. WIDTH	45.0 AS SQ.IN	45.0 P(1) KSF	45.0 P(2) KSF	45.0 SAFETY FACTOR
5	8	1.66	10.00	3.00	0.300M	1.13	0.07	2.93	10.00	3.50	0.300M	0.99	0.07	2.57
6	8	2.08	11.00	3.50	0.340M	1.40	0.03	2.81	11.00	4.25	0.340M	1.10	0.12	2.65
7	8	2.50	11.00	4.00	0.340M	1.58	0.02	2.76	11.00	4.75	0.340M	1.31	0.08	2.50
8	8	2.91	12.00	4.75	0.380M	1.60	0.15	2.97	12.00	5.25	0.380M	1.60	0.01	2.37
9	8	3.33	12.00	5.25	0.380M	1.76	0.15	2.91	12.00	6.00	0.380M	1.62	0.09	2.45
10	8	3.75	13.00	5.75	0.420M	2.03	0.12	2.84	13.00	6.50	0.420M	1.92	0.02	2.35
11	8	4.25	13.00	6.25	0.420M	2.25	0.10	2.81	13.00	7.25	0.420M	1.97	0.09	2.44
12	8	4.66	14.00	6.75	0.460M	2.51	0.06	2.76	14.00	7.75	0.460M	2.27	0.02	2.35
13	10	5.08	14.00	7.25	0.460M	2.70	0.06	2.74	14.00	8.50	0.569	2.30	0.11	2.42
14	10	5.50	15.00	7.75	0.500M	2.97	0.03	2.71	15.00	9.00	0.656	2.59	0.04	2.35
15	10	5.91	15.00	8.25	0.532	3.13	0.03	2.69	15.00	9.75	0.817	2.61	0.12	2.41
16	10	6.41	16.00	9.00	0.599	3.20	0.14	2.81	16.00	10.25	0.921	2.94	0.04	2.36
17	10	6.83	16.00	9.50	0.726	3.36	0.14	2.78	16.00	10.75	1.123	3.15	0.00	2.31
18	10	7.25	17.00	10.00	0.803	3.63	0.11	2.75	17.00	11.50	1.244	3.25	0.06	2.35
19	10	7.66	17.00	10.50	0.955	3.79	0.11	2.74	17.00	12.00	1.490	3.46	0.01	2.31
20	10	8.08	18.00	11.00	1.044	4.06	0.08	2.71	18.00	12.75	1.631	3.56	0.07	2.35

HORIZONTAL BACKFILL WITH 300 PSF SURCHARGE

H FT	WALL TOP IN.	HEEL WIDTH FT.	30.0 LBS/CU.FT WALL BOT.	30.0 FTG. WIDTH	30.0 AS SQ.IN	30.0 P(1) KSF	30.0 P(2) KSF	30.0 SAFETY FACTOR	45.0 LBS/CU.FT WALL BOT.	45.0 FTG. WIDTH	45.0 AS SQ.IN	45.0 P(1) KSF	45.0 P(2) KSF	45.0 SAFETY FACTOR
5	8	2.25	10.00	4.50	0.300M	0.95	0.05	2.44	10.00	5.50	0.300M	0.81	0.05	2.24
6	8	2.66	11.00	5.00	0.340M	1.20	0.02	2.37	11.00	6.00	0.340M	1.06	0.00	2.13
7	8	3.00	11.00	5.50	0.340M	1.35	0.02	2.32	11.00	6.75	0.340M	1.12	0.04	2.15
8	8	3.41	12.00	6.25	0.380M	1.43	0.10	2.46	12.00	7.50	0.380M	1.26	0.07	2.20
9	8	3.75	12.00	6.75	0.380M	1.57	0.10	2.42	12.00	8.00	0.456	1.42	0.03	2.12
10	8	4.08	13.00	7.25	0.420M	1.78	0.08	2.39	13.00	8.75	0.540	1.54	0.06	2.15
11	8	4.50	13.00	7.75	0.455	1.95	0.08	2.37	13.00	9.25	0.699	1.72	0.03	2.10
12	8	4.83	14.00	8.25	0.520	2.16	0.06	2.35	14.00	10.00	0.800	1.83	0.06	2.13
13	10	5.25	14.00	8.75	0.646	2.34	0.06	2.35	14.00	10.50	1.002	2.02	0.03	2.10
14	10	5.58	15.00	9.25	0.722	2.55	0.04	2.33	15.00	11.25	1.120	2.13	0.07	2.13
15	10	5.91	15.00	9.75	0.876	2.68	0.05	2.32	15.00	11.75	1.372	2.29	0.04	2.09
16	10	6.33	16.00	10.25	0.963	2.93	0.02	2.32	16.00	12.50	1.508	2.42	0.08	2.13
17	10	6.66	16.00	10.75	1.147	3.06	0.03	2.31	16.00	13.00	1.817	2.58	0.04	2.09
18	10	7.08	17.00	11.25	1.244	3.31	0.00	2.31	18.00	13.75	1.805	2.70	0.08	2.13
19	10	7.41	17.00	11.75	1.461	3.44	0.02	2.30	19.00	14.25	1.965	2.87	0.05	2.09
20	10	7.83	18.00	12.50	1.569	3.49	0.12	2.38	21.00	14.75	1.985	3.13	0.00	2.07

RET. WALLS

TABLE 11-1 CANTILEVERED RETAINING WALLS F'C = 4.00 KSI FY = 60.0 KSI
REMARKS -- M = MIN. STEEL

HORIZONTAL BACKFILL

H	WALL	HEEL	-----EQUIV. FLUID PRESSURE 30.0 LBS/CU.FT-----						-----EQUIV. FLUID PRESSURE 45.0 LBS/CU.FT-----					
FT	TOP	WIDTH	WALL	FTG.	AS	P(1)	P(2)	SAFETY	WALL	FTG.	AS	P(1)	P(2)	SAFETY
	IN.	FT.	BOT.	WIDTH	SQ.IN	KSF	KSF	FACTOR	BOT.	WIDTH	SQ.IN	KSF	KSF	FACTOR
5	8	1.66	10.00	3.00	0.300M	1.13	0.07	2.93	10.00	3.50	0.300M	0.99	0.07	2.57
6	8	2.08	11.00	3.50	0.340M	1.40	0.03	2.81	11.00	4.25	0.340M	1.10	0.12	2.65
7	8	2.50	11.00	4.00	0.340M	1.58	0.02	2.76	11.00	4.75	0.340M	1.31	0.08	2.50
8	8	2.91	12.00	4.75	0.380M	1.60	0.15	2.97	12.00	5.25	0.380M	1.60	0.01	2.37
9	8	3.33	12.00	5.25	0.380M	1.76	0.15	2.91	12.00	6.00	0.380M	1.62	0.09	2.45
10	8	3.75	13.00	5.75	0.420M	2.03	0.12	2.84	13.00	6.50	0.420M	1.92	0.02	2.35
11	8	4.25	13.00	6.25	0.420M	2.25	0.10	2.81	13.00	7.25	0.420M	1.97	0.09	2.44
12	8	4.66	14.00	6.75	0.460M	2.51	0.06	2.76	14.00	7.75	0.460M	2.27	0.02	2.35
13	10	5.08	14.00	7.25	0.460M	2.70	0.06	2.74	14.00	8.50	0.561	2.30	0.11	2.42
14	10	5.50	15.00	7.75	0.500M	2.97	0.03	2.71	15.00	9.00	0.647	2.59	0.04	2.35
15	10	5.91	15.00	8.25	0.526	3.13	0.03	2.69	15.00	9.75	0.803	2.61	0.12	2.41
16	10	6.41	16.00	9.00	0.592	3.20	0.14	2.81	16.00	10.25	0.904	2.94	0.04	2.36
17	10	6.83	16.00	9.50	0.715	3.36	0.14	2.78	16.00	10.75	1.097	3.15	0.00	2.31
18	10	7.25	17.00	10.00	0.791	3.63	0.11	2.75	17.00	11.50	1.214	3.25	0.06	2.35
19	10	7.66	17.00	10.50	0.938	3.79	0.11	2.74	17.00	12.00	1.446	3.46	0.01	2.31
20	10	8.08	18.00	11.00	1.025	4.06	0.08	2.71	18.00	12.75	1.581	3.56	0.07	2.35

HORIZONTAL BACKFILL WITH 300 PSF SURCHARGE

H	WALL	HEEL	-----EQUIV. FLUID PRESSURE 30.0 LBS/CU.FT-----						-----EQUIV. FLUID PRESSURE 45.0 LBS/CU.FT-----					
FT	TOP	WIDTH	WALL	FTG.	AS	P(1)	P(2)	SAFETY	WALL	FTG.	AS	P(1)	P(2)	SAFETY
	IN.	FT.	BOT.	WIDTH	SQ.IN	KSF	KSF	FACTOR	BOT.	WIDTH	SQ.IN	KSF	KSF	FACTOR
5	8	2.25	10.00	4.50	0.300M	0.95	0.05	2.44	10.00	5.50	0.300M	0.81	0.05	2.24
6	8	2.66	11.00	5.00	0.340M	1.20	0.02	2.37	11.00	6.00	0.340M	1.06	0.00	2.13
7	8	3.00	11.00	5.50	0.340M	1.35	0.02	2.32	11.00	6.75	0.340M	1.12	0.04	2.15
8	8	3.41	12.00	6.25	0.380M	1.43	0.10	2.46	12.00	7.50	0.380M	1.26	0.07	2.20
9	8	3.75	12.00	6.75	0.380M	1.57	0.10	2.42	12.00	8.00	0.451	1.42	0.03	2.12
10	8	4.08	13.00	7.25	0.420M	1.78	0.08	2.39	13.00	8.75	0.533	1.54	0.06	2.15
11	8	4.50	13.00	7.75	0.449	1.95	0.08	2.37	13.00	9.25	0.686	1.72	0.03	2.10
12	8	4.83	14.00	8.25	0.514	2.16	0.06	2.35	14.00	10.00	0.784	1.83	0.06	2.13
13	10	5.25	14.00	8.75	0.637	2.34	0.06	2.35	14.00	10.50	0.977	2.02	0.03	2.10
14	10	5.58	15.00	9.25	0.711	2.55	0.04	2.33	15.00	11.25	1.092	2.13	0.07	2.13
15	10	5.91	15.00	9.75	0.859	2.68	0.05	2.32	15.00	11.75	1.328	2.29	0.04	2.09
16	10	6.33	16.00	10.25	0.944	2.93	0.02	2.32	16.00	12.50	1.459	2.42	0.08	2.13
17	10	6.66	16.00	10.75	1.120	3.06	0.03	2.31	16.00	13.00	1.742	2.58	0.04	2.09
18	10	7.08	17.00	11.25	1.214	3.31	0.00	2.31	17.00	13.75	1.891	2.70	0.08	2.13
19	10	7.41	17.00	11.75	1.419	3.44	0.02	2.30	17.00	14.25	2.227	2.87	0.05	2.09
20	10	7.83	18.00	12.50	1.524	3.49	0.12	2.38	18.00	14.75	2.392	3.13	0.00	2.07

cu ft equivalent fluid pressure is computed. Concrete with $f_c' = 3000$ psi and steel with $f_y = 60{,}000$ psi will be used (see Fig. 11-1).

Moments due to horizontal forces:

$$\begin{aligned} 0.135(21.75)^2/2 &= & -\ 31.93 \text{ ft-kips} \\ 0.979(21.75)^2/6 &= & -\ 77.19 \\ & & = -109.12 \text{ ft-kips} \end{aligned}$$

Moments due to vertical forces:

$$\left.\begin{aligned} 0.83\,(20.0)\ 0.15 &= 2.49\ (5.586) = & +\ 13.91 \\ 0.92\,(20.0)\ 0.05/2 &= 0.46\ (6.336) = & +\ 2.92 \\ 8.75\,(20.0)\ 0.10 &= 17.50(10.376) = & +181.58 \\ 14.75\ (1.75)0.15 &= 3.87\ (7.375) = & 28.55 \end{aligned}\right\} 226.96$$

$$\Sigma V = 24.32 \text{ kips} \qquad \Sigma M = +117.84 \text{ ft-kips}$$

The distance of the resultant to the front edge is:

$$a = 117.84/24.32 = 4.85 \text{ ft}$$

eccentricity is:

$$e = \frac{14.75}{2} - 4.85 = 2.53 \text{ ft}$$

the corresponding maximum and minimum soil pressures are:

$$f_1 = \frac{24.32}{14.75} + \frac{24.32(2.53)}{(14.75)^2/6} = 1.65 + 1.65 = 3.30 \text{ ksf}$$

$$f_2 = 1.65 - 1.65 = 0.00 \text{ ksf}$$

The factor of safety against overturning:

$$226.96/109.12 = 2.08$$

is ample. The resisting force against sliding: (use 0.58 as friction coeff)

$$\begin{aligned} &\text{Due to vertical load: } 24.32 \text{ kips } (0.58) = & 14.10 \rightarrow \\ &\text{Lateral forces: } 0.135(21.75) = & 2.94 \leftarrow \\ &\qquad 0.979(21.75)/2 = & 10.65 \leftarrow \\ & & \Sigma = 13.59 \leftarrow \end{aligned}$$

The factor of safety against sliding is $14.10/13.59 = 1.04 < 1.5$, N.G. Use an adequate key to increase passive earth pressure.

The external stability of the wall has now been ascertained, and it remains to determine the required reinforcement and to check the internal stresses.

Stem. The moment at the bottom section of the stem is:

$$
\begin{aligned}
0.135(20.0)^2/2 &= \quad 27.00 \text{ ft-kips}\\
0.900(20.0)^2/6 &= \quad \underline{60.00}\\
M &= \quad 87.00 \text{ ft-kips}
\end{aligned}
$$

Design moment:

$$M_{u_{\text{required}}} = 1.7(87.00) = 147.90 \text{ ft-kips}$$

$$d = 21 \text{ in.} - 2.5 \text{ in.} = 18.5 \text{ in.}$$

From Table 3-7:

$$\overline{M}_{u_{\text{concr.}}} = 240.60 \text{ ft-kips}$$

$$\text{Ratio of } M_{u_{\text{required}}} \text{ to } \overline{M}_{u_{\text{concr.}}} = 147.90/240.6 = 0.614$$

From Fig. 3-8, $a_u = 4.03$

$$A_s = \frac{M_{u_{\text{required}}}}{a_u(d)} = \frac{147.9}{4.03(18.5)} = 1.98 \text{ sq in.}$$

The required area is provided by #9 at 6 in. o.c.

In order to facilitate construction, the footing is placed first and the construction joint provided at the base of the stem, as shown in Fig. 11-1. The main bars of the stem, therefore, end at the top of the base slab, and dowels are placed in the latter and spliced with them. Since the entire strength of the wall depends on the strength of this splice, using Class C splices, with lap length of $1.7\, l_d$, is recommended. For l_d see Chapter 9.

One part of each dowel must be extended into the toe portion of the footing and provide reinforcement, and the other portion into the key, to produce the necessary length of development for embedment.

Base slab. Toe and heel are designed for moments and shears caused by the bearing pressure acting upward, and counteracted by the superimposed weight of fill, including the weight of the footing proper.

(1) Heel: for bearing pressure diagram see Fig. 11-2.

On the heel, the combined weights are:

$$20.0(0.100) + 1.75(0.150) = 2.26 \text{ kips/ft}$$

The moment and shear produced by this load are counteracted by those from the bearing pressure. Hence,

$$M = 2.26(7.83)^2/2 - 1.75(7.83)^2/6 = 51.40 \text{ ft-kips}$$

Design moment will be:

$$M_{u_{\text{required}}} = 1.7\,(51.40) = 87.38 \text{ ft-kips}$$

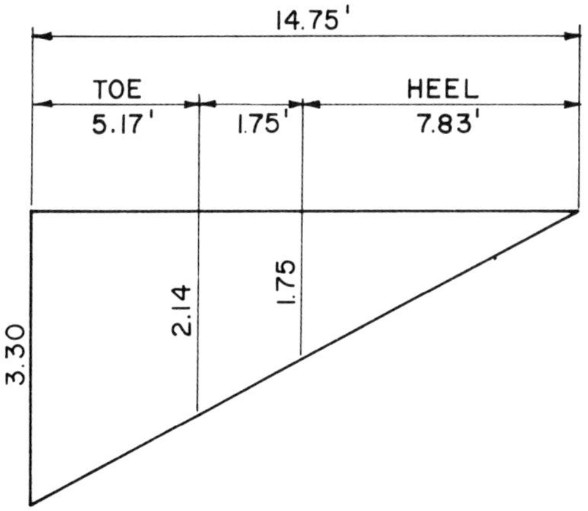

Fig. 11-2.

Using $d = 18.5$ in., from Table 3-7:

$$\overline{M}_{u_{\text{concr.}}} = 240.60 \text{ ft-kips}$$

Ration of $M_{u_{\text{required}}}$ to $\overline{M}_{u_{\text{concr.}}} = 87.38/240.60 = 0.363$. From Fig. 3-8, $a_u =$ 4.25

$$A_s = \frac{M_{u_{\text{required}}}}{a_u(d)} = \frac{87.38}{4.25\,(18.5)} = 1.11 \text{ sq in./ft}$$

The required area is provided by #8 at 9 in. o.c.

$$V = 2.26(7.83) - \frac{1.75(7.83)}{2} = 17.69 - 6.83 = 10.86$$

Design shear is:

$$V_u = 1.7(10.86) = 18.46 \text{ kips}$$

From Table 3-7:

$$\overline{V}_{c_{\text{concr.}}} = 20.70 \text{ kips} > 18.46 \text{ kips (O.K.)}$$

No shear reinforcement is required.

(2) Toe: For the toe, the fill above it will be neglected as being subject to possible erosion. Hence, with only the weight of the footing,

$$1.75\,(0.150) = 0.26 \text{ kips/ft}$$

acting downward,

$$M = 2.14\,\frac{5.17^2}{2} + (3.30 - 2.14)\,\frac{5.17^2}{3} - 0.26\,\frac{5.17^2}{2}$$

$$M = 28.59 + 10.33 - 3.47 = 35.45 \text{ ft-kips}$$

Design moment will be:

$$M_{u_{\text{required}}} = 1.7(35.45) = 60.26 \text{ ft-kips}$$

Using $d = 21 - 4 = 17$ in., from Table 3-7:

$$\overline{M}_{u_{\text{concr.}}} = 203.20 \text{ ft-kips}$$

Ratio of $M_{u_{\text{required}}}$ to $\overline{M}_{u_{\text{concr.}}} = 60.26/203.20 = 0.296$. From Fig. 3-8, a_u = 4.28

$$A_s = \frac{M_{u_{\text{required}}}}{a_u(d)} = 60.26/4.28(17) = 0.83 \text{ sq in.}$$

The required steel area is provided by #9 at 12 in. o.c.

$$V = \frac{3.30 + 2.30}{2}(3.75) - 0.26(3.75) = 9.52 \text{ kips}$$

Design shear is:

$$V_u = 1.7(9.52) = 16.18 \text{ kips}$$

From Table 3-7:

$$\overline{V}_{c_{\text{concr.}}} = 19.00 \text{ kips} > 16.18 \text{ kips (O.K.)}$$

No shear reinforcement is required.

Table 11-1 gives the dimensions of the concrete structure and the reinforcement area at the location of maximum stress, i.e., at the bottom of the wall for cantilevered retaining walls varying from 5 to 20 ft in height above the top of the footing, for two loading conditions: horizontal back fill and horizontal back fill with 300 psf surcharge. Two values of equivalent fluid pressure of 30 and 45 pcf were used in the tabulation. Two heel widths are provided: the first part of the table shows minimum heel width, the second part utilizes moderate heel width.

The tabulation gives, for varying wall height, the wall thickness at top, the width of heel, measured from the inside face of the wall, the maximum wall thickness at the bottom of the wall, the total width of footing in feet, and the reinforcement required at the bottom of the wall for a 12 in. length of wall. Where the minimum reinforcement, per equation (10-1), governs, M is printed adjacent to the reinforcement area. The earth pressure under the footing at the heel, as well as the safety factor, which is the ratio of resisting moment to the overturning moment, are also printed.

Load factors were used for computing the lateral pressure, the soil pressure under the footing, and the safety factor. However, 1.7 load factor was used for computing the bending moments and shear forces for the design of the concrete wall.

Horizontal lateral force was used in the design. The neglected vertical component of the lateral pressures results in a conservative design. The reaction is limited to the middle third of the footing; therefore the soil pressure is zero at the heel.

The thickness of wall, used in the tabulations, is less than the minimum $h/12$ thickness or the $h/10$ limitation by the old 1963 Code, since the deflection of the wall does not control the design.

The requirements of the minimum steel area govern the reinforcement selection generally up to 14 feet wall height, depending on the loading conditions and materials. It is assumed that the thickness of foundation slab equals the maximum thickness of the wall. The assumed effective thickness of the wall is 2.5 in. less than the total thickness.

Water Tower Inn, Chicago, Ill. Hausner & Macsai, Architects. Paul Rogers & Associates, Structural Engineers.

12

Special Provisions for Seismic Design, in Accordance with ACI 318-71 and Recommendations of SEAC

Californians are constantly aware of seismic forces. It was only a short time ago that the San Fernando earthquake jolted this area; the rest of the country, however, is much less earthquake conscious, in spite of the fact that many areas of the United States and its territories are seismic zones.

With the advent of nuclear power stations, a strong impetus was given to seismic design, because the Atomic Energy Commission insists that all nuclear plants be seismic resistant irrespective of their locations.

The Structural Engineers Association of California (SEAC) introduced seismic recommendations several decades ago and these, as a rule, became officially adopted by the Uniform Building Code and other regulatory agencies and building officials. The American Concrete Institute's Standard Building Code 318 has, in the past, ignored the influence of seismic effects. Its inclusion in the Code could not be postponed, however, and the ACI 318-71 contains provisions similar to the SEAC requirements.

It should be mentioned, at this time, that the theory and application of dynamic response analysis, both elastic and nonlinear, based on time history and spectra, has advanced to the point that its application is not now difficult. It is expected that virtually all important structures in the future will be analyzed by dynamic response analysis.

SPECIAL DUCTILE FRAMES, ACI 318-71
DESIGN PHILOSOPHIES

Regions of low seismicity	1. Design for actual (unreduced) forces 2. Ductility as provided by main body of Code
Regions of high seismicity	1. Design to reduced forces 2. Special provisions for ductility to allow inelastic deformation (Appendix A)

Fig. 12-1.

GENERAL PRINCIPLES

The guiding principle for seismic design is to produce a ductile, energy-absorbing, structural system which develops extra strength through formation of plastic hinges instead of relying on elastic flexural, shear, and axial stresses. Proper ductility is aimed at preventing brittle or abrupt types of failures associated with "overreinforced" members.

In actual design practice the ACI 318-71 Strength Design (formerly USD) provisions are used. Loads are increased by load factors (U), and by capacity reduction factors (ϕ), and sized up to ultimate capacities. Thus the post-yielding extra capacities are not, as a rule, directly recognizable.

As far as design philosophies are concerned, one should differentiate between regions of low and high seismicity, as shown in Fig. 12-1. Performance criteria are shown in Fig. 12-2. In order to employ Strength Design, the actual loads are multiplied by the load factors. Since there is substantial difference between the ACI and SEAC requirements, both are featured in Fig. 12-3.

The author considers the SEAC requirements more acceptable, particularly in truly earthquake prone areas. Particular attention is directed to the reversibility of the seismic forces; thus the last equation is very significant.

Neither of the present Codes considers, at this time, the effect of vertical accelerations, and the simultaneous two-directional effects of an earthquake. Both phenomena were evident during the San Fernando earthquake.

It should also be remembered that the actual maximum lateral displacements during a real earthquake may be several times larger than the maximum com-

SPECIAL DUCTILE FRAMES, ACI 318-71
PERFORMANCE CRITERIA

1. Resist minor earthquakes without damage
2. Resist moderate earthquakes without structural damage but with some nonstructural damage
3. Resist major earthquakes without collapse

Fig. 12-2.

SPECIAL DUCTILE FRAMES, ACI & SEAC
LOADING COMBINATIONS
New ACI 318-71

$U = 1.4D + 1.7L$
$U = 0.75\ (1.4D + 1.7L + 1.87E)$
$U = 0.9D \pm 1.43E$

New SEAC-71

100% Ductile frames	Shear walls
$U = 1.5D + 1.8L$	$U = 1.5D + 1.8L$
$U = 1.4\ (D + L + E)$	$U = 1.4\ (D + L) + 2.8E$
$U = 0.9D \pm 1.4E$	$U = 0.9D \pm 2.8E$

where D = Dead load
L = Live load
E = Earthquake load

Fig. 12-3.

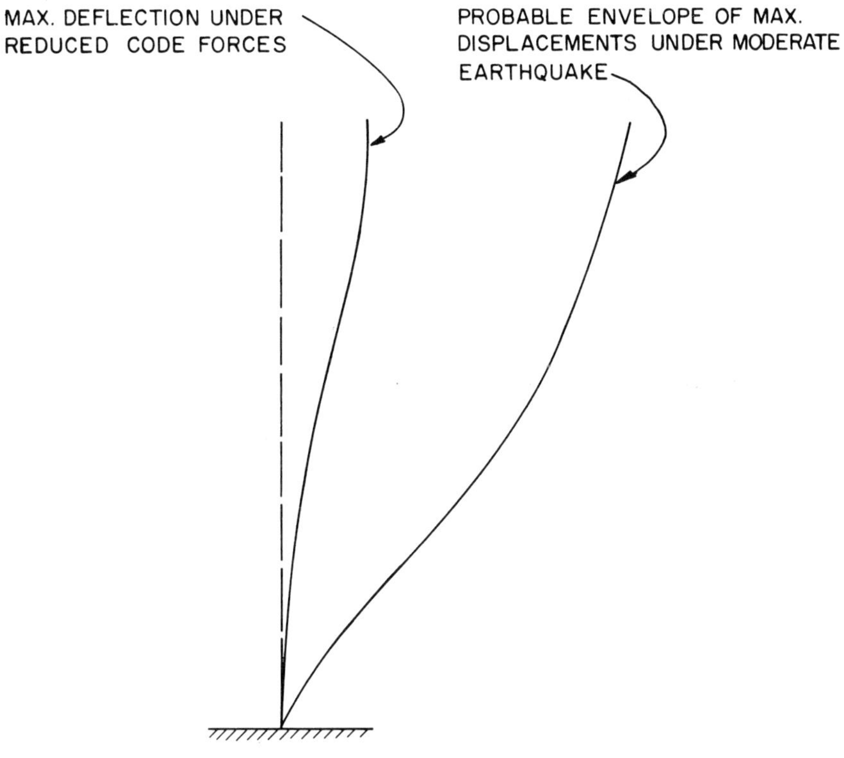

Fig. 12-4.

puted Code displacements. While this is within the ductile design philosophy, the stability of the structure should be considered. There are also indications that high-rise structures go into harmonic vibrations *after* the first violent shaking of the structure. Resonance of natural and forced vibrations should be avoided. Until recently, only the column reinforcement of ductile frames could be of ASTM A-615-60 steel, while the girder and beam reinforcement was limited to ASTM A-615-40 steel. The explanation is given in Fig. 12-5. It is obvious from this diagram that 40 ksi yield point steel has a long yield plateau necessary for ductility. The 60 ksi yield point steel has a shorter plateau, but it is now considered safe for ductile design. Again, the SEAC requirements are more stringent since these prescribe that the yield stress, f_y, must not exceed the specified one by more than 18 ksi. Too, the ultimate tensile stress must not be less than 1.33 times the *actual* yield stress.

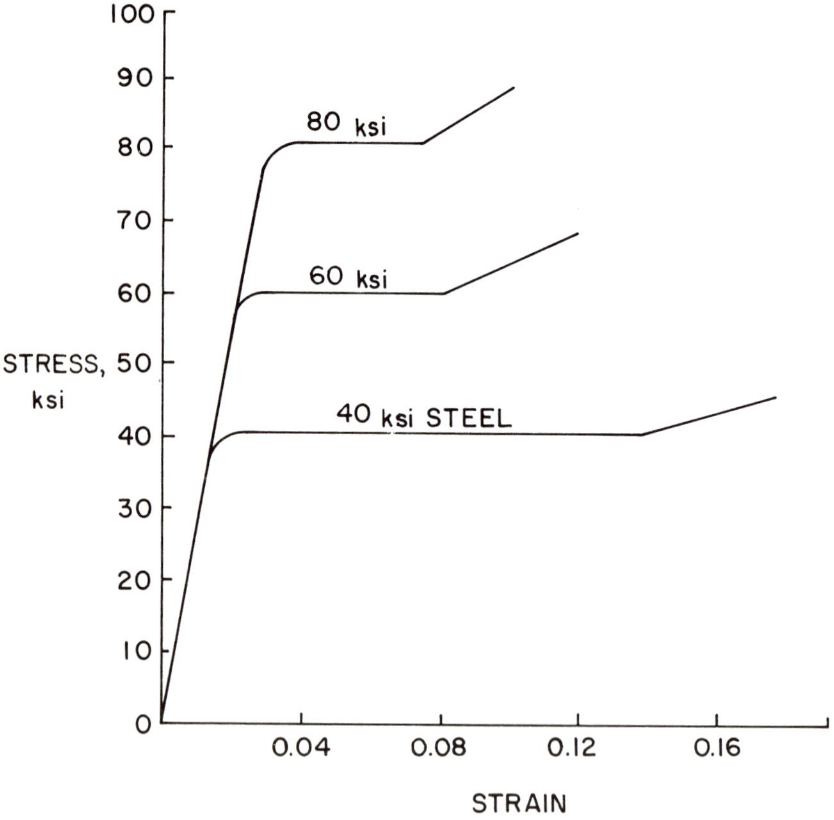

Fig. 12-5 Typical stress–strain curves for various steels.

Flexural members are subject to the following restrictions:

1) The ratio ρ_{max} of flexural reinforcement shall not exceed 0.50 of the ratio necessary for balanced condition. According to SEAC, ρ_{max} shall not exceed 0.025.
2) At supports, positive (bottom) moment capacity reinforcement equal to at least one half of the corresponding negative (top) moment capacity reinforcement to be provided for.
3) All flexural reinforcement, top or bottom, shall consist of two or more bars. The minimum reinforcement ratio, ρ_{min}, is to be $200/f_y$. A minimum of one fourth of the larger negative reinforcing to continue for the full length of the beam.
4) Since stress reversals may occur, neither top nor bottom reinforcement should be lap-spliced or terminated within the support (except at ends). Splices must be located not closer than twice the beam depth from the face of the column. At ends, both top and bottom reinforcements must terminate in standard 90° hooks at the far end of the column.

If the beam continues, the anchorage length to be as shown in Fig. 12-7. The anchorage length is again different for ACI and SEAC:

ACI	SEAC
As per Chapter 12, reduced to 2/3 in confined region, but not less than 16 in.	$L = \dfrac{A_s f_y}{1.5\, u_u \Sigma_0} \geqslant 24$ in.

Note that ACI eliminated the bond stress calculations.

Anchorage is to start from the extreme position of the point of inflection and be not less than 0.25 l_n from the face of support. Where the confined end of a

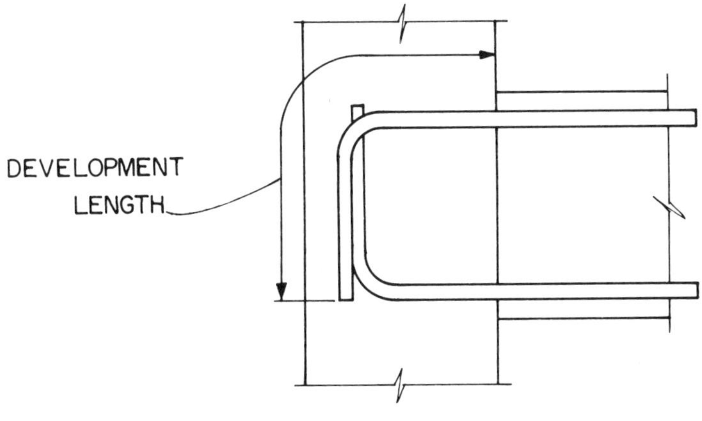

Fig. 12-6 Anchorage of beam reinforcement within confined region of exterior column.

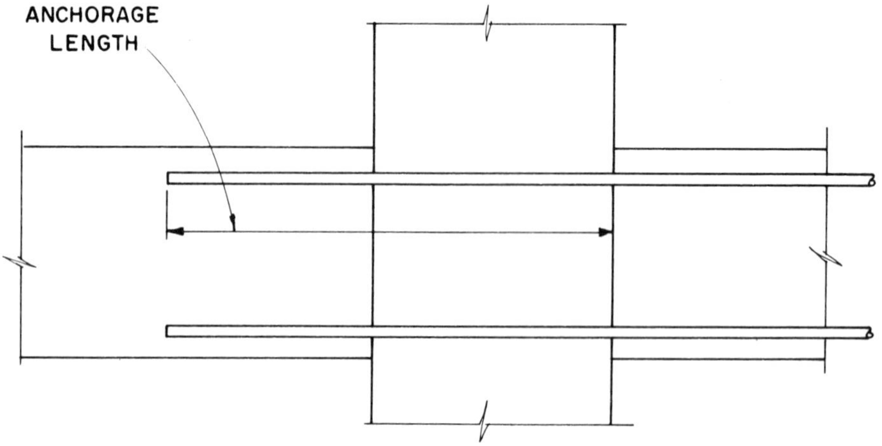

Fig. 12-7 Anchorage of beam reinforcement within column and opposite beam.

beam is wider than the column, one half of the top and bottom reinforcement may be anchored outside of the column core.

Compact sections. The SEAC requires compact sections for columns, girders, and beams that are part of a ductile frame. This implies that flat slabs and flat plates may not be part of the ductile framing system. *The ACI does not require compact sections.*

Web reinforcement of beams. Ductile design requires substantial web reinforcement in order to guarantee the development of plastic hinges. Since the differences between ACI and SEAC are major, both are presented in Fig. 12-9.

According to SEAC, the ultimate moment capacities M_{p_1} and M_{p_2} should be computed without the ϕ factor but with 125 percent of f_y. Use the ϕ factor for V_u, however.

Spacing of stirrup ties is shown in Fig. 12-10. Wherever compression reinforcement is required, the members are to be treated as tied columns with single or overlapping stirrup ties and, perhaps, with supplementary cross-ties.

Columns, in ductile frames are, as a rule, subject to axial loads and bending. The following restrictions apply (see also Fig. 12-8):

1) Columns shall have a compact section:

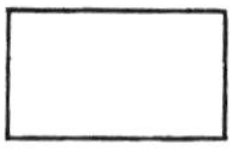

$b \geqslant 0.4h \geqslant 12$ in. (SEAC only)

$h \geqslant 12$ in.

2) Ratio of vertical reinforcement is $\rho = 0.06$. Except for small-size columns, however, the upper limit of 6 percent, as a rule, is only possible if bundled bars are employed.

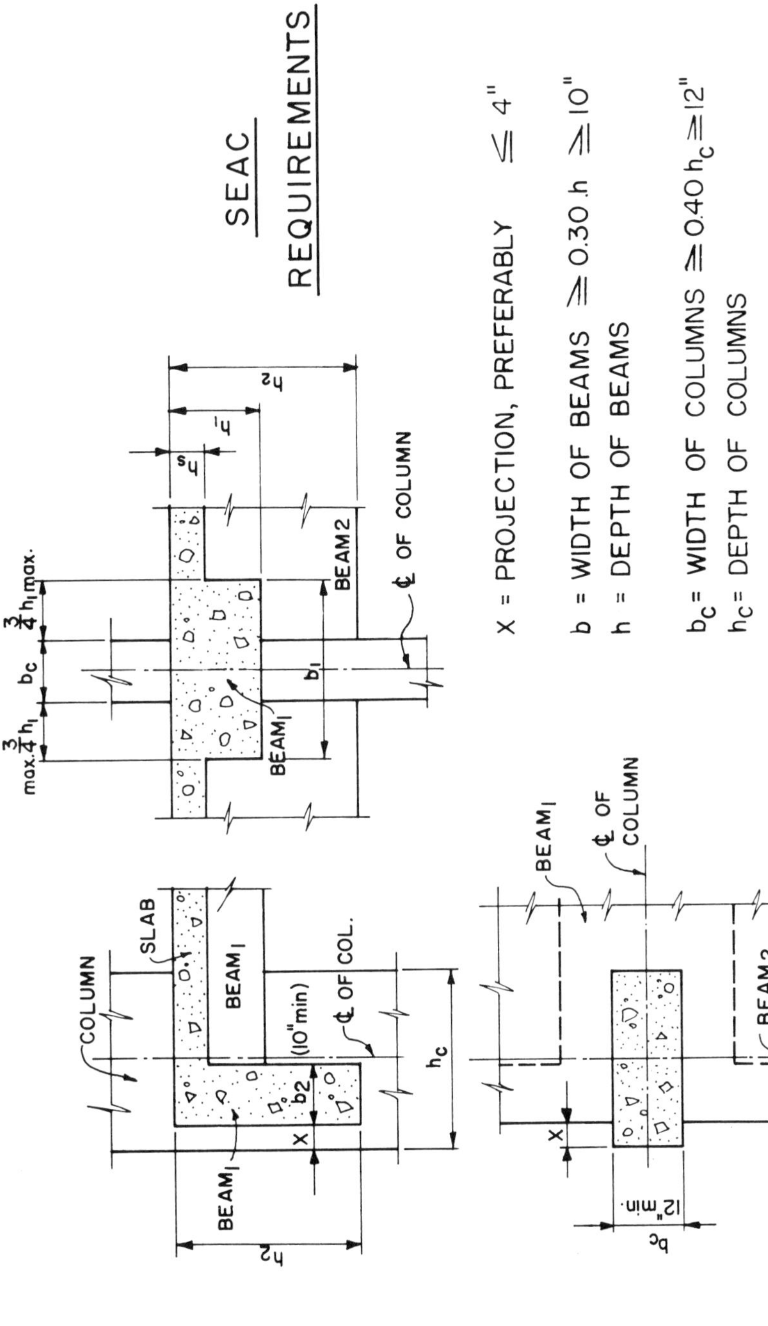

Fig. 12-8 Compact sections.

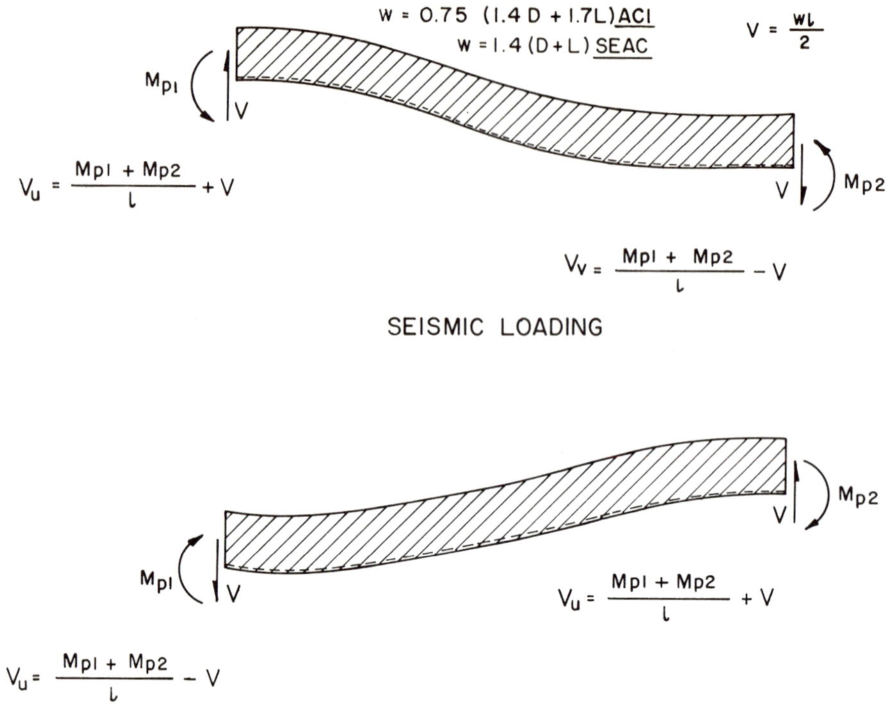

Fig. 12-9 Reverse seismic loading.

3) In order to ensure stability, it is required, generally, that plastic hinges should form only in beams.
4) Lightly loaded columns ($P_e \leqslant 0.40\ P_b$) must comply with flexural design. ACI has relaxed the confinement reinforcement for such columns, although SEAC has not done so.
5) Longitudinal bars are to be lap-spliced near midheight and the splice length should not be less than 30 bar diameters (SEAC).
6) Columns with axial loads greater than $0.40\ P_b$ are to be provided with special transverse reinforcement at the ends to protect the cores. Minimum length of this protection is 18 in. or one-sixth of the column height. Spiral reinforcement is most effective but rectangular hoops may be used with reduced capacity.

The ratio of spiral reinforcement, $\rho_s = 0.45 \left(\frac{A_g}{A_c} - 1 \right) \frac{f'_c}{f_{ys}}$ or $\rho_s = 0.12 \frac{f'_c}{f_{ys}}$; however, if rectangular hoops are used, the ratio has to double unless advantage is taken of supplementary cross-ties, if any. The area of

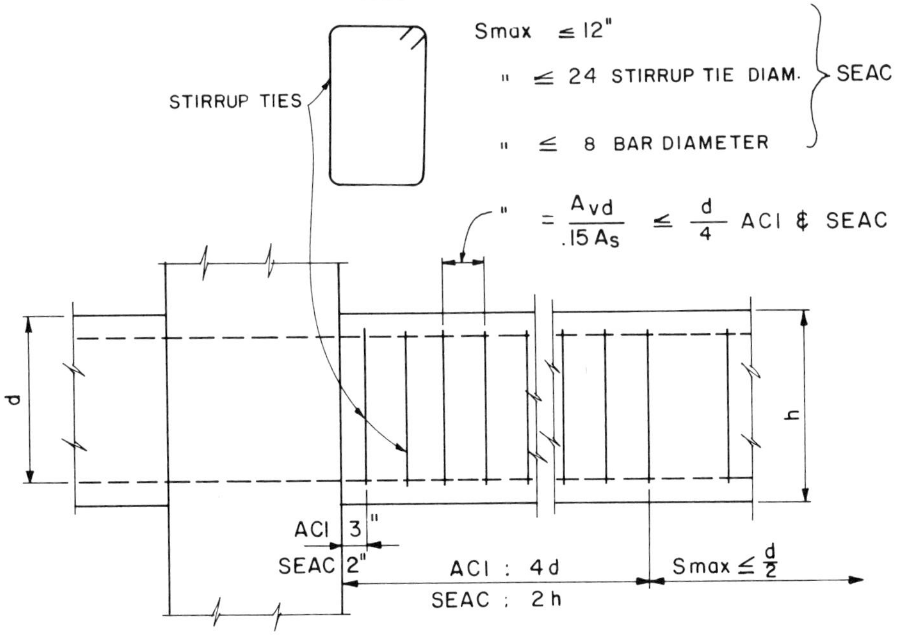

Fig. 12-10 Beam web reinforcement limitations, min. bar size #3.

transverse hoop bar (one leg) is:

$$A_{sh} = \frac{l_h \rho_s s_h}{2} \text{ (substitute } A_{ch} \text{ for } A_c\text{)}$$

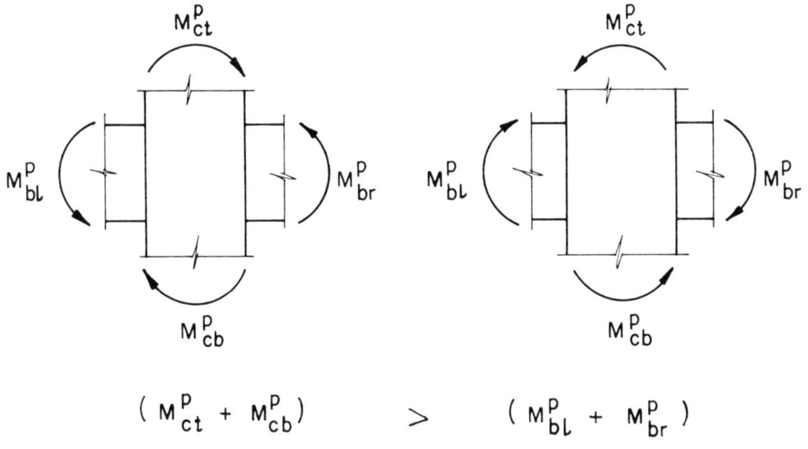

Fig. 12-11 "Strong column-weak beam" frame.

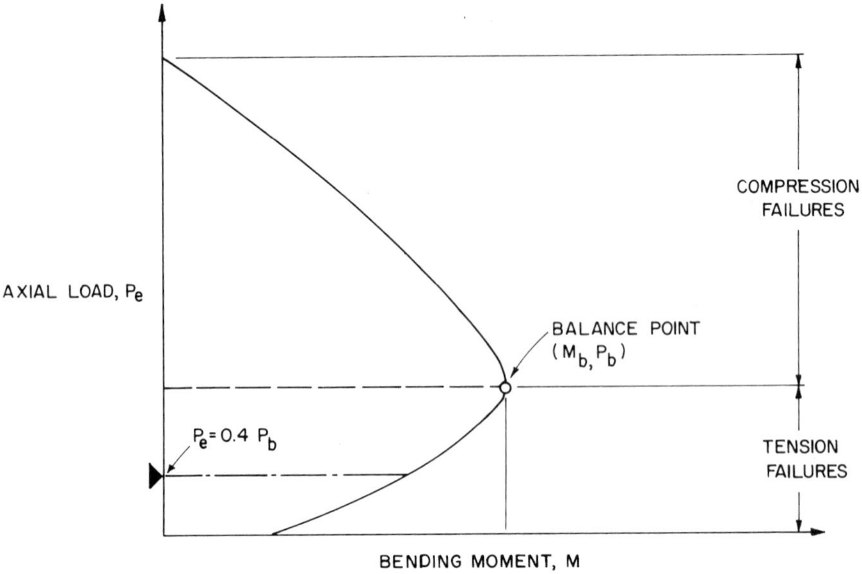

Fig. 12-12 Typical interaction diagram.

where l_h = unsupported length of rectangular hoop
s_h = center-to-center spacing of hoops $\leqslant$ 4 in.
Note: All hooks to be 135°; transverse spacing of supplementary cross-ties $\leqslant$ 14 in.

7) The special transverse reinforcement described in (6) is to continue for full height for columns supporting discontinuous members such as shear walls, braced frames, or other rigid elements.
8) Transverse column reinforcement subject to bending and axial compression shall satisfy:

$$A_v f_y \frac{d}{s} = V_u - V_c, \text{ where } V_u = \frac{M_u^B + \frac{1}{2} M_b}{l_u} \leqslant \frac{M_u^T + M_u^B}{l_u}; M_u^T \text{ and } M_u^B$$

are ultimate moment capacities of column at top and bottom, under design earthquake axial load.

M_b = maximum sum of moment capacities of the beams framing into top connection. Omit the factor $\frac{1}{2}$ for only one column.
l_u = clear height of column
V_c = $v_c bd$, where v_c as per ACI 318-71, except that $V_c = 0$ when $P/A_g < 0.12 f'_c$
d = effective depth

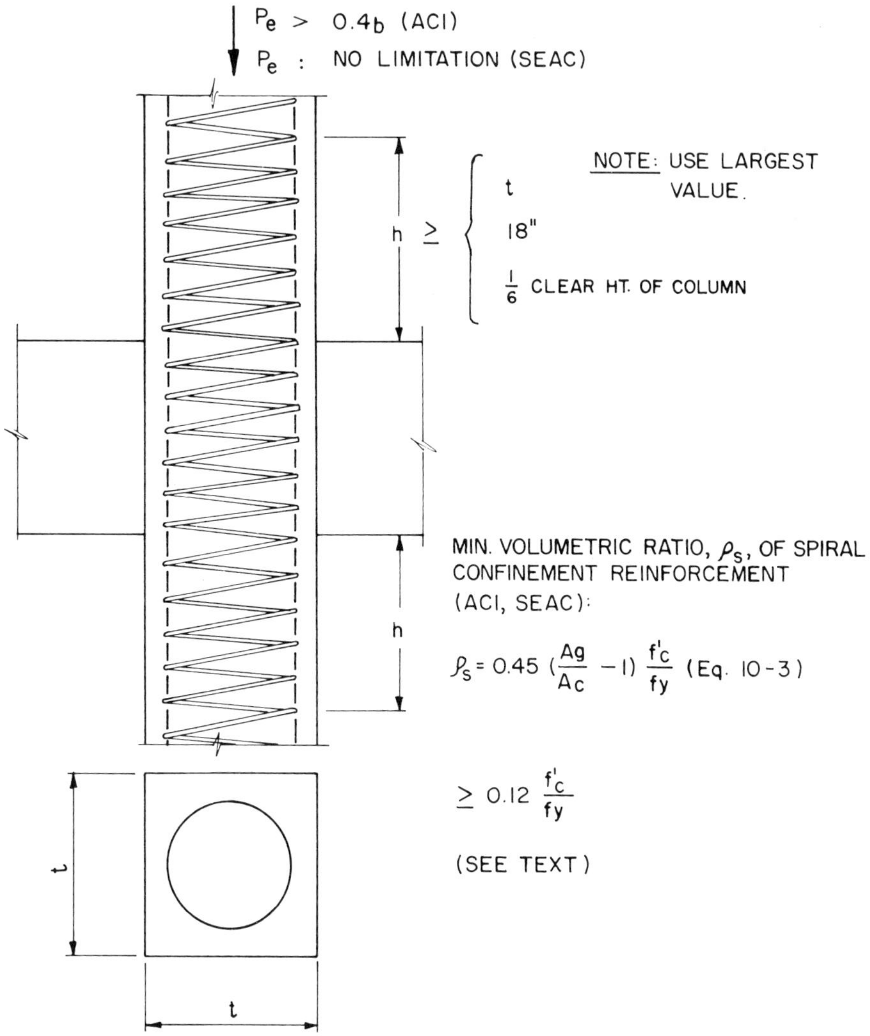

Fig. 12-13 Confinement reinforcement at column ends–spirals.

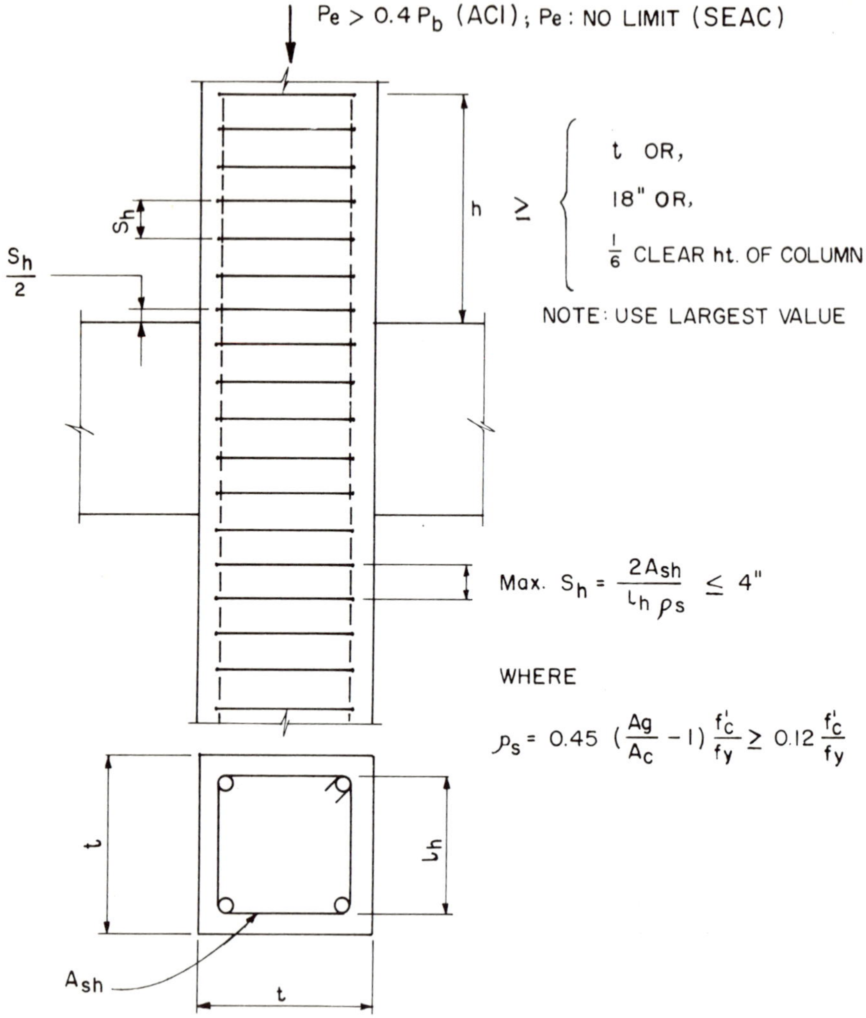

Fig. 12-14. Confinement reinforcement at column ends–rectangular hoop.

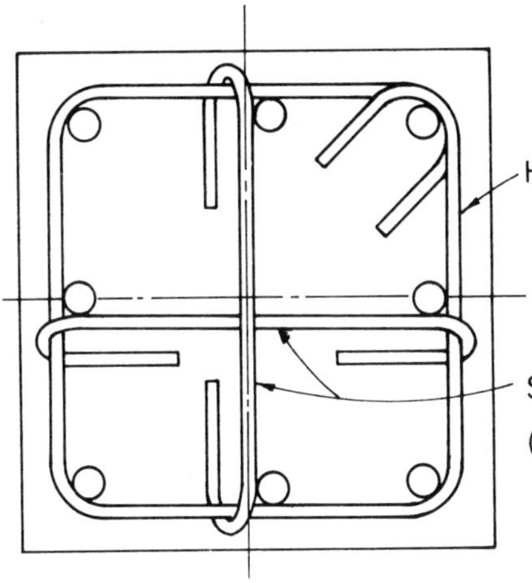

Fig. 12-15 Column hoops with supplementary crossties.

s = spacing $\leqslant \frac{1}{2}$ of minimum column dimension

A_v = total cross-sectional area of special transverse reinforcement within a distance s, reduced to two-thirds for circular spirals.

9) Special transverse column reinforcement may be reduced by 50 percent if beams frame into it on all four sides (SEAC).
10) Special transverse reinforcement is to continue through the connection for all end and corner columns, and for interior columns with unprotected corners larger than 4 in (SEAC).

Shear walls, in connection with ductile frames have to satisfy certain prerequisites. Again the SEAC is more conservative than the ACI. The load factors for special shear walls as part of ductile frames are:

$$U = 1.4\,(D + L) + 2.8E, \text{ or}$$
$$U = 0.9\,D \qquad \pm 2.8E \text{ (SEAC)}$$

The ACI uses the regular load factors except that the horizontal force factor $K = 1$.

It is to be mentioned that the overturning forces ($2.8E$) are a function of each individual shear wall length and are *not* dependent on the overall width or length of the structure. Thus, in case of short shear walls, the overturning forces can be very substantial.

The SEAC also insists on boundary members for all special shear walls, whereas

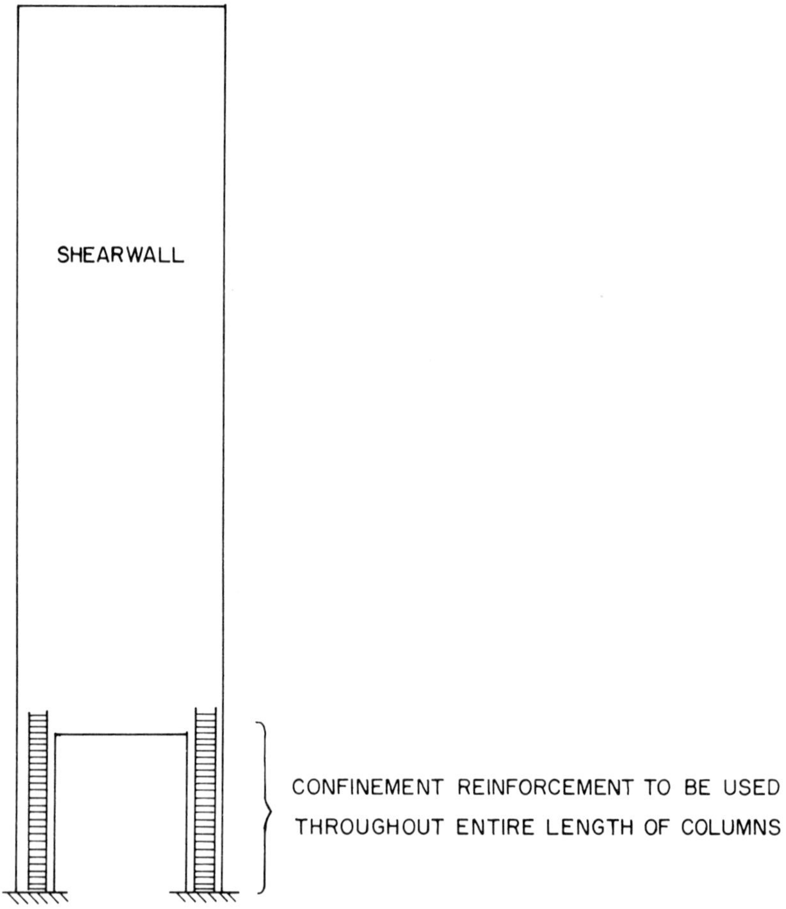

Fig. 12-16 Columns supporting discontinued shearwall.

the ACI limits this requirement to cases where:

$$P_e < 0.4P_b$$

The boundary members are to carry all vertical loads from all sources and shall have confinement reinforcement for full height. When $P_e < 0.4P_b$ and the maximum factored tensile stress exceeds 1.5 f_r (modulus of rupture), then the ACI (but not SEAC) permits the omission of the boundary members, but the shear wall is to be designed as a vertical cantilever and both edges to have a minimum flexural reinforcement of

$$A_s = (200/f_y)hd$$

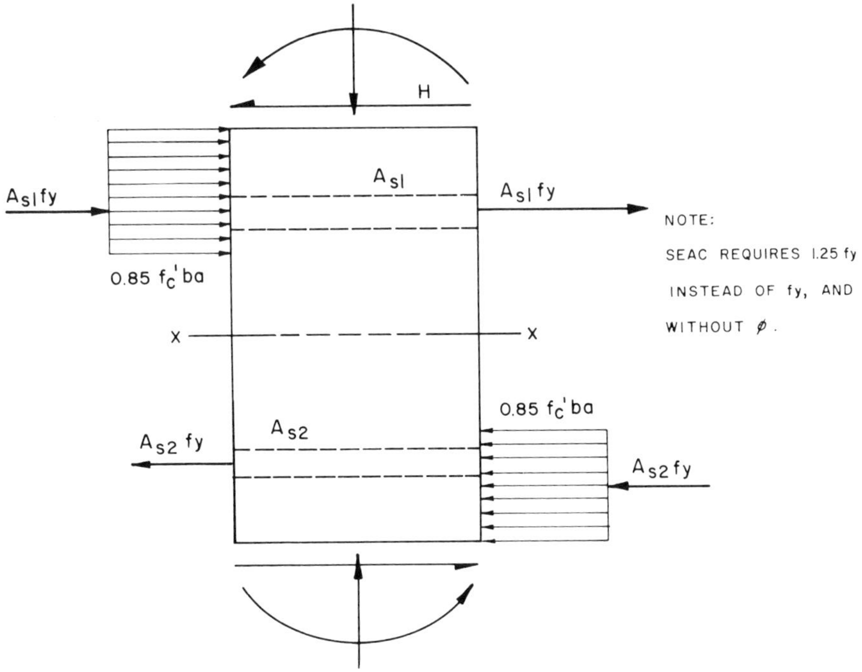

Fig. 12-17 Beam-column connection.

The earthquake load (E) mentioned in this chapter is created by the lateral forces due to an earthquake. The magnitude of the lateral forces and their distribution along the building is a function of the base shear. This is the total seismic design force on the building structure in the particular direction that is being considered, but which generally is normal to a principal axis (in plan) of the structure. The base shear is the horizontal force transmitted from the ground into the building. Inversely, the base shear, or the shear at any level, is the summation of the individual lateral forces from the top downward to the base or to the level in question.

The magnitude of the base shear and its distribution along the height of the building has been covered, until recent times, by formulas developed through static analyses. More recently, dynamic response analyses have been introduced which furnish forces, moments, deformations, etc., representing, to a substantial degree, the real conditions and stresses that occur under complex earthquake motions.

For the calculations of lateral forces by static approach, as per the SEAC recommendations, see Figs. 12-22, 12-23 and 12-24.

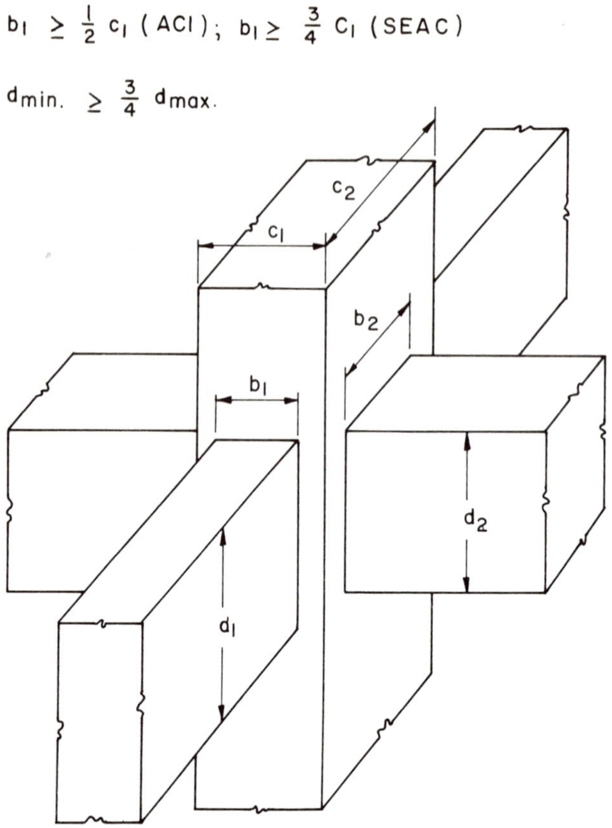

Fig. 12-18 Size limitations on framing beams for effective confinement of joint.

The dynamic response approach, which is now proposed for the City of Los Angeles Building Code, is described, briefly, on page 262.

The City of Los Angeles is about to modify its building code relative to earthquake-resistant design, introducing dynamic analyses and adopting the same requirements as outlined by SEAC. In accordance with the proposed modifications, every structure shall have structural and deflectional capacities sufficient to resist the effects of earthquake upon the structure as determined by dynamic analyses. These analyses shall combine the natural periods and corresponding significant mode shapes of the structure with the ground shaking prescribed for the site in a Geology-Seismology Report. Every Geology-Seismology Report shall be subject to review and approval by the Department of Building and Safety.

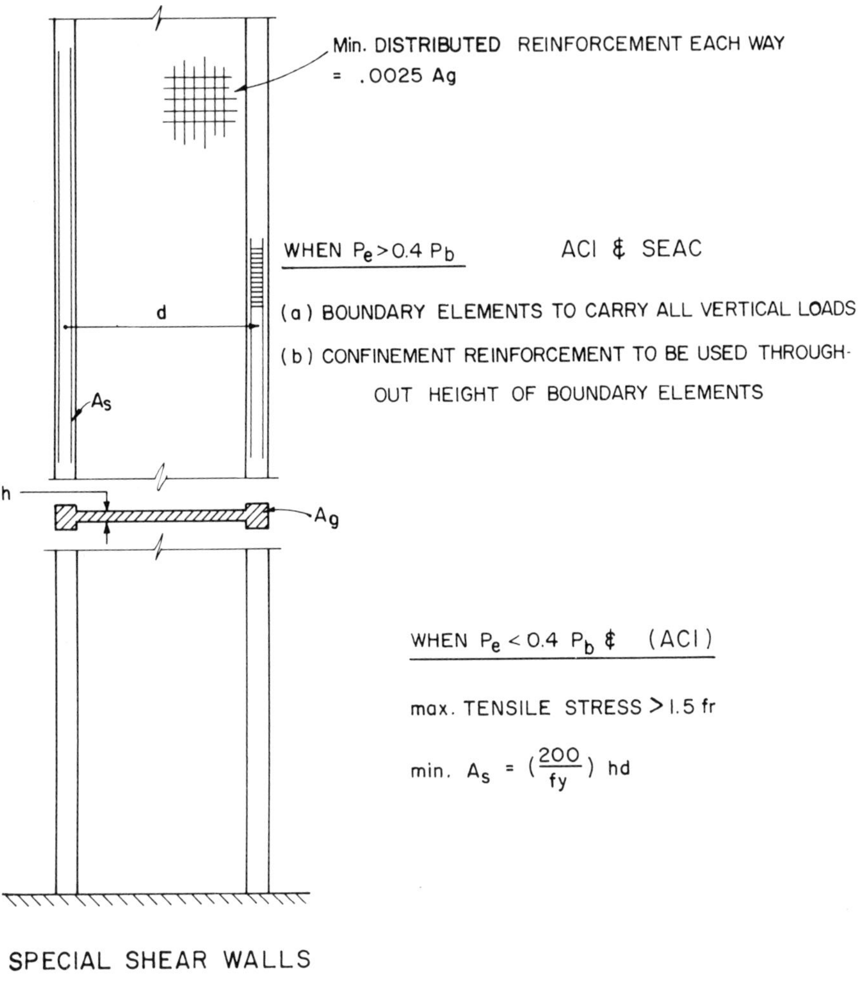

Fig. 12-19.

Exceptions are:

1) Structures 160 feet or less in height, other than essential facilities, may be designed for earthquake forces as developed by static analyses.
2) Essential facilities 160 feet or less in height may be designed for one and a half times the earthquake forces given by the static analyses.

Essential Facilities are those structures or buildings which must be safe and usable for emergency purposes after an earthquake in order to preserve the

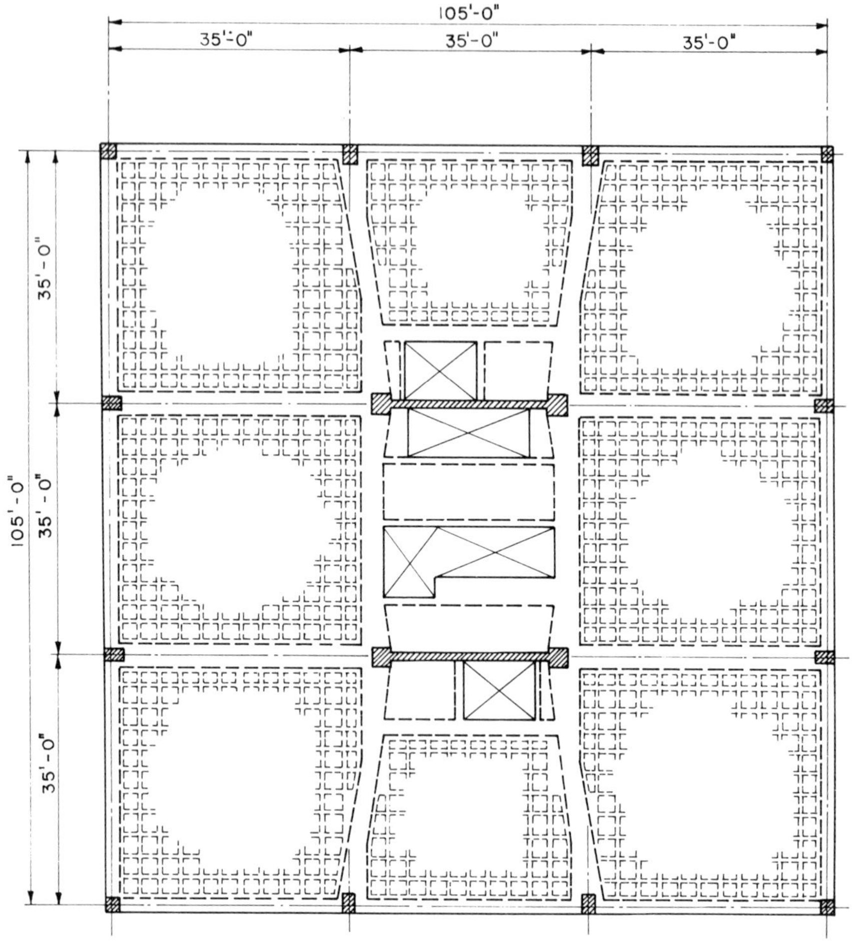

Fig. 12-20 Typical floor framing plan, Century City Medical Plaza.

peace, health, and safety of the general public. Such facilities shall include, but are not limited to:

1) Hospitals and other medical facilities having surgery or emergency treatment areas
2) Fire and police stations
3) Municipal government centers

Every structure 160 feet or less in height may be designed and constructed to withstand minimum total lateral seismic forces assumed to act nonconcurrently

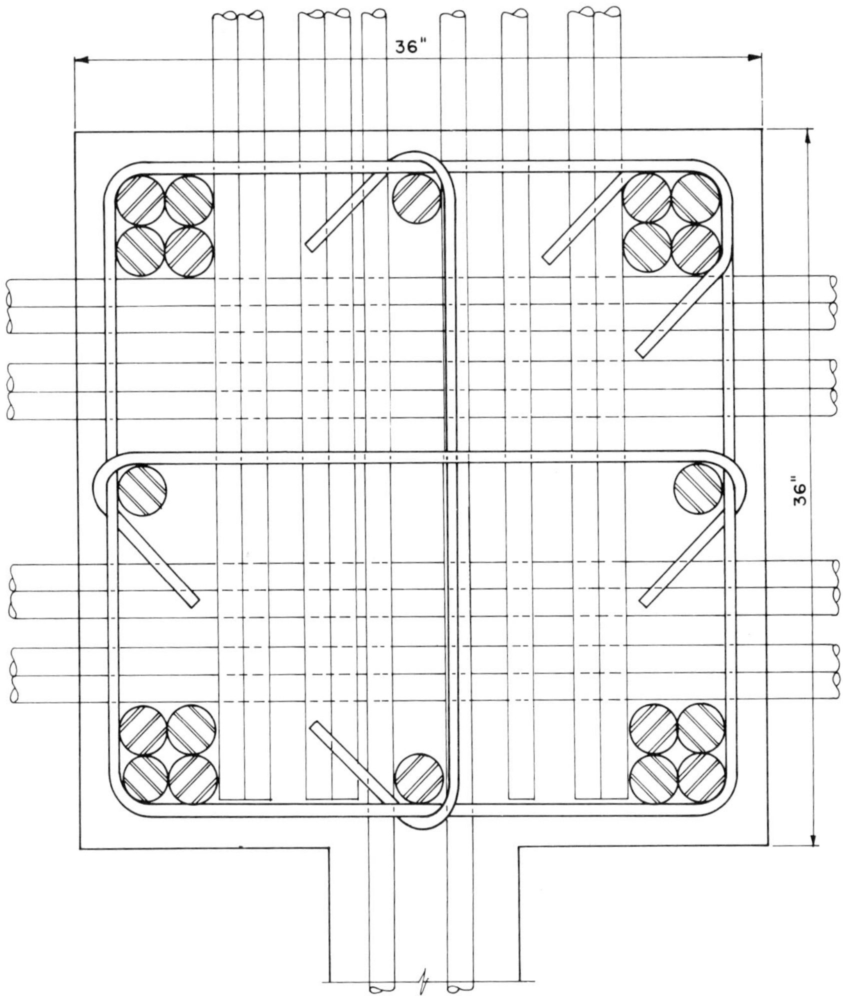

Fig. 12-21 Girder-column intersection.

in the direction of each of the main axes of the structure, in accordance with the following formula:

$$V = KCW$$

For the values of K, C, and W, see Fig. 12-23. The total lateral force V must be distributed over the height of the structure, according to Fig. 12-24.

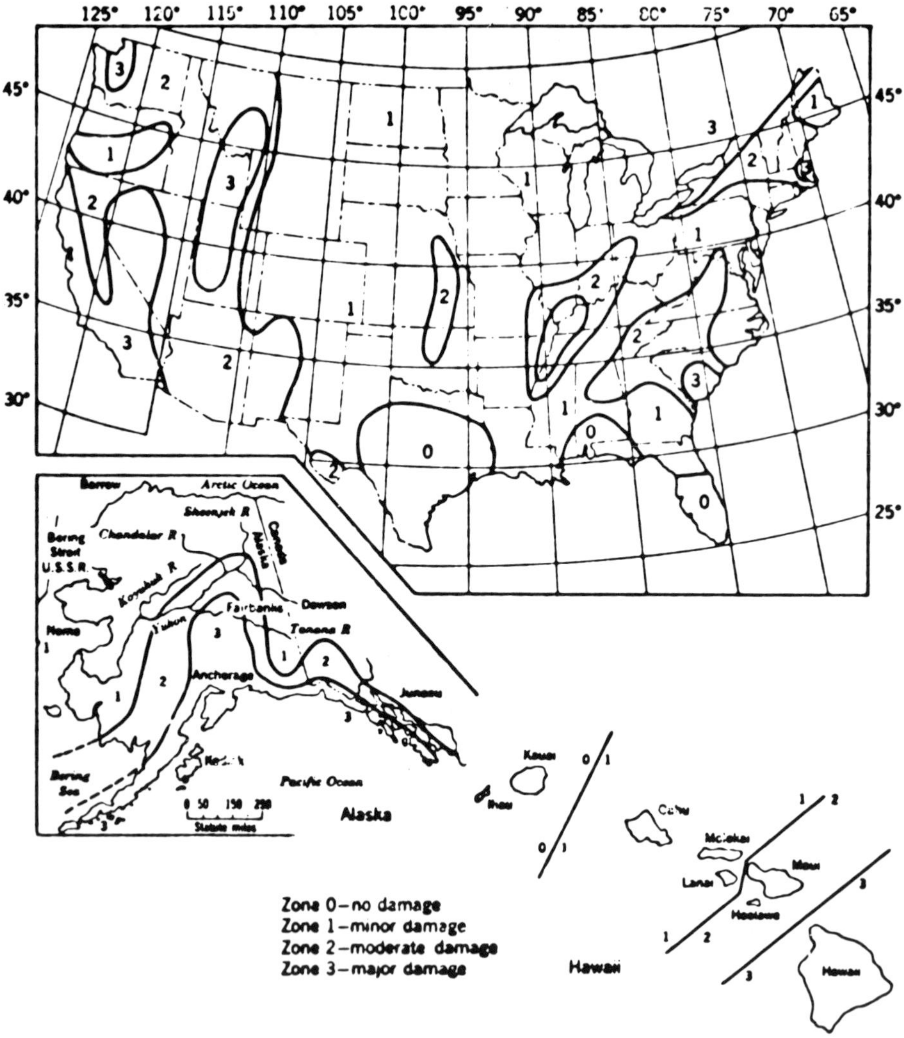

Fig. 12-22 Seismic risk map of the United States.

In accordance with the proposed modifications the structural systems must comply with the following special requirements.

1) All buildings designed with a horizontal force factor $K = 0.67$ or 0.80 shall have space frames-ductile moment resisting.
2) Buildings more than 160 feet in height shall have space frames-ductile moment resisting capable of resisting not less than 25 percent of the re-

quired seismic forces for the structure as a whole, if shear walls are also used.

3) All concrete space frames required by design to be part of the lateral force resisting system and all concrete frames located in the perimeter line of vertical support shall be space frames-ductile moment resisting. *Exception:* Frames in the perimeter line of vertical support of buildings designed with shear walls taking 100 percent of the design lateral forces need only be checked for conformance with the following subitem (3a).
   3a) All framing elements not required by design to be part of the lateral force resisting system shall be investigated for adequacy for vertical load and induced moment due to four times the distortions resulting from the code-required lateral forces. The rigidity of other elements shall be considered in accordance with the provisions of the existing code.
4) Moment-resisting space frames and ductile moment-resisting space frames may be enclosed by or adjoined by more rigid elements which would tend to prevent the space frame from resisting lateral forces where it can be shown that the action or failure of the more rigid elements will not impair the vertical and lateral load-resisting ability of the space frame.

The major portion of the proposed modifications concerns the earthquake resistant design of structures by a dynamic method of analyses and would require one of the following methods:

1) Time-dependent response method. (Currently being used for most modern high-rise structures.)
2) Spectral modal analysis method. (A dynamic response method currently used.)
3) Pseudodynamic analyses for regular buildings not exceeding six stories in height. (Simplified manual calculation method.)
4) Simplified static application for minor structures such as all one- and two-story buildings and three-story wood-frame buildings.

There are several publications available which present the types of dynamic analyses described in numerous research studies by educational institutions, professional associations, and individuals.

SEISMIC BASE SHEAR BY STATIC ANALYSIS

V = Minimum design base shear = $ZKCW$
Z = Seismic coefficient dependent on the seismic probability zone (Fig. 12-22)
W = The total dead load, including permanent installations, kips
K (Horizontal force factor) = 0.67 for ductile frame buildings without shear walls
= 0.80 for ductile frame buildings with shear walls
= 1.33 for buildings with a box system as defined by SEAC
T (fundamental period of vibration of the building in seconds in the direction under consideration)
= $0.05h_n/D^{1/2} = 0.10N$
C = $0.05/T^{1/3} = 0.10$ (for main frame)
D = The dimension of the building in feet in a direction parallel to the applied forces
N = The total number of stories above exterior grade to level n
h_n = The total height of the building above the base, in feet

Distribution of base shear V to be made as illustrated in Fig. 12-24.

Fig. 12-23. Seismic base shear by static analysis.

DISTRIBUTION OF LATERAL FORCES BY STATIC ANALYSIS

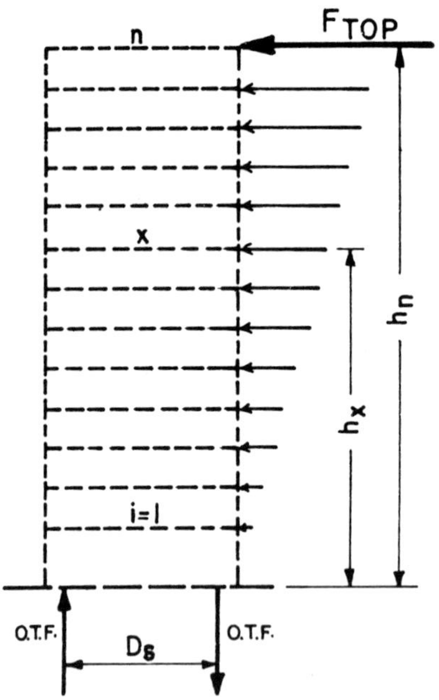

V (base shear) = $ZKCW$

$F_{top} = 0.004\,V(h_n/D_s)^2 = 0.15\,V$

$$F_x = \frac{(V - F_{top})\,w_x h_x}{\sum_{i=1}^{n} w_i h_i}$$

D_s = The plan dimension of the vertical resisting system in feet
w_i, w_x = That portion of W which is located at or is assigned to level i or x respectively
i = 1, designates the first level above the base
OTF = Overturning force (kips)

Exception: One- and two-story buildings shall have uniform distribution.

Fig. 12-24 Distribution of lateral forces by static analysis.

MINIMUM WALL REINFORCEMENT.

t	WELDED WIRE FABRIC REINF. (ASTM A 185)			
	(ACI 318-71 L.A., U.B.C. 1970		RECOMMENDED	
t	HORIZONTAL	VERTICAL	HORIZONTAL	VERTICAL
%	.0020	.0012	.0027	.0015
6"	(.137) 5/0 @ 12 ON ℄	(.0865) 2/0 @ 12 ON ℄	(.195) 6/0 @ 10 ON ℄	(.108) 4/0 @ 12 ON ℄
8"	(.183) 7/0 @ 12 ON ℄	(.1153) 4/0 @ 12 ON ℄	(.260) 7/0 @ 8 ON ℄	(.144) 5/0 @ 12 ON ℄
8"	(.183) 3/0 @ 12 E.F.	(.1153) # 1 @ 12 E.F.	(.260) 2/0 @ 8 E.F.	(.144) # 0 @ 12 E.F.
10"	(.228) 3/0 @ 10 "	(.144) # 0 @ 12 "	(.324) 3/0 @ 8 "	(.180) 2/0 @ 12 "
12"	(.274) 4/0 @ 10 "	(.173) 2/0 @ 12 "	(.389) 4/0 @ 8 "	(.216) 4/0 @ 12 "
14"	(.319) 5/0 @ 10 "	(.202) 3/0 @ 12 "	(.453) 4/0 @ 6 "	(.252) 4/0 @ 12 "
16"	(.365) 6/0 @ 10 "	(.230) 4/0 @ 12 "	(.518) 5/0 @ 6 "	(.288) 5/0 @ 12 "
18"	(.410) 5/0 @ 8 "	(.260) 5/0 @ 12 "	(.583) 5/0 @ 6 "	(.324) 6/0 @ 12 "
20"	(.456) 6/0 @ 8 "	(.288) 5/0 @ 12 "	(.647) 6/0 @ 6 "	(.360) 7/0 @ 12 "
22"	(.502) 6/0 @ 8 "	(.317) 6/0 @ 12 "	(.713) 7/0 @ 6 "	(.396) 6/0 @ 10 "
24"	(.547) 7/0 @ 8 "	(.346) 5/0 @ 10 "	(.777) 1/2" φ @ 6 "	(.432) 7/0 @ 10 "

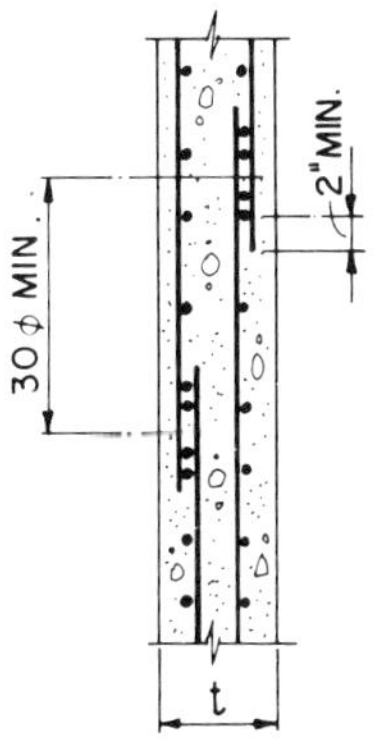

PLAN

Typical concrete wall.

MINIMUM WALL REINFORCEMENT.				
THICKNESS	DEFORMED BAR REINFORCEMENT (ASTM A15 OR A432)			
t	(ACI 318-71, L.A., U.B.C. 1970		RECOMMENDED	
	HORIZONTAL	VERTICAL	HORIZONTAL	VERTICAL
%	.0025	.0015	.0035	.0020
6"	#4 @ 12 ON ℄	#4 @ 18 ON ℄	#4 @ 9 ON ℄	#4 @ 16 ON ℄
8"	#4 @ 10 ON ℄	#4 @ 16 ON ℄	#4 @ 7 ON ℄	#4 @ 12 ON ℄
8"	#4 @ 18 E. F.	#4 @ 18 E. F.	#4 @ 14 E. F.	#4 @ 18 E. F.
10"	#4 @ 16 "	#4 @ 18 "	#4 @ 11 "	#4 @ 18 "
12"	#4 @ 12 "	#4 @ 18 "	#4 @ 9 "	#4 @ 16 "
14"	#4 @ 11 "	#4 @ 18 "	#5 @ 12 "	#4 @ 14 "
16"	#4 @ 10 "	#4 @ 16 "	#5 @ 11 "	#4 @ 12 "
18"	#5 @ 14 "	#4 @ 12 "	#5 @ 10 "	#5 @ 16 "
20"	#5 @ 12 "	#4 @ 12 "	#5 @ 9 "	#5 @ 14 "
22"	#5 @ 11 "	#4 @ 10 "	#6 @ 11 "	#5 @ 13 "
24"	#5 @ 10 "	#4 @ 10 "	#6 @ 10 "	#5 @ 12 "

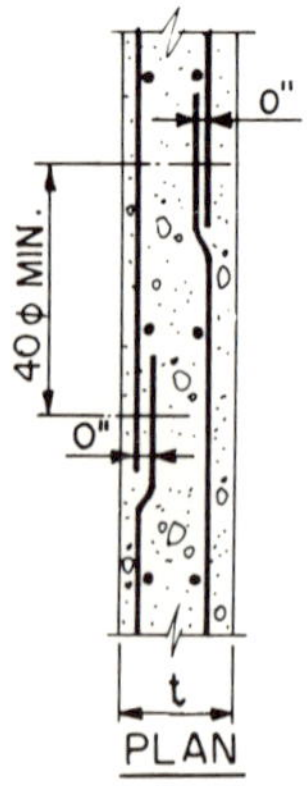

NOTE:

1. EITHER VERT. OR HORIZ. STEEL MAY BE OUTSIDE UNLESS OTHERWISE SHOWN OR NOTED.
2. ALL HORIZONTAL BAR SPLICES SHALL BE STAGGERED A MIN. OF 40 DIAMETERS FROM SPLICES IN ADJACENT HORIZONTAL BARS IN THE SAME OR OPPOSITE FACE.
3. IF WALL IS DESIGNED FOR COMPUTED FLEXURE, CALCULATED REINFORCEMENT SHALL BE USED, BUT NOT LESS THAN SHOWN IN THE TABLE ABOVE.

Typical concrete wall.

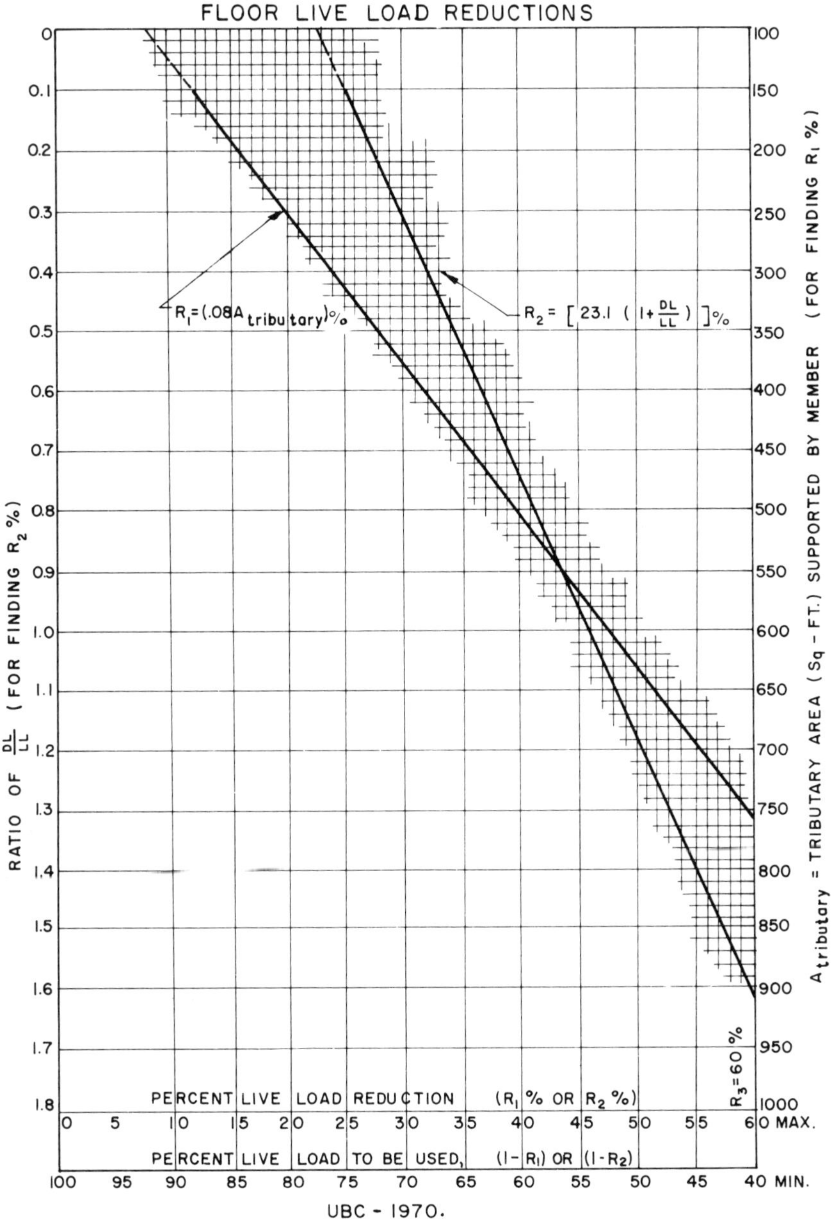
FLOOR LIVE LOAD REDUCTIONS
RATIO OF DL/LL (FOR FINDING R2 %)
0
0.1
0.2
0.3
0.4
0.5
0.6
0.7
0.8
0.9
1.0
1.1
1.2
1.3
1.4
1.5
1.6
1.7
1.8
R1=(.08A tributary)%
R2= [23.1 (1+DL/LL)]%
R3=60%
PERCENT LIVE LOAD REDUCTION (R1 % OR R2 %)
0 5 10 15 20 25 30 35 40 45 50 55 60 MAX.
PERCENT LIVE LOAD TO BE USED, (1-R1) OR (1-R2)
100 95 90 85 80 75 70 65 60 55 50 45 40 MIN.
A tributary = TRIBUTARY AREA (Sq - FT.) SUPPORTED BY MEMBER (FOR FINDING R1 %)
100
150
200
250
300
350
400
450
500
550
600
650
700
750
800
850
900
950
1000
UBC - 1970.

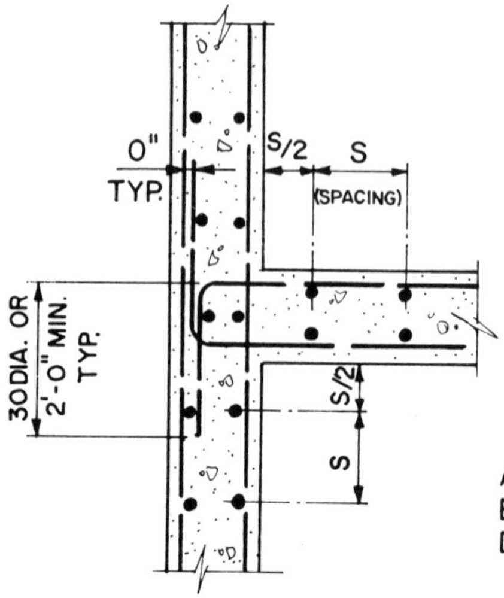

DOUBLE CURTAIN OF REINFORCEMENT

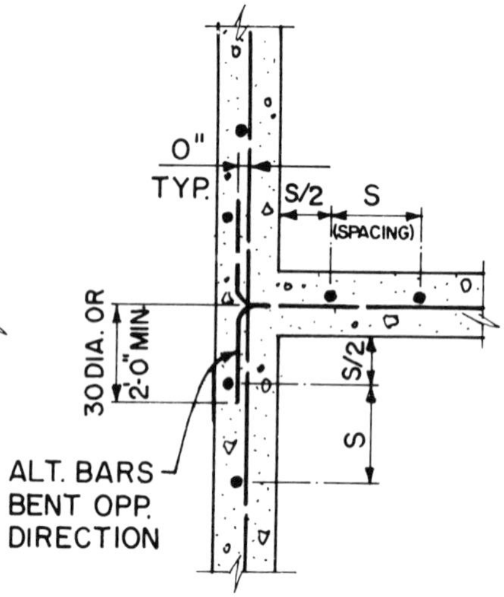

SINGLE CURTAIN OF REINFORCEMENT

Concrete wall intersections.

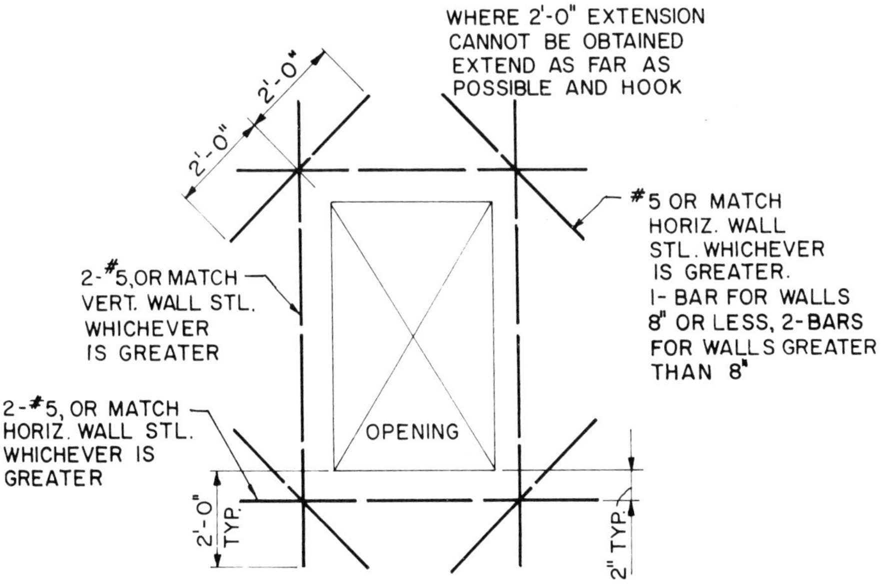

NOTE: FOR WALL OPENINGS LESS THAN 24"x24" NO ADDITIONAL STEEL REQUIRED.

DESIGNERS NOTE:

2-#5 TRIM BAR REQUIRED BY A.C.I.

1-#4 TRIM BAR REQUIRED BY U.B.C.

Typ. reinforcing at openings in concrete wall.

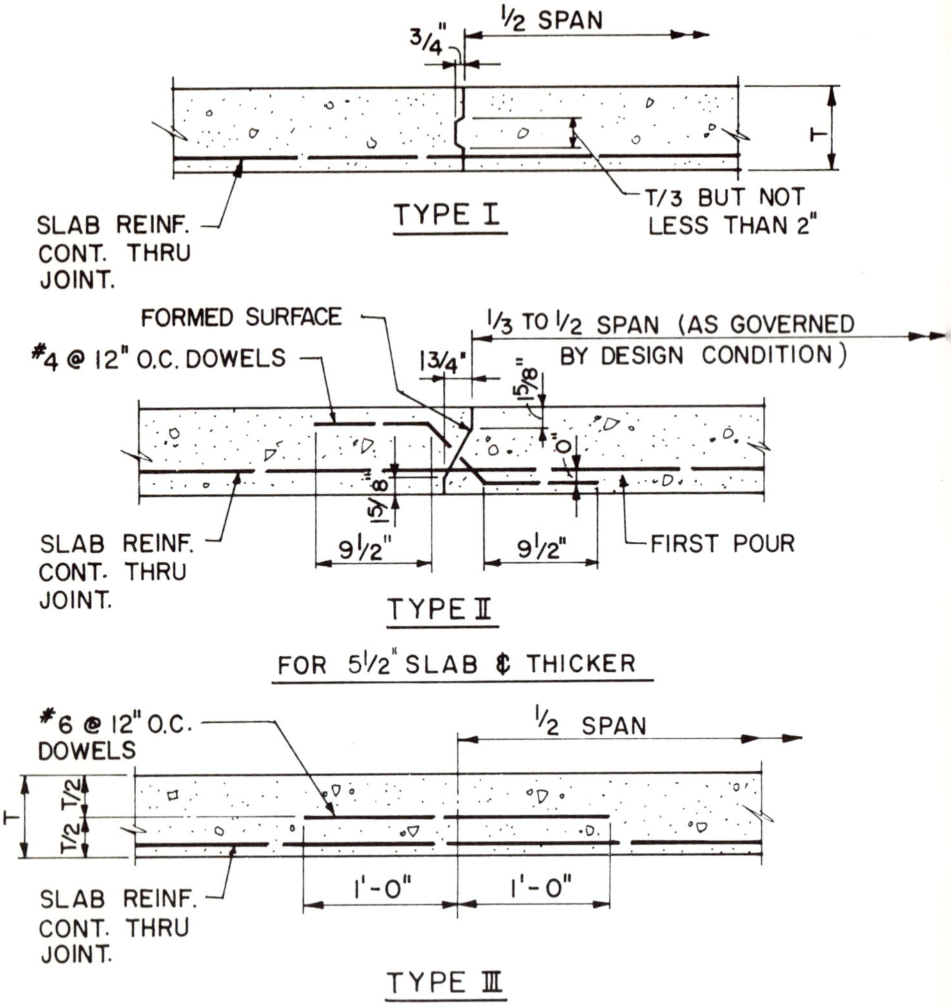

Construction joints for structural slabs.

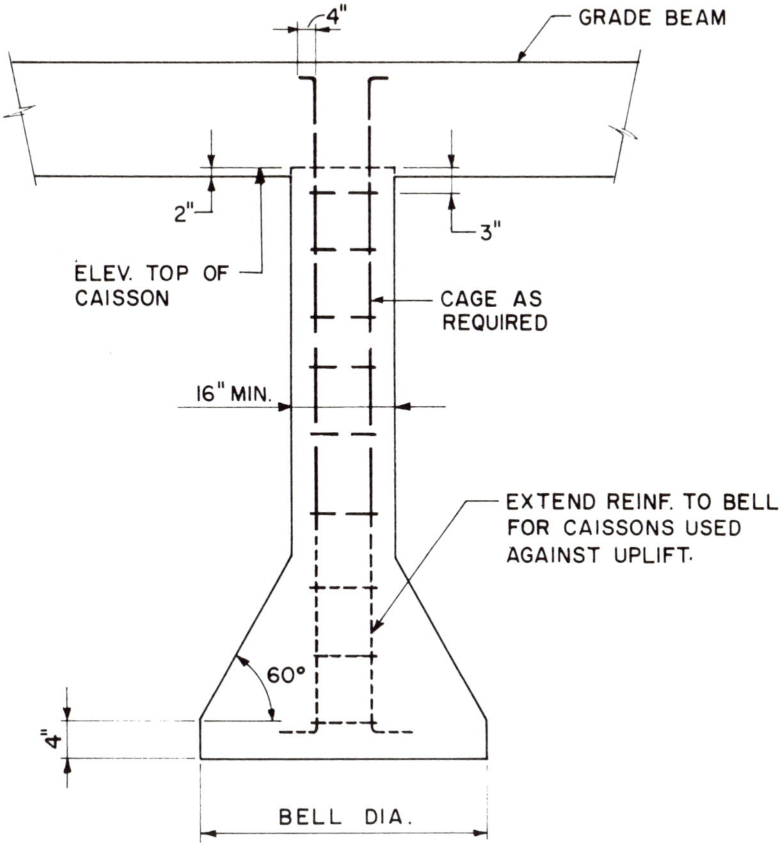

Typical belled caisson.

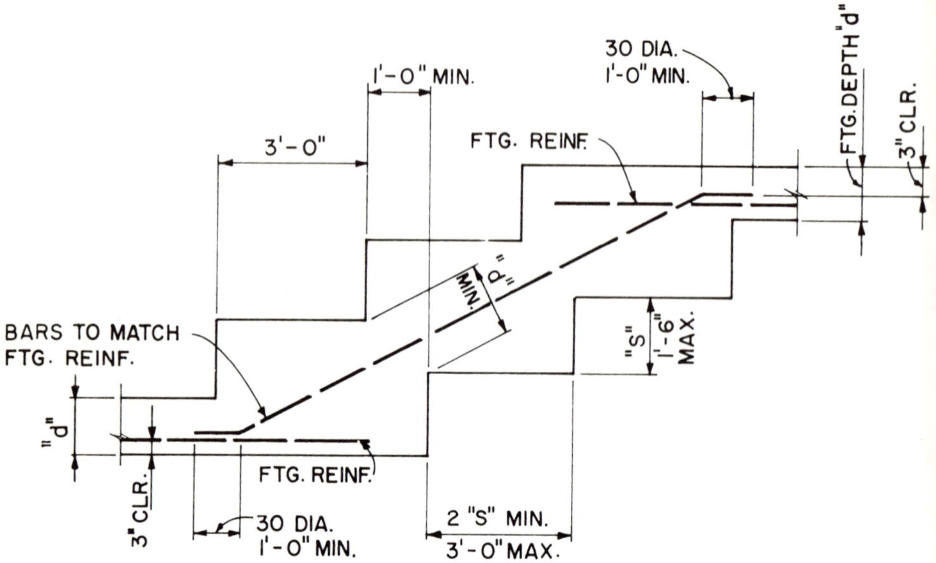

Stepped footing detail.

Index